RARE EARTH FRONT

RARE EARTH FRONTIERS

From Terrestrial Subsoils to Lunar Landscapes

Julie Michelle Klinger

CORNELL UNIVERSITY PRESS   ITHACA AND LONDON

First published 2017 by Cornell University Press

Printed in the United States of America

Library of Congress Cataloging-in-Publication Data

Names: Klinger, Julie Michelle, 1983– author.
Title: Rare earth frontiers : from terrestrial subsoils to lunar landscapes /
Julie Michelle Klinger.
Description: Ithaca : Cornell University Press, 2017. | Includes bibliographical
   references and index.
Identifiers: LCCN 2017012446 (print) | LCCN 2017014409 (ebook) |
   ISBN 9781501714603 (epub/mobi) | ISBN 9781501714610 (pdf) |
   ISBN 9781501714580 (cloth : alk. paper) | ISBN 9781501714597
   (pbk. : alk. paper)
Subjects: LCSH: Rare earth metals—Social aspects. | Rare earth metals—Political
   aspects. | Rare earth metals—China—Inner Mongolia. | Rare earth metals—
   Brazil—Amazonas. | Lunar mining.
Classification: LCC TN490.A2 (ebook) | LCC TN490.A2 K56 2017 (print) |
   DDC 553.4/94—dc23
LC record available at https://lccn.loc.gov/2017012446

Cornell University Press strives to use environmentally responsible suppliers and materials to the fullest extent possible in the publishing of its books. Such materials include vegetable-based, low-VOC inks and acid-free papers that are recycled, totally chlorine-free, or partly composed of nonwood fibers. For further information, visit our website at cornellpress.cornell.edu.

To the people of Bayan Obo, and Baotou, whose waters, soils, and bodies have borne a burden few outside of rare earth mining regions can imagine; to the people of São Gabriel da Cachoeira; to those devoting their lives to more just and sustainable regimes of rare earth production and consumption; this work is humbly dedicated to you.

Contents

Illustrations

Acknowledgments

The creative and often surprising ways that people get involved in a project of this scope would, I suspect, be worthy of its own novel, but with these few words I wish to acknowledge those principal characters without whose support this book would not be.

To Michael Watts, Nathan Sayre, Harley Shaiken, Paola Bacchetta, Richard Walker, and You-tien Hsing at the University of California, Berkeley, and to Kenneth Pomeranz, at the University of Chicago. Thank you for your kind mentorship.

To those who hosted me at various research institutions throughout the duration of this project, a special thank you to Alexandre Barbosa at the University of São Paulo Institute of Brazil Studies; to Wu Baiyi at the China Academy of Social Science Institute of Latin American Studies; to Liu Weidong at China Academy of Science Institute of Geographic Sciences and Natural Resources Research; to Bi Aonan at the China Academy of Social Science Research Center for Chinese Borderland History and Geography; to Yang Tengyuan at the Inner Mongolia Autonomous Region University; to Ingo Richter, Sabine Berking, and Stefanie Schafer of the Irmgard Coninx Stiftung at the Wissenschaftszentrum Berlin. Without the 'home away from home' that you provided, this work could not have been accomplished.

For funding, it has been an honor to have this research supported by the East Asian Career Development Professorship Award at Boston University, the National Science Foundation Graduate Research Fellowship Program, the University of California, Berkeley Department of Geography, and the Irmgard Coninx Stiftung in Berlin.

To those who provided the critical administrative support to keep this enterprise running, thank you to Elaine Bidianos, Christian Estrella, and Noorjehan Khan at Boston University, and thank you to Marjorie Ensor and Natalia Vonnegut at University of California, Berkeley.

To my excellent research assistants, thank you to Tara Moore at Boston University and Wang Xingchen in Inner Mongolia and Beijing for helping me track down all manner of archival sources. For typing support while I recovered from an injury to both of my arms, thanks to Jenna Hornbuckle and Nick Scheepers. For excellent graphic design work, thank you to Molly Roy. For all manner of technical support, thank you to Nick Bojda.

To my circle of writers, thank you for your companionship and solidarity, especially Noora Lori, Jessica Stern, Manjari Miller, Saida Grundy, Ashley Farmer, Cornel Ban, Kaija Schilde, and Renata Keller. To my dear friends and fellow geographers, for your camaraderie and companionship over the years, especially Shaina Potts, Zoë Friedman-Cohen, Meleiza Figueroa, Annie Shattuck, Mary Whelan, Aharon de Grassi, Anne Bitsch, and Danny Bednar.

To those whose love and hospitality provided a warm respite amid the more intense periods of fieldwork, thank you. To Ward Lynds and Zhang Yazhou, my dear friends of fifteen years in Changchun. To Brendan and Angela Acord in Beijing, my beloved *bon vivants*. To Jeffrey Warner, Yang Weina, and Zhao Qiuwan in Shanghai. To Gustavo Oliveira in Brasília, and to my dear Brazilian family, whom I had the good fortune of encountering in both Brazil and China. our long conversations, celebrations, and laughter sustain me. Renato, Dulcinea, Daniela, Guilherme, Kika, Manuela, Anahi, and most of all, José Renato Peneluppi Jr. To my intellectual godparents, mentors, and dear friends, Joshua Muldavin and Monica Varsanyi. Thank you for lighting my way and sharing so much of your lives with me, from Beijing to New York City.

To my family, thank you for your love, encouragement, and zest for adventure. As long as I can remember, you've told me to go far, be brave, have fun, and do the right thing. I will always do my best to make you proud.

To my dear spouse, life partner, and best friend, Nick Bojda. Thank you for every single conversation, for circling the globe with me, and for not only enduring long absences but cheering me through them. Your love and support mean the world to me.

Portions of chapters 1 and 2 appeared in earlier form in "Historical Geography of Rare Earth Elements. From Discovery to the Atomic Age," *The Extractive Industries and Society* 2, no. 3 (2015): 572–80. Portions of chapters 1 and 3 are reprinted from "The Environment-Security Nexus in Contemporary Rare Earth Politics," in *The Political Economy of Rare Earth Elements. Rising Powers and Technological Change,* edited by Ryan David Kiggins (New York: Macmillan, 2015), 133–55. Thank you for supporting my scholarly work.

Finally, I am immensely grateful for the support provided by my editor, Jim Lance, and the publishing team at Cornell University Press. Thank you for your stewardship over this project.

All translations in the text are my own, as are any remaining errors or shortcomings. Except where indicated in the text, this book uses metric measurements. Large volumes are measured in *tonnes,* the singular of which is a unit of mass equal to 1,000 kilograms. These are not to be confused with the American *ton*, which is a unit of mass equal to 907.2 kilograms, or 2,000 pounds.

RARE EARTH FRONTIERS

WELCOME TO THE RARE EARTH FRONTIER

> **Non-availability means that resource conflict is an immediate threat with negative short- and long-term geostrategic consequences.**
>
> —Rare Earth Elements World Report (June 21, 2012)

> **The problem we face on earth is that beyond their scarcity, these elements are not evenly distributed throughout the world. We need to disrupt this market. By finally being able to reach the Moon and harvest the resources that are there, we can overcome the scarcity of rare earth elements and create the infrastructure necessary for innovation to continue.**
>
> —Naveen Jain, Founder of Moon Express (May 24, 2012)

> **Unfortunately, "strategic metals" are among those perennially misunderstood policy issues with strange lives of their own. The myth of shortage simply refuses to die.**
>
> —Russell Seitz and Jerry Taylor (July 28, 2005)

Rare earths are not rare. Because they were unknown at the time of their discovery—as most things are—they were presumed to be rare. Such faulty thinking would shape the political life of these elements from the moment they were first identified in 1794 until the present.

It is true that rare earths are so thoroughly integrated into our everyday lives that just about everything would grind to a halt without them. They enable both the hardware and the software of contemporary life to be lighter, faster, stronger, and longer ranging. The incredible array of essential applications will be discussed later, but the good news is that rare earth elements are not at all rare on earth. These seventeen chemically similar elements, distinguished by their exceptional magnetic and conductive properties, abound in Earth's crust. The bad news is that minable rare earth deposits coincide with all sorts of other hazardous elements: uranium, thorium, arsenic, fluoride, and other heavy metals.

Yet even this is insufficient to explain our contemporary circumstances, wherein 97 percent of global production concentrated in China in 2010 (see figure 1).

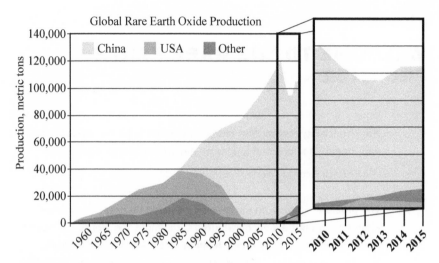

FIGURE 1. Global rare earth oxide production.

Sources: Data compiled from Information Office of the State Council (2010); Orris and Grauch (2013); and United States Geological Survey (2016). Image by Molly Roy.

FIGURE 2. This map indicates research sites visited in 2010–2015, with the exception of the Moon.

Source: Image by Molly Roy.

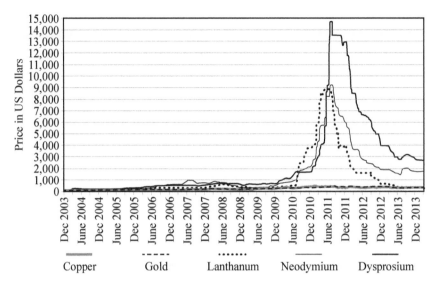

FIGURE 3. Price index USD/kg for selected rare earth oxides, 2003–2013, compared to copper and gold. Prices rebased to 100 for fourth quarter of 2003.

Source: Bartekova (2014). Image by Molly Roy.

Against this situation, unexpected alliances have emerged to attempt mining rare earths in impossible places. The second decade of the twenty-first century saw campaigns to mine rare earths in the most forbidding of frontiers: in ecologically sensitive indigenous lands in the Amazon, in war-torn Afghanistan, in protected areas of Greenland, in the depths of the world's oceans, and even on the Moon.

Currently, rare earths elements are mined and processed in ways that generate tremendous harm to surrounding environs and their inhabitants. In my primary research sites, the abandoned homes and noxious waterways provided visible evidence of the costs of mining for global consumption without regard for local landscapes and lives. Less visible but more profound were the devastating effects on the bodies of people living nearby and downstream of these operations. Cancers, birth defects, and the decomposition of living people's musculoskeletal systems: these constitute an epidemiological crisis affecting some two million people in northern China and many others living former rare earth production sites in Southern California, Malaysia, and Central Asia.

There are readily available alternatives to this devastating state of affairs. Since 2010, some firms have invested in building the industrial capacity to expand more sustainable production practices. Researchers on three continents have launched initiatives to improve recycling techniques. All of this is supported by growing movements of people committed to cleaning up the lifecycles of our everyday technologies. With unprecedented public awareness of these elements and their importance, the time would seem ripe to make our systems of rare earth production and consumption greener, safer, and more reliable.

But we have not yet risen to the occasion. In fact, until relatively recently, few knew what rare earths were or why they mattered. Even fewer demonstrated concern over the tremendous harms generated by contemporary rare earth production practices. Fewer still were those contemplating how we are all implicated and endangered not only by the devastation wrought by their production but also by alarmist reactions to our contemporary situation. In 2010, this began to change.

What happened?

The Situation and the Questions

In late September 2010, China's military blocked a routine shipment of rare earth elements to Japan. What was initially an independent maneuver at a single port facility by the People's Liberation Army in the ongoing tensions between the two countries came to be interpreted by the international community as China flexing its geoeconomic muscle. China's foreign ministry intervened to resume shipments in November 2010 and later denied that such a disruption had taken place. But the rude awakening had already happened (Areddy, Fickling, and Shirouzu 2010; Bradsher 2010; Hur 2010). China then provided over 97 percent of the global supply of rare earth elements on which nearly every industrial country depends, and for which there were no synthetic alternatives.

Although annual global consumption remains at a relatively modest 120,000 tonnes (Castilloux 2014),[1] rare earth elements define modern life. Without them, the technologies on which we rely for global communication, transportation, medicine, and militarism, as well as nuclear, petroleum-based, and renewable energy production would not exist. Sudden supply disruptions had never occurred since the elements had become so thoroughly embedded in contemporary life.

Over a decade prior to the 2010 incident, China's central government began implementing policies to curb rare earth production in response to alarming environmental crises in mining regions, and to enforce export quotas to mitigate against the perceived threat of resource exhaustion (Chen 2010). The first discernible effect of these policies occurred in 2008, marking the first year in which exports decreased relative to the previous year (Zepf 2013). Then between 2008 and 2011, prices increased as much as 2,000 percent for some elements. For example, dysprosium, an element used in commercial lighting, lasers, and hard drives, rose from US$110/kg in 2008 to US$2,031/kg in 2011 (see figure 3). As an indicator of how this affected downstream industry, one study found that the price increases between July and September 2011 reduced the net income of a major hard drive producer in the United States by 37 percent (Monahan 2012). Others reported "a chilling effect" on renewable energy start-ups in the Euro-American world (Bradsher 2011a), while still others claimed that dependence on China for materials used in critical defense applications posed a "national security threat" to the United States and allied countries (Coppel 2011).

Prices began rising in 2008, but not until late 2010 did China's monopoly come to be seen as a global threat, prompting market panic and unleashing waves of speculation, prospecting, and bellicose political discourse across the world (Caramenico 2012; Z. Chen 2011; Fulp 2011). It was no longer invisible within the global economic status quo prevailing since the 1980s. Rare earth mining and processing, and increasingly, the production of critical technological components, followed the trajectory of many global industries as they concentrated in China in search of cheaper labor and fewer environmental regulations. During the latter decades of the twentieth century, the deindustrialization of much of the West intersected with its mirror opposite in China: massive state-directed initiatives of integrated scientific and industrial development in certain strategic sectors. The result of these intersecting processes is the contemporary East Asian dominance in heavy industry and manufacturing that defines the present. But rather than address the issue in a substantive or historically informed manner, Anglophone commentators unleashed a sensationalist maelstrom describing the concentration of rare earth production in China as a "stranglehold" (Evans-Pritchard 2013) that threw the world into "crisis" (Bourzac 2011) and constituted a "threat" (Hannis 2012) to the national security and economic stability of downstream countries (Portales 2011). In such a framing, the actual origins of China's rare earth monopoly were obscured by accusations of conspiracy and geopolitical posturing.

The situation prompted a flurry of dramatic responses and counterstrategies across the globe.[2] For example, since the crisis of 2010, economic officials within the United States, European Union (EU), and Japan acknowledged that their dependence on China's rare earths arose from longer-term shifts in the global division of labor, whereby dirty industry relocated to China and then undersold Western firms to the point of their bankruptcy. Several elected officials publicly advocated for national plans to revive domestic industries in the Americas (Bennett 2010; Clancy and Banner 2012). Although such initiatives would have required significant political and technological capital, the urgency of the period between 2010 and 2013 inspired efforts to restore domestic capacity through rather creative means, as illustrated by the Brazil and US cases examined in chapter 4. But at the same time, the United States, EU, and Japan filed a WTO suit against China's production and export quotas in order to preserve the very global division of labor that had brought about the demise of rare earth mining and processing industries in the West.

The extent of the actual shortage of rare earth oxides in 2010 is debatable. But the very possibility drew together diverse currents circulating in different parts of the world. Growing international anxieties with respect to China's rise, creeping resource nationalism, and frustrated bids for geopolitical power: these collided with the shock of sudden awareness of dependence on China into a perfect storm that drove mining interests into previously protected places across the globe and even beyond. China's contemporary rare earth production dominance—or more precisely, the delayed international response to the central government's decade-old decisions to curb output—impelled the opening of vast new horizons on the global rare earth frontier: stimulating new investments in prospecting and mining activities while renewing struggles over who bears the staggering environmental costs of production. Each of these developments are driven by the strategic value with which these elements are imbued, underscored by the specter of price volatility, perceived global supply shortages, and the stubborn fiction that rare earths are, in fact, rare (Brown 2013; Lima 2012; Ting and Seaman 2013; Wang 2010).

The importance and relative ubiquity of rare earth elements would seem more likely to drive exploitation "closer to home," so to speak, in well-connected regions within major consuming economies. However, this is not the case. Instead, in the race to open up rich new extraction points, less remote, apparently easier to access deposits have been overlooked in favor of the far northwestern Amazon and the Moon.

In 2011, Brazil's Rousseff administration issued a public call to mining firms to evaluate the feasibility of exploiting rare earth deposits on indigenous territory in a sensitive border region of the far northern Amazon, while the National Aeronautics and Space Administration (NASA) and the US Department of Defense partnered with Silicon Valley firms to develop the required technological and legal infrastructure to extract these elements from the Moon. US entities have not been the only ones seeking these elements beyond Earth: China successfully landed the *Jade Rabbit* lunar rover on the Moon on December 14, 2013, with the purpose of gathering scientific data and exploring for minable minerals, including rare earths (Radio 2013; Shefa 2014; Wang 2013).

Perhaps such agitation would make sense if rare earth elements were, in fact, rare. But many are more abundant than copper, as common as lead, and as of late 2015 there were more than 800 known minable land-based deposits[3] on Earth. In the years since the crisis, herculean efforts to open up new production sites have modestly reduced China's share of global production. In 2015, China produced 85 percent of all rare earth elements consumed worldwide. Meanwhile, hundreds of new mining initiatives have failed. But some—such as the far-flung campaigns examined herein—persist in defiance of familiar market logics.

Unfortunately, this hyperbolic behavior has been supported by hyperbolic commentary. A few popular nonfiction books on rare earths emerged in the aftermath of 2010 (Veronese 2015; Wang 2010), and tellingly, even more works of fiction on the same theme have been published (Asher 2015; Besson and Weiner 2016; Bunn 2012; Mason 2012; Sellers 2016). If books—both fiction and nonfiction—on the topic of resource scarcity that make only brief mention of rare earths are included, the list grows prodigiously. Without exception, all emphasize the threat posed by global dependence on China and paint apocalyptic pictures of urgently intensifying geopolitical contest in a fictitious context of disappearing global resources. In describing the putative "race for what's left" (Klare 2013), their objectives are to illustrate rather than deconstruct the status quo as perceived from the more paranoid segments of the English, French, or Chinese speaking world. This has had the result of amplifying bellicose discourses at the expense of opening new lines of inquiry toward more collaborative solutions. By reciting powerful tropes that have emerged in the confusion surrounding rare earths, these works obscure far more than they clarify. None of these works are supported by in-depth site-specific research, much less engage the multiple stakeholders or volumes of primary source materials in languages local to the mining sites. The sole exception is the 2015 *The Elements of Power: Guns, Gadgets, and the Struggle for a Sustainable Future* by David S. Abraham, which describes the processes by which rare earths and other metals travel from various mines to some of our everyday technologies. Although Abraham still relies on the mischaracterization of certain metals as "rare," he insightfully noted: "It's not hyperbole to state that the fate of the planet and our ability to live a sustainable future . . . depends on our understanding and production of rare metals and our avoidance of conflict over them" (2015, xiv). The key point here is that it is not the elements that pose the danger, but how we source and consume them. Given these high stakes, it is unfortunate that so much fiction abounds.[4]

The first step to understanding the politics surrounding rare earth elements is getting the story straight. Rare earths are not rare. Abundant geological and mineralogical research shows that we are nowhere near exhausting Earth of potential mining sites, for rare earths or otherwise. If we are interested in resolving rather than aggravating the contentious politics surrounding rare earth mining and processing, we need to understand why certain places have emerged on the global rare earth frontier in the first place. Despite the abundance of post-2010 books on the topic, fundamental questions remain wholly or partially unanswered: What, precisely, are rare earths, how did they come to be so important, and why, given their relative ubiquity in Earth's crust and overwhelming importance to everyday life, is the geography of their production confined to so few places?

This book answers these questions. It investigates the global geopolitical superstorm raging around rare earths by diving into several of its constitutive cells, from the Mongolian Steppe to the High Amazon to the Moon. This is, first and foremost, a work of geography—literally, "writing the world." Geography is concerned with how particular spaces are produced by people interacting with the environment in specific times and places. In this approach, nothing is a priori or determined, but rather, written into being through a host of interacting, aggregate, and (un)intentional actions unfolding over time. There are no "externalities" in geography: the very word "externality" reflects a way of thinking that does not match reality. As residents in an integrated biophysical Earth system, there is no part of the Earth that is external to our affairs. Pollutants do not respect boundaries, nor do our efforts to acquire the elements essential to contemporary life. In geographical research, the biophysical, historical, political, cultural, and economic dimensions all matter. Lave et al. (2014) put it best with the statement that "specific modes, strategies, and institutions of governance and development interact with stochastic, contingent physical processes to shape the earth; racism, the movement of global capital, and the history of colonialism are as fundamental as the hydrologic cycle, atmospheric circulation, and plate tectonics" in producing the present (7). Such an approach is both explicitly political and deeply attuned to environmental conditions: if we eschew facile determinism, then we can see that much that defines our contemporary world is of our own making. To a far greater extent than we might generally acknowledge, we are responsible for the outcomes of our actions in a world defined by tremendous diversity and possibility.

This is no less true in a socially necessary enterprise such as rare earth mining. It may seem strange that the sourcing of the elements on which contemporary life depends is characterized by unusual geographies of production. It may even seem paradoxical that mining the elements so necessary to our greenest and greatest technologies generates immense environmental and epidemiological devastation. Such a situation is far from natural or inevitable, but how we arrived at the crisis of 2010 and what continues to drive the strange geography of the global rare earth frontier can be historically understood. When we take history and politics into account, we see that the geography of the global rare earth frontier is geologically contingent, rather than geologically determined. Hence the present work examines how the global rare earth frontier, in selected diverse places, is written into being by specific actors, events, and institutions. And how, in these remote places, rare earths sometimes serve as a mere pretext for broader geopolitical and economic struggles.

If we are concerned with sourcing rare earths in a manner that is not only more reliable and less crisis-prone but also socially and environmentally just, then we must first understand what it is that drives production to certain places. Navigating this far-flung terrain requires a grounded approach. Thus the book begins with the first chapter devoted to the deceptively simple question: What are rare earths? In addition to their defining role in contemporary life, rare earths are also heavily mythologized. Myths serve a purpose: propagating narratives, norms, and forms of social control. To unpack rare earth myths in their local forms at several points across the globe and beyond, I examine the local landscapes and transnational histories with which rare earths are literally and figuratively intertwined and from which the myths draw their potency. This book is therefore about much more than the elements. It is about the way they are given meaning, and how those meanings reconfigure space in specific far-flung places. The actual spaces to which these myths are addressed are those over which their propagators seek control. This is the frontier space, defined in the next section.

In light of the indisputable importance of rare earths and the abundance of accessible deposits, the scattered geography of the global rare earth frontier cannot simply be taken for granted, attributed to geological determinism or the "free hand" of the market. Chapters 2 and 3 examine how and why China's Bayan Obo mine in Baotou, Inner Mongolia Autonomous Region emerged as the single greatest source of rare earth elements worldwide, showing that neither China's grand strategy nor Western lassitude provide sufficient explanation for our present global arrangements, as has been repeatedly alleged. Chapter 3 delves into the local environmental and epidemiological problems in northern China that prompted the change in China's political economic priorities from export dominance to conservation. Chapters 4, 5, and 6 follow the ramifications of these shifts as they unfold across several—but certainly not all—new sites on the global rare earth frontier.

The contemporary global geography of rare earth prospecting and mining is of course linked to the story of China's contemporary dominance in rare earth exploitation in particular, as well China's global integration and attendant geopolitical developments more generally. But this is not simply another story of how China might be taking over the world. In order to identify what drives our destructive and conflict-prone practices of sourcing rare earths, this analysis goes deeper to investigate the particular dynamics at work in three types of sites: established, explored, and prospective. The *established* site is China's Bayan Obo mine in Baotou Municipality in Inner Mongolia Autonomous Region, which borders Mongolia to the north and is the source of roughly half of all rare earth elements consumed worldwide. The *explored* but unindustrialized site is São Gabriel da Cachoeira, the northwesternmost municipality of Amazonas state in Brazil, which borders Venezuela to the north and Colombia to the northwest and holds some of the highest concentrations of rare earths identified to date. The *prospective* site is located on the western lunar highlands, on the Moon, which is currently enshrined in robust international treaty regimes as the indivisible patrimony of all humankind. As presently discussed, each of these sites exemplify the (un)making of the frontier by connecting the political economy and political ecology of rare earths to territorial contests preceding and emergent from the 2010 crisis. To support the in-depth analysis of each of these sites, several others are briefly discussed throughout the book (see figure 2).

Although this analysis is global in scope, it is not exhaustive. Rather, it delves into the multilayered significance of rare earths, exposing their roles in everyday life and exploring how they illuminate ongoing territorial struggles in some of Earth's most iconic places. The emergence of Baotou, São Gabriel da Cachoeira, and the Moon in particular as key points along the global rare earth frontier cannot be explained by mere accidents of geology or exercises of economic rationality. This is precisely what makes these sites key to understanding contemporary strategic resource geopolitics: by examining the dynamics that brought these three far-flung places into the global rare earth frontier, we can identify some of the more intractable obstacles to sensible sustainable resource use globally.

Therefore this work is structured around an inquiry into the spatial politics with which specific rare earth deposits are entangled at local, national, and international scales in Bayan Obo, São Gabriel da Cachoeira, and the Moon to explain why these specific sites have emerged on the global rare earth frontier. Spatial politics are concerned with the concrete, material processes of (re)production, power, governance, and everyday life that define human-environment relations within and across specific places.[5] This inquiry is concerned with precisely which practices, ideas, and environmental factors shape our global arrangements of rare earth exploration and production. To wit: Through what processes were the deposits at Bayan Obo transformed into the single greatest contemporary source of rare earth elements worldwide? What local material conditions prompted the shift in China's production priorities from export dominance to becoming a net importer? How has this ramified across the global rare earth frontier? Specifically: Toward what end is the Brazilian government undoing its own indigenous and ecological protection laws to mine São Gabriel da Cachoeira, a historically contested border region shared with Venezuela and Colombia, when there are more easily accessible deposits under production in existing mining sites elsewhere in the country? And why have NASA and the US Department of Defense chosen to partner with Silicon Valley start-ups to mine these elements from the Moon, while the United States throws away hundreds of tonnes of rare earths annually in mine tailings and e-waste?

All of these questions are posed at the intersection of specific local and broader global processes, thus conceived in order to illuminate contemporary rare earth politics the way a lighting strike reveals, for an instant, the shape and order of myriad things whipping about in the darkness and confusion of a superstorm. No storm is caused by the wrath of the gods—we need not be mystified by them. Nor should we be mystified by the state of global rare earth politics. They emerge from discernible and knowable phenomena that, if understood, can inform reasonable action despite perpetual fears of impending crisis or prognostications of "mineral eschatology" (Bardi 2014, 241). Conversely, if storms are left to the stuff of myths, if we take the fear and hyperbole surrounding rare earth elements at face value, then any number of fictions of impending disaster might be leveraged to force people to accept things they would otherwise quite sensibly reject. By demystifying rare earths and laying bare the operations of the global rare earth frontier in specific places, this book takes aim at those who would have us undo hard won progress in environmental and social protection and peaceful international cooperation in the name of impending resource apocalypse.

There are two short answers to the series of questions posed above. The first is that mining these places is about more than rare earths. It is about demonstrating, through high-tech militarized means, the capacity to stake a claim to historically contested and geopolitically significant spaces rather than about the actual practicalities of establishing mining operations. As the history of Bayan Obo illustrates, the ethno-national and geopolitical ends served by establishing mining operations in a hostile and historically contested terrain justified the immense multinational undertaking to build an industrial base on the southern Mongolian steppe. The second answer is perhaps more fundamental: the social and environmental hazards involved in producing rare earths exert an outward (or inward, depending on your perspective) pressure on the placement of rare earth mining and production. The twin desires to isolate the hazards while capturing the geopolitical benefits of keeping the production of these strategically vital elements within a particular set of borders drives production to the frontiers of empire, state, and capital.

This tension explains why the rare earth frontier is found in borderlands and hinterlands, in places where local landscapes and lives are deemed sacrificable in the name of some greater good (Campbell 2000; Hecht 2005; Johnson and Lewis 2007). The "greater good" refers to the utilitarian principle in economic, political, and philosophical discourse that views the best possible outcome as whatever brings the greatest possible benefit for the greatest possible number of people. This principle is often used to justify some measure of harm or sacrifice concentrated somewhere. These places where the toxic enterprises and their ill effects ultimately land are known as "sacrifice zones" because their destruction is considered indispensable to achieving the greater good. Sacrifice zones are where the so-called negative externalities are located. They are not ephemeral or intangible: they have a specific geography that can be mapped. The destruction of landscapes and lives in pursuit of rare earth mining has generally been considered a fair price to pay, generally by those who do not live in the sacrifice zone.

The greater good operates transnationally to temporarily resolve the otherwise impossible tension in which rare earths must be procured by industrialized countries, but for which very few wish to assume the risk of extracting them from their own subsoils or investing in greener production practices. This tension also drives the dynamism of the global division of toxic labor, which is never settled, but resolved only through periodic fixes as toxic industry moves from place to place, seeking out new locales where local landscapes and lives are imagined to be worth less. As shown in subsequent chapters, Euro-American production migrated almost entirely to China by the late 1990s, following a series of extensive— and expensive—environmental disasters at Western production sites. Over the first decade of the twenty-first century, as ecological and epidemiological crises deepened in China's rare earth mining regions, the central government formulated a long-term policy portfolio aimed at shifting the country's position from a net exporter to net importer of basic rare earth commodities. This domestic fix operates by driving the environmental burden of rare earth mining and processing beyond China's borders through multiple trade, investment, and aid partnerships. Efforts to transnationalize China's rare earth hinterland do not end there. Seizing on the inability of the global market to support greener rare earths, private sector firms and military planners across the globe have leveraged scarcity myths to advance a campaign to enclose outer space resources.

Neither the price nor the actual availability of rare earth elements is sufficient to explain such extreme measures. While there is a clear need to isolate and contain the toxic wastes generated by rare earth mining and processing, the strange geography of the global rare earth frontier is driven by the desire to capture the geopolitical benefits of establishing rare earth production in certain places. This is a key feature of our contemporary global arrangement of rare earth production, which must be examined in global perspective so that we might identify workable global solutions. Toward that end, this investigation uncovers shared historical experiences across vast distances that have, until now, been overlooked due to entrenched Orientalist and Cold War-era assumptions that "East" and "West" are mutually unintelligible. The fact that a compelling link between Inner Mongolia and Amazonas, or between extractive frontiers on Earth in outer space, might seem far-fetched or counterintuitive highlights the limits of the received wisdom with which we seek to understand contemporary global issues.

In fact, Bayan Obo, São Gabriel da Cachoeira, and the Moon have much in common. They are each frontiers for the extractive aspirations of states and empires, and they are each current and historical sites of struggle against the imposition of sacrifice or destruction. While their geological endowments provide some logical basis for their potential as mining sites, this is not sufficient explanation for their emergence on the global rare earth frontier since there are so many other less controversial, or at the very least, operational sites around the world. Were the global geographies of rare earth prospecting and mining the mere result of the practical organization of global resource provision, Inner Mongolia, the Amazon, and not to mention the Moon, would be unlikely sites. Instead, they are definitive of contemporary global resource geopolitics, in which geographies of extraction are dictated by racially charged territorial ambitions intertwined with geological and economic circumstance.

In these three instances separated by immense spatial and temporal distance, the production of geological knowledge has been used to advance broader colonial, imperial, national, and private sector projects to control the landscapes and lives under which rare earth deposits are situated. For decades, even centuries, conflicting land use regimes and competing claims over local resources thwarted large-scale mining operations. In each of the sites examined herein, frustrated territorial ambitions leveraged broader shifts in global resource geopolitics to frame mining rare earths as vital to some greater good, whether that is defined as national development, economic security, or all humanity. In short, Baotou, São Gabriel da Cachoeira, and the Moon emerged as key points along the contemporary rare earth frontier in part because they had been frontiers of another sort, and continue to pose a frontier problem to multiple territorial powers.

What Is the Frontier?

We cannot begin to understand our present rare earth situation without critically examining the sorts of spaces in which rare earths are mined: frontiers. Precision is necessary when invoking the term "frontier." As a spatial, temporal, cultural, political, and scientific signifier, the word is used so broadly that it must be carefully defined to serve any useful analytical purpose. There is a sense in which the frontier is an ideation, or used figuratively to convey a sense of civilizational progress, as in the frontiers of research or technology. But even these figurative uses bear implicit spatial politics: the actions taken to reach an imagined or figurative destination are material, and therefore place-based and spatial insofar as they involve specific people and resources. The frontier is also used literally to refer to a place, whether that place is an ambiguous zone or a Cartesian line. Such places are also ideations, but they are ideations to which people give meaning through enforcement mechanisms in specific places. A frontier always refers to a real and imagined place that is specific to times, places, and cultures in which they are invoked. In both literal and figurative senses, frontiers change over time. In this book, the frontier refers to the more or less vaguely specified zones over which multiple actors and institutions compete for control, both over the place and over the extraction of its strategically valued rare earths. Secondarily, the frontier refers to the manners in which such places are described, imagined, and problematized.

These dialectical characteristics of the frontier—literal and figurative, real and imagined, material and meaningful—are co-constituted with the exercise of state and corporate power. Most basically, the frontier implies a limit: the limits of state power and rule of law, of the known and disciplined, and of a set of particular social relations or identities. Therefore the frontier narrative, when invoked, represents a set of spatialized intentions to transform a place that is unknown and ungoverned into the known and disciplined: to penetrate the impenetrable, to transform untapped minerals into wealth and power. For our purposes, the use of the term implies a project to turn the space in question into something else. The desired outcome of that project is to enclose the space containing strategically valued resources. The act of enclosure transforms that space from a frontier beyond the reach of state or corporate power into a hinterland, the (re)productive activities of which are reoriented from sustaining local economies to enriching extralocal actors. Thus frontiers are not objective facts existing in any a priori sense. The frontier is conjured in order to be spectacularly destroyed (Tsing 2005); its environments mythologized in order to be pillaged or policed; its inhabitants exoticized or dehumanized in order to be minoritized or murdered.

A conspicuous feature of this project is the tendency on the part of extralocal actors—states, firms, strategists of all kinds—to view frontiers as zones of legal ambiguity or lawlessness (Evans 2009; Haynes 2014). While this is sometimes the case, it can also be the case that local social relations, property regimes, and governance structures, by virtue of being independent of or contrary to the ambitions cultivated in distant metropoles, are simply ignored by state, corporate, or imperial agents. The reason for this is straightforward: accumulation by extralocal actors cannot occur if those same actors do not possess orchestrative control over the land, property relations, and authoritative institutions local to particular resources.

In mining sites, the contradiction between local livelihoods and extralocal impositions is absolute. Minerals are for the most part located beneath the surface. Large-scale mining operations cannot proceed without annihilating the landscapes and lives atop the deposits, which is an activity of such upheaval that it requires the exercise or invocation of legal exceptionalism in order to proceed. Therefore the frontier represents both a limit and a possibility where the exercise of extraterritorial, extrajudicial, and extraordinary state power is concerned. By expressing a limit, the state, corporate, and military actors can conjure a space in which power can be exercised with fewer restraints. Hence, as the diverse cases in this book show, the frontier narrative reconstructs local identities as underdeveloped, unproductive, or even nonexistent for the purpose of extralocal exploitation.

But there was always a before, a time when local landscapes and lives were not pathologized in a frontier narrative. This is true for the sites comprising the global rare earth frontier, even those beyond the scope of this book. Frontiers are often imagined as empty of (indispensable) people yet full of the particular variety of riches fancied by extralocal actors. For example, to early explorers and planners, Bayan Obo was a "wasteland"; for centuries, surveyors, missionaries, and federal officials described São Gabriel da Cachoeira as "the end of the world"; the Moon has become, to aspiring miners and colonizers, "a treasure trove" that "belongs to no one." But such imaginings are false. Where the sites are not populated with communities, they are filled with collectively held significance. The key is that spaces must be literally or discursively made empty in order to be coherently reimagined as a frontier.

Therefore the frontier project is supported by forcefully enacting a reality that matches the narrative. Erasing or problematizing local territorial orders whose very existence contradicts the frontier narrative becomes the first and ongoing order of business. A territorial order refers to the social organization of the landscapes and lives rooted in a particular place. As people build their lives, they transform the spaces in which they live according to collective social logics: they build and sustain territorial orders such as villages, productive ecosystems, or cities. People do not exist without territorializing space, without making it into something that supports their existence. To (attempt to) erase these deep social logics is not just a matter of "impacting" local populations, but also of transforming the landscape. In critical social theory, this (attempted) erasure is known as deterritorialization (Deleuze and Guattari 1987). What follows is not empty space but a new territorial order. *Re*territorialization is the (attempted) imposition of a new order, which necessarily follows *de*territorialization as two aspects of connected yet spontaneous processes. The preceding territorial order cannot be wholly or partially destroyed without another order taking its place, even if that so-called order is violent disorder.

Once deterritorialized, frontiers can be reconstituted by the state, reterritorialized through settlement campaigns, infrastructure construction, or military fortification. Ambiguous zones can be carved up with borders and zoning ordinances to remake frontiers into the hinterlands from which the "cores" draw their resources (Lefebvre 2009; Pomeranz 1993). The purpose of imposing a border is to assert a sovereign claim over a given space and the resources therein, while the purpose of zoning is to orchestrate local (re)productive activities according to broader governance objectives. Borders represent inward assertions of the physical boundaries of "imagined communities" (Anderson 1982) and outward assertions of geopolitical power. In the case of commercial mining enterprises, a clear border and an exclusive zone are essential to securing investments. Ambiguous zones are the bane of nation building and commercial enterprises alike, therefore imposing borders is a critical component of the frontier project insofar as it represents an effort at greater precision: the frontier may be a vaguely specified zone, but the border is a line. Commercial forays into new spaces rely heavily on the force and backing of the state to maintain conditions amenable to exclusive large-scale extraction. The very absence of borders in "the final frontier" is a central problem for private outer space mining enterprises, which have enlisted the expertise of legal scholars since the mid-1990s in order to formulate a property-rights regime for the privatized exploitation of lunar resources (Al-Rodhan 2012; Lewis 1996).

Because both frontiers and borders tend to be produced through a gaze from elsewhere, their very production is an act of both incorporation and peripheralization. By imposing a border, people and places are simultaneously zoned into (or out of) the national geo-body (Winichakul 1988) and also placed at its periphery. Hence the boundaries between (un)known and (un)governed tend to cleave along racial and cultural lines. The project of concretizing this line in the form of a border generates ambiguities and violence as competing sovereignties are divided or suppressed.

In the northwestern Amazon as well as on the Inner Mongolian Steppe, the history of border-marking and resource extraction has been "written in blood" (Mote 1999; Pinheiro 1995), characterized by struggles between Euro-Brazilian and Han Chinese colonizers on the one hand, and indigenous peoples on the other. As the cases of China and Brazil show, it takes multiple generations of violence and massive in-migration campaigns to inculcate a frontier subjectivity in local inhabitants. The "frontier subjectivity" refers to that profoundly alienated sensibility that the native soil on which one stands is somehow distant or peripheral as well as fundamentally different from the soils nearby, occupied by one's kin, but nevertheless on the other side of the border. This cognitive-cultural sleight demands that local inhabitants reconceptualize their local territorial orders as anachronistic and subjugated to a larger territorial order, defined in the cases of São Gabriel da Cachoeira and Baotou according to the priorities of a distant yet locally proliferated central state.

An ideal frontier subject is sufficiently self-reliant to minimize state obligations to the local population, but sufficiently dependent on the state so as not to directly challenge state sovereignty. Understandably, people tend to resist or redefine the imposition of such a condescending subjectivity. It is atop deep histories of contestation that rare earth reserves have been be framed as part of the national patrimony in China and Brazil, while on the Moon, where assertions of national sovereignty are forbidden, resource exploitation is framed as a necessity for "all humankind" (Moon Express 2013). The invocation of national collective interest is hardly peculiar to rare earth extraction; the justification for sacrificing local landscapes and lives is consistently framed in terms of a "greater good" in extractive and other toxic enterprises (Fox 1999). But as the cases examined herein show, extralocal actors tend to define the greater good according to extralocal interests. Livelihood activities that inhibit state coercion of local groups, such as small scale, informal, or family mining enterprises, are criminalized. Generally, when commercial large-scale mining is framed as essential to the nation, artisanal or clandestine exploitation is framed as a crime against the citizenry (as in China), or as evidence of an ineffective state (as in Brazil).

Like frontiers and borders, the hinterland designation implicates a vantage point from elsewhere: if not a self-proclaimed center, then an urban or commercial space. As populations condense within (sub)urban centers of spectacular consumption, glittering with flat screens, ringing with smartphones, surveilled with evermore sensitive technologies, and ensconced within durable steel-alloy architecture—all of which depend on rare earth elements—the land and resource area needed to sustain high-technology consumption expands. Resource hinterlands are not contiguous features of metropoles: they are defined by their distance, measured in miles and otherness, from those who claim to uphold the center of human civilization. They are scattered across the globe and beyond, driven by that fundamental tension between the need to sequester the hazards of mining and processing as well as the desire to capture potential economic and geopolitical benefits associated with controlling rare earth production.

This tension defines the global rare earth frontier. For each site that could be included in an exhaustive global catalogue, there are stories of enclosure, contestation, and violence as states and firms work to confine the hazards of mining and processing in places deemed sacrificable to the greater good of rare earth production. The case of Australia's Lynas Corporation, Ltd splitting its operations between mining Mount Weld in Australia and ore processing in Kuantan, Malaysia is a case in point.

The Mount Weld deposits were discovered in 1988. In 2001, the twenty-year-old gold mining company, Yilgangi Gold NL, sold off it gold division and rebranded itself as Lynas to focus on developing this particular rare earth deposit following the closure of what had been the last remaining rare earth mine outside of China: the Mountain Pass facility in the United States (discussed in chapters 1 and 4). For several years, Lynas worked on raising capital to develop the deposit.

In 2009, China's State-Owned Non-Ferrous Metal Mining Group offered to purchase a controlling 51.6 percent stake in the company. Australia's Foreign

Investment Review Board blocked the transaction (Bloomberg News 2011). Later that year, the company received approximately $330 million from JP Morgan to keep the company afloat by inducing Australian and New Zealand shareholders to purchase additional shares[6] (Lynas Corporation 2009). In the heat of the late 2010 tensions between China and Japan, Lynas signed an agreement with Sojitz, a Japanese rare earth trading company, to export three thousand tonnes to Japan beginning in late 2011 (Tabuchi 2010). Meanwhile, the company worked to open the offshore Lynas Advanced Materials Processing (LAMP) facility in Gebeng, Kuantan, Malaysia. Former CEO Nicholas Curtis explained the site selection as a matter of good business. The Gebeng Industrial Estate is a Malaysian state-owned initiative designed to attract chemical industries to an area with robust industrial infrastructure, abundant water, and a skilled, lower cost workforce. According to company press releases, locating the processing facilities in Malaysia would be good for the company, and good for Malaysia's economic development.

Local residents and officials disagreed on several counts. First, a twelve-year tax break offered to Lynas meant the Malaysian government would not collect revenues from the operation for the foreseeable future. Second, residents, officials, and observers were alarmed at reports of structural and engineering flaws in the facility (Bradsher 2011b; Butler 2012). Galvanized by the cancers and birth defects suffered by workers and community members at the Japanese-run Bukit Merah Mitsubishi rare earth processing facility in the 1980s (Consumers Association of Penang 2011), an international coalition organized under the banner of *Stop Lynas Save Malaysia* (2014) to halt construction of the LAMP facility. As protests peaked in 2011 and 2014, the Malaysian police imprisoned demonstrators from Malaysia, Australia, and New Zealand (Bradsher 2011b; Davey 2014a, 2014b; Lee 2011).

Despite public outcry, validated in part by findings published in a June 2011 report by the International Atomic Energy Agency (IAEA) concerning inadequate waste disposal and containment measures, the facility became operational in 2012. The company reports that it is meeting all of its environmental and safety requirements (Lynas Corporation 2016), while mining researchers, investors, and officials in Australia and Malaysia characterize public concerns as disproportionate to the actual levels of radioactivity contained in the ores brought to the facility (Ali 2014; Matich 2015). International environmental activists countered that if in fact the materials are so safe, the wastes should be reimported to Australia for value-added processing there. The Australian government refused to accept any responsibility for waste material produced by Lynas' Malaysia plant. Western Australian Minister for Mines and Petroleum, Norman Moore, stated "national legislation stipulates that Australia will not accept responsibility for any waste product produced from offshore processing of resources purchased in Australia such as iron ore, mineral sands, and the rare earths produced by Lynas Corporation" (Sta Maria 2012). The IAEA follow-up report in 2014 found that many of its recommendations presented in 2011 had been wholly or partially satisfied, but recommended greater monitoring and transparency concerning the ecological and public health effects of radioactive effluent in the Balok River and other bodies of water (IAEA 2015).

While public pressure was insufficient to prevent *on*shoring some of the more hazardous aspects of rare earth processing in Malaysia, it has been crucial in drawing attention to the need for greater transparency and accountability in the sourcing of rare earth elements. In 2015, the new CEO of Lynas, Amanda Lacaze, moved to Kuantan to oversee the operation more closely. A celebratory May 2016 celebratory piece in *The Australian* reported that Lacaze's husband currently does local charity work for disabled children in the community while she turns "a company heading for collapse into an efficient, viable business" (quoted in Korporaal 2016, 21). The case of Lynas can be, and has been, analyzed as one of environmental violence enacted by an Australian company on the peoples in the developing country of Malaysia. Opponents to the LAMP facility argued repeatedly that Malaysian lives were as important as Australian lives (ABC Radio Australia 2014). It has also been described as a straightforward cost-savings measure undertaken by former CEO Nicholas Curtis who wished to take advantage of the industrial capacity, cheaper skilled workforce, and access to abundant water and regional markets afforded by the Malaysia location. Whether the driving factors are understood as environmental racism or simply smart business, the central issue has been the spatial allocation of sacrifice and the management of the harms generated by rare earth processing. In this case, differing social and regulatory thresholds scattered the rare earth hinterland between Australia and Malaysia, between a developed and a developing country. During its brief revival in 2012–15, the US Mountain Pass mine resorted to similar measures, discussed in chapter 3.

Where the matter is not primarily the spatial allocation of sacrifice, the geography of rare earth mining and processing is driven to some places and not others by the desire to extend corporate, national, or imperial power. In some cases, building an industrial mining operation serves as a method of territorial control. Rare earth ventures in Afghanistan and Kyrgyzstan illustrate this point. The Ak-Tyuz mine is located in the northern Tien Shan Mountains in Kyrgyzstan and was opened by the USSR in 1942. This was an important source of strategic elements—from antimony to rare earths and uranium—for the Soviet military industrial complex through the late 1970s (Djenchuraev 1999). Construction of the facilities at Ak-Tyuz was part of the larger project of incorporating new territories by developing military-industrial bases in Soviet republics and satellites across inland Eurasia. The broader objectives were to quite literally build a world communist empire directed from Moscow. As detailed in chapter 2, the Baotou facility was also built with extensive Soviet support under the aegis of provisioning the world communist revolution.

In post–Cold War, post-9/11 Afghanistan, the territorial dividends of establishing industrial rare earth mining and processing complexes have caught hold of the imaginations of some planners and geologists among the US-led forces still operating in the country. Chapter 4 discusses the Coalition Forces' promotion of rare earth deposits in Helmand, the Afghani province with the greatest opium production, highest volume of refugee flows, and strongest militant presence, as an "exciting" investment opportunity for an "enterprising company" (Coats 2006; Tucker 2014). The extensive efforts around the globe—from southeast China and India to Afghanistan and the Amazon—to criminalize or eliminate artisanal mining further demonstrates that only certain forms of extraction are welcome on the global rare earth frontier. Aside from the risks faced by miners of direct exposure to the hazards of small scale mining and processing, which states have a history of tolerating for economic and political reasons,[7] the primary issue is that small-scale operations are very difficult for national governments to tax. As in the cases of Brazil, Colombia, and Afghanistan, where revenue streams sometimes support groups with territorial claims contrary to existing governments, small-scale mining is an important source of autonomy.

Each rare earth mining site, past and present, is worthy of its own book. No two sites are identical in history, geology, technology, or in the social organization of extractive labor. The present work is not concerned with cataloguing all the rare earth ventures in the world, but with examining the role of rare earths in nation building, geopolitical contests, and global political economy in world historical perspective. As noted, this is done by examining an established, an emergent, and a prospective mining venture, and inspired by renewed awareness of our dependence on these elements. The broader appreciation of the importance of rare earth elements has not, however, rendered production and consumption more rational or sustainable. This suggests that a more complex and intractable set of spatial politics are at work.

If we look at rare earths devoid of their spatial politics, we see naught but a market mysteriously dominated by China and a few other minor players. If we look at only a few rare earth frontiers devoid of their world-historical contexts, we risk reinforcing certain (post)colonial assumptions that some people and places are simply more appropriate for waste than others. Or, denying such racism, we risk relying on facile geological determinism. Such analyses are not only despatialized and ahistorical, they are also dehumanized. This is why it is important to study the global rare earth frontier in its historical and political complexity: to make sense of where, how, and why diverse actors and institutions across the globe continue to invest in a destructive and crisis-prone system of production to sustain the globally integrated life most of us know today. We must understand where, precisely, the elements of everyday life come from, and what places, people, and struggles are involved in producing this state of affairs.

Therefore further precision is necessary when referring to the *global rare earth frontier* that is the topic of this book, because there is a sense in which rare earth-enabled innovations drive the frontiers of everything: technology, consumerism, surveillance, warfare, postpetroleum possibilities, human exploration of subatomic and outer space, and indeed, contemporary geopolitics, power, and accumulation. The term "global rare earth frontier" could, and should, refer to the multiple sites and situations in which rare earth elements are used; where researchers engage in concerted efforts to force technology beyond current limits of possibility; where ores are processed and to what extents toxic by-products penetrate surface and subsurface environments as industrial run-off; how we incorporate rare earth-bearing products into everyday life in evermore intimate and mundane ways; and where these products land when the larger machine of which they are part is discarded as waste. These questions cry out for further research. This present work is concerned with three primary and several secondary examples from the frontier that is fundamental to all others: the mined and to-be-mined.

The central drivers of this frontier process are evolving sets of geological knowledge about particular rare earth deposits that have been vested with strategic significance. Not just any deposit—there are, after all, hundreds scattered across the globe—but those defined as strategic by state, military, public, and private actors in order to reorganize the surrounding environs.

This lies at the heart of the geopolitical projects driving the geography of the global rare earth frontier, masquerading in context-specific discourses of security or development. Two valances of geopolitics are useful here: conventional and critical. Conventional geopolitics are concerned with the balance of power among states, for which exercising orchestrative control over an internationally recognized territory is a precondition for enforceable sovereignty, and as noted above, necessary for industrial extraction to proceed. For China, Brazil, and other state actors in the contemporary rare earth game, erstwhile desires for (inter)national power and recognition have been pinned to gaining control over a critical share of rare earth production by exploiting a problematic place. Transforming regions that have historically evaded state control into mining and industrial hinterlands serves important geopolitical purposes of disciplining historically autonomous regions and projecting power internationally.

There is, of course, more to the story. While useful, conventional conceptions of geopolitics have limitations: they can fall into the "territorial trap" (Agnew 1994), which occurs when states are taken as primary actors in international affairs, rather than examined as one set of nodes in a network of multiscalar social relations. Such approaches often assume that power is organized in discreet national containers, which erases differing interests within states and ignores common interests across transnational space. Representing nation states as discreet monolithic entities must be treated as such: a representation that, however compelling, is fundamentally different from the complex realities unfolding on the ground. Taking the simplifications of state discourses at their word limits the extent to which conventional geopolitics can help us understand the global rare earth frontier.

The second valence, that of critical geopolitics, is concerned with "examining politics at scales other than that of the nation-state; by challenging the public/private divide at a global scale; and by analyzing the politics of mobility" (Hyndmann 2001, 210). This book examines the state as one of many actors working at multiple scales to mine impossible places. Accordingly, how private enterprises and public oversight mechanisms are differentiated is interrogated rather than taken for granted in each of the sites. The public/private divide is a legal artifice and a historical contrivance whose precise definition varies across time and space. Legally, it is used to orchestrate state jurisdiction over the actions, resources, and (re)productive capacities of people and firms in a selective designation of what is subject to public oversight, and what falls within "a separate private realm free from public power" (Horwitz 1982, 1424). Historically, the public/private divide refers to the violently enforced difference between paid and unpaid labor. Because reproductive work is historically devalued, considered a private matter and gendered female, the public/private divide is gendered (Federici 2004). Where the Westphalian state has consolidated power, this has historically translated into marking men and women respectively as legitimate and illegitimate agents of political economic life. This division of labor in society does not stop at gender. Such a division is also characterized by the imposition of racially coded difference between hegemonic self and subaltern "ethnic" other (Guillamin 1995), with its spatial corollaries of core and periphery, capital and frontier, metropole and hinterland, the "right" and "wrong" side of the border. These raced and gendered divides are manifest in practices of extractivism. "'Ethnic" women engage in much higher numbers in artisanal operations that are more vulnerable to expropriation and criminalization. Conversely, men are over-represented in industrial mining enterprises that proceed at the behest of the state. The manner in which multiple actors invoke rare earth mining and prospecting as a matter of national significance contrary to competing local interests is an object of analysis because it shows how the frontier project works to redefine notions of accountable public rights and unaccountable private matters. The content of public/

private categories shifts over time and space while the medieval fiction of the divide between the two remains instrumental to geographies of extraction, production, profit, and sacrifice.

The lens of critical geopolitics allows us to see that the rare earth frontier cuts literally and figuratively across human bodies. How we organize the risks and benefits of rare earth production across space—and the inevitable measure of destruction that entails—is intricately interwoven with (un)conscious evaluations of the worth of some human lives relative to others. Which people incorporate the sacrifice zone into their bodily tissues is not a matter of random happenstance. These evaluations are riddled with raced and gendered notions of difference, where some landscapes and lives are considered more expendable than others. The intimate politics of sacrifice assume many forms: uprooted communities, the impositions of gendered labor practices at production sites, the shortened lives and miscarried pregnancies that result from prolonged exposure to the air, soil, and water contaminated by rare earth industries.

None of this is to suggest that possessing certain markers of privilege is sufficient to protect an individual from the toxic effects of rare earth production. However, it is notable that in the primary cases examined herein as well as several others beyond the scope of this book, those setting out to conquer the rare earth frontier are vested with multiple forms of power—including relatively privileged race, class, and/or gender power—over those who live atop the deposits or downstream from the processing site. That race and waste coincide, that poorer communities comprised of differentially raced or ethnicized "others" are disproportionately burdened with toxic industries and effluents is well established by hundreds of case studies from across the globe, revealed in research that has reinterpreted the past several centuries of global change (Bryant and Mohai 1992; Dillon and Sze 2016; Harvey 1996; Merchant 2003; Pulido 1996). The distribution of harms generated by rare earth mining and processing is not exempt from broader historical practices that sort labor and sacrifice according to race, gender, and class.

These (un)intentional sorting mechanisms that shape the spatial politics of the rare earth frontier have not been resolved by legal measures. Legislative attempts to mandate greener, nonviolent mineral sourcing have had mixed results at best. Often, mining activities cannot be easily sorted into "legal" or "illegal" categories. There is a complicated coexistence between the two. Generally, the difference depends on whether a particular operation lends itself to taxation or other forms of surplus extraction. Much has been made of campaigns in both Brazil and China to crack down on clandestine mining and the black market mineral trade (Li 2017; Taylor 2016). This is not necessarily a new phenomenon, but rather an episodic censure of customary practices. In truth, mining companies have historically depended on artisanal miners to assume the risks of prospecting and exploration (Hecht and Cockburn 1990). Despite government and NGO initiatives in many parts of the world, contemporary downstream industries in practice do not discriminate between legally and illegally sourced materials—even those from known conflict zones—if the price is right.[8] Therefore the abundance of illegally mined rare earths estimated to be in circulation[9] is hardly novel or peculiar to the rare earth sector.

When considering issues of legality in resource extraction, it is crucial to heed critical legal geographers' call to view law as a set of "lived institutions and relations" in order to examine "how and why law works to perpetuate particular relations of social authority, power, exploitation and oppression" (Chouinard 1994, 415). Legal mining is often synonymous with corporate mining to the exclusion or attempted erasure of small-scale and artisanal mining. Within frontier narratives, artisanal mining is too often marshaled as evidence to demonstrate the putative backwardness or lawlessness of a given place. As a term that is meant to provoke or problematize, the frontier is necessarily a temporary signifier, invoked precisely because it is meant to be transformed, for our purposes, into an enclosed rare earth hinterland from which the rest of the world might draw these essential elements.

The particular geography of rare earth mining and prospecting is not a straightforward matter of meeting global needs. The persistent myth of rare earths' rarity skews their politics and possibilities, reviving bold, even mercenary, mandates to explore far-flung and perhaps dangerous places in pursuit of treasure.[10] Myths of rarity work together with the myths of the frontier: while the myth of rare earth scarcity may legitimate far-out prospecting activities, the frontier signifier works to recast a given space in terms of riches to be captured and a problem to be solved, regardless of what may already be there. Hence rare earths entangle with broader territorial ambitions to conquer all manner of frontiers so that one myth drives the other in service of territorial control and accumulation. Without a historically informed analysis of why certain unlikely sites emerge and persist on the global rare earth frontier when there are hundreds of known deposits, and hundreds of ventures to exploit less remote sites have failed, we cannot formulate meaningful proposals for a more just and sustainable production paradigm.

The Primary Sites

Naming a place a frontier recasts its qualities as problems requiring a certain kind of intervention. Baotou, São Gabriel da Cachoeira, and the Moon have been frontiers many times over, for many purposes, among them the construction of colonial and Cold War empires, the pursuit of collaborative international scientific research, and nation building in the name of development and security. The objectives of these various initiatives have been to rationalize space, to transform the unknown into the known. The rationalization of space is a social project: it is the ongoing project of states and firms to reorganize people and places in ways that serve extralocal governance and accumulation. In the cases examined herein, rare earth prospecting, mining, and processing serve these objectives in their contemporary forms.

Baotou, São Gabriel da Cachoeira, and the Moon respectively illustrate established, explored, and prospective aspects of the global rare earth frontier. Historically speaking, they reveal the historical, contemporary, and future-present of global development and resource geopolitics refracted through rare earth prospecting, mining, and processing. The historical processes through which these sites have been configured into key points along the global rare earth frontier are crucial to understanding contemporary global resource geopolitics.

Baotou was surveyed and territorialized by the contemporary state, and industrialized sixty years ago. By contrast, struggles in São Gabriel da Cachoeira and the Moon are unfolding as I write in anticipation of ramped up and regularized rare earth mining. This means that there were very different qualities and quantities of data available for each of these cases, which is reflected in the organization of the book. What is striking is that despite the enormous distances in time and space among these three frontiers, analogous and linked processes draw them into this common concern.

While the differences among the three sites are many and important, it must be said that the Moon is less unlike Baotou or Sao Gabriel de Cachoeira than one might expect. From the point of view of rare earth mining, at least, these distinctions primarily concern the technology required to bring the Moon within the purview of material engagement[11] and to reduce transport costs.[12] A further distinction is that the Moon is entirely unpopulated and is governed solely by international legal conventions, which prohibit claims of sovereignty or acts of enclosure. In this way, the Moon is similar to the deep seabed and the Arctic Circle, both of which figure significantly in global resource and territorial geopolitics (Macdonald 2007).

But from the point of view of rare earth extractors and attendant prospectors, territorial agents, and financial speculators, the commonalities between the Moon, Baotou, and São Gabriel da Cachoeira exceed the differences. No single mining site is exempt from the fundamental contradiction between extracting subterranean ores and preserving the landscapes and livelihoods unfolding on the surface. Even those sites inhospitable to humans such as the Moon or the ocean floor are imbued with competing valuations of the same space, whether it is scientific research or biodiversity and heritage conservation. This complicates discourses that describe the frontier as "empty." In many, but not all, populated sites, national and international norms now compel industry actors to exercise some form of social and environmental responsibility. This often means consulting "impacted" communities to define the terms of compensation for the obliteration of the lands on which their livelihoods depend, and/or seeking a "social license to operate" from those impacted communities. Although these forms of engagement are important, they remain problematic insofar as they are often a rehearsal rather than a deliberative forum, generally conducted by paid subordinates of the mining company whose job is to secure a minimal baseline of assent rather than determine whether the project should proceed at all.

As a multibillion-dollar industry, the move to drive mining off-Earth has gathered scientific and moral momentum precisely because of the assumption that there are no social or environmental hurdles to space mining activities. Such ventures are endorsed by ethical and scientific leaders such as Neil deGrasse Tyson[13] and the Vatican Astronomer Brother Guy J. Consalmango. These developments demand that we look beyond our Earthly provincialisms that anachronously place the Moon and near-Earth outer space beyond the purview of global political economy and hence, of critical concern.

Many aspects of the frontier project that drove rare earth prospecting interests to Bayan Obo and São Gabriel da Cachoeira are operative with respect to the Moon as well. The technology and infrastructure requirements are perhaps the most apparent example of the temporal and technological relationality among these sites that emerge over time with changes in the global division of toxic labor. Bayan Obo was opened by rail and unpaved road built by mobilized migrant labor in post–World War II and postrevolution China. For mining to be feasible in São Gabriel da Cachoeira, reliable road or rail transport would have to be built, which has proven impossible over sixty years of campaigns to develop and integrate the Amazon into regional economies. For mining to be feasible on the Moon, transport costs would have to be dramatically reduced in order to bring elements back to Earth from outer space, so space-faring states as well as private firms are developing multiple strategies to creatively work around these costly logistical hurdles.

In each of the cases, the logistical challenges leading up to rare earth extraction require significant capital, expertise, and political will to overcome. The transport technology situations are indicative of the broader world-historical contexts in which these sites emerged on the global rare earth frontier. In other words, the contemporary global prominence of Bayan Obo is one outcome of postrevolution, postreform, and neoliberal development between China and the West since 1950. The contest over the future of exploitation of São Gabriel da Cachoeira is one aspect of shifts in the global division of toxic labor intertwined with Brazilian expansionism and geopolitical ambition as a major BRICS (Brazil, Russia, India, China, and South Africa) power, against which indigenous peoples seek to both maintain and modify statutory protections won in the last three decades. The latest race to the Moon is an outcome of four decades of neoliberal policies resulting in an immense concentration of wealth in the hands of an extreme minority,[14] the subcontracting of state research and military operations to the private sector, and the resurgent geopolitical anxieties provoked by perceptions of intensifying resource scarcity dramatized by China's rare earth production and export controls.

In addition to these world-historical relationalities, there are several material and meaningful processes defining the sites respectively and collectively. Here, I highlight five specific points of comparison, selected precisely because they are so often framed in exceptional terms in the histories of each place.

First, as explored in-depth in chapter 1, these three sites share notable geological similarities. Elements of the crisis and the broader incidence of rare earth deposits may be fictive, but the basic dynamics of their geological formation are not. Rare earth deposits are formed during repeated cycles of small degrees of partial heating and cooling within Earth's mantle. This gradual process produces alkaline magmas which, when they ascend and cool in Earth's crust undergo further changes stimulated by local temperature, pressure, and chemical variations (Long 2010). Too much melting, or too rapid cooling, and rare earth deposits fail to coalesce. The deposits of both São Gabriel da Cachoeira and Baotou are situated along accretionary orogenic belts between major intracontinental cratons (Lujan and Armbruster 2011; Nutman, Windley, and Xiao 2007), which formed after intense periods of the subduction of the Mongolian and Guianan oceanic plates beneath the northern cratons of their respective continents (Chao et al. 1997; Voiçu, Bardoux, and Stevenson 2001).

On the Moon, those gradual cycles of heating and cooling occurred under a very different set of conditions, which are hypothesized to have formed the KREEP[15] deposits (Shervais and McGee 1999). It is believed that the Moon formed after a Mars-sized object smashed into Earth and broke off debris that eventually consolidated into the Moon we recognize today. The power of the collision liquefied much of the debris tossed into space, which formed lunar magma. The lower density of the Moon left an ocean of this magma trapped between the mantle and the crust, which cooled very gradually, thereby enabling high concentrations of rare earths to coalesce (Heiken, Vaniman, and French 1991). Although the characterization of rare earth deposits in new sites along the global rare earth frontier tends to be prone to hyperbole, there is, nevertheless, a geological basis for the emergence of these three particular sites.

Second, deposits alone are not reason enough for some places to be mined and others not. The existence of rare earth deposits does not, in itself, explain why a particular place emerges on the global rare earth frontier. In each of the sites, deposits were valorized into coherent bodies of geological knowledge. That knowledge circulated in a way that enabled political and economic actors to envision mining operations in these particular places as a means to achieve broader territorial objectives. Professional prospecting in remote regions is a high-cost and high-risk endeavor, often requiring the backing of the state, military, or major research institutions to catalogue the attributes of vertical territory. In all cases examined in this book, the discovery of rare earths generated little immediate interest within the larger territorial project of which geological surveying was one crucial component.

The timeline of exploration tells a story of successive territorial and scientific ambitions to conquer new frontiers. Baotou, São Gabriel da Cachoeira, and the Moon were sites of early twentieth century international geological prospecting as multiple states and empires sought knowledge, wealth, and territorial control. Dutch, Swiss, Japanese, and Soviet geologists visited Baotou, but the Chinese geologist Ding Daoheng is credited with discovering the iron reserves at Bayan Obo in 1927 and publishing his results in 1933. In 1935, a team of chemists discovered that the iron was associated with rare earths, which at that point played a marginal role in technological applications. Due to the Japanese invasion and the civil war, Bayan Obo remained undeveloped until the 1950s, although the Japanese imperial forces conducted some exploratory work in this area (Ding 1933; Ma 1995). In 1972, the Brazilian Geological Service identified sizable niobium-coltan-REE deposits at Morro dos Seis Lagos in São Gabriel da Cachoeira, which some analysts claim are the largest in the world (Gomes, Ruberti, and Morbidelli 1990; Orris and Grauch 2013). This deposit was further explored and mapped in 1975 by the Brazilian Federal Mineral Resources Research Company (Companhia de Pesquisa de Recursos Minerais) (Cuadros Justo and de Souza 1986). Other deposits in the region identified by indigenous and small-scale miners are located in alluvial deposits close to the surface, requiring minimal blasting which some claim would make production relatively cheaper provided infrastructural constraints could be overcome (Jacobi 2009). On the Moon, rare earth elements were identified in samples from the western lunar highlands brought back from the Apollo 12 and 14 missions and the Soviet Luna 16, 20, and 24 missions (Shervais and McGee 1999; Zak 2013). Lunar samples are legally available for scientific research to all countries (United Nations 1967). These rocks have been examined by international geologists visiting NASA and affiliated planetary science institutes, the Russian Space Agency, the European Space Agency, and the China National Space Administration.

Initial international geological prospecting proceeded in these sites under broad mandates of scientific knowledge acquisition in the "peripheries" for global centers of power (Braun 2000). The precise geographical locations of "centers" and "peripheries" change over time. In the early twentieth century, these centers of power and calculation relative to Inner Mongolia were located in colonial European, imperial Japanese and Soviet states, and only toward the mid-twentieth century in Beijing. In the mid-twentieth century, the centers of calculation and power relative to the Moon were located in the space agencies of the Soviet Union and the United States. But this has shifted to include start-ups in the San Francisco Bay Area, and the dozens of space agencies and hundreds of contractors across the globe pursuing lunar mining. In the latter quarter of the twentieth century, the centers of power and calculation relative to the northwestern Amazon were located in Brasília, Manaus, and Washington DC, and since the turn of the millennium, China. Put another way, the expansion of rare earth technology and awareness of their potential economic and geopolitical importance revived dormant bodies of geological knowledge in global metropoles. This has reanimated the frontier imaginary among planners, financiers, inventors, and politicians who imagine personal and national prosperity built on exploiting rare earths in these "peripheries."

This is how frontiers are produced by extralocal powers reaching across immense distances to configure the space in question into something else (Tsing 2005). It is in the metropoles that political, economic, scientific, and military actors jockey, in oft-unexamined abstractions, to transform the unknown into the known. The raison d'être, in all cases, is the pursuit, production, and differentially controlled circulation of geological knowledge. Geological knowledge that is strategically valued imbues the territory in question with sense of verticality. This vertical sense of territory assigns different values to different layers of substrata. Under these conditions, it is possible to value subterranean resources in such a manner that devalues the lives and norms that define the surface. Each chapter illustrates how the history of rare earths mining is also a history of contestation over the value of landscapes and lives relative to the elements underfoot.

Which geological knowledge is recognized as legitimate and who is empowered to wield it is contested. In China, much geological data is treated as a state secret, a practice informed by a living memory of foreign occupation. With respect to the Moon, selenological knowledge has constituted part of the patrimony of all humankind since 1967. Any data gathered from lunar missions must be made available to any researcher regardless of nationality. In Brazil, there is a dynamic and at times violent interplay between local indigenous knowledge and federally recognized geological data used to grant mining concessions. In China and Brazil, state geological knowledge coexists and draws from local vernacular knowledge held by artisanal and black market miners. Far from objective, apolitical catalogues of vertical territory, the collection and circulation of geological data is crucial to multiple competing territorial orders. This is especially so when geological knowledge is mobilized to justify the enclosure and sacrifice of other peoples' lands.

Third, all three sites are characterized by partial, overlapping, or contested legal regimes. In Inner Mongolia, mining activities intersect with economic development, minority protection, and conservation laws. The first concerns the strategy to "Open Up the West" and to better utilize the autonomous region's mineral resources in service of the national mandate of "All-Around Scientific Development" (He 2009). The second requires that local pastoralists' livelihood interests be protected, meaning that mining cannot proceed without consent from and compensation for impacted pastoralists. Enforcement lies with both the local bureau of land and resources, and the head of the village collective, which often have conflicting interests. The third concerns a set of environmental protection laws containing quotas on the proportion of grassland that can be used for purposes other than conservation or grazing. Coordination or harmonization of these laws in practice has been partial at best; local, autonomous region and national leaders concede that economic resource needs (i.e., mining)[16] supersede other concerns. The contradiction is material and immediate. Inner Mongolia's mineral wealth lies beneath the grassland resources fundamental to pastoralist livelihoods. These overlapping laws result in periodic violence between local leaders, mining bosses and their employees, and local pastoralists (Agence-France Presse 2015; Jacobs 2011; Reuters 2011). But because rare earth elements are classified as strategic national resources, the Bayan Obo mining district has been re-zoned to legally erase possible conflicts: officially, there are no farmers or pastoralists in Bayan Obo. This is of course not true—during fieldwork in 2013 I noted at least thirty agropastoral homesteads in the vicinity of the mine. But there is no longer a local census category for farmers. Land is zoned according to three uses: mining, urbanization, or wind energy generation. Livelihoods that do not fit within these zoning codes—whether agropastoralism or artisanal mining—are not counted and therefore exist in precarious relation to the local state.

In Brazil, ecological and indigenous legal protections overlap with Amazonian border security laws, which designate all land within sixty kilometers of the national border to be part of the Calha Norte (northern trench) under the exclusive domain of Brazil's military. The deposits at Morro dos Seis Lagos lie sixty kilometers south from the borders of Colombia and Venezuela within two national parks. One, Neblina Peak National Park, was established by executive decree in 1979, by the military-appointed President João Figueiredo following failed migration and construction campaigns. The second, Morro de Seis Lagos State Biological Reserve, was established by the state of Amazonas in 1990 as part a land management and zoning plan, which ostensibly provided employment and land use entitlements for indigenous inhabitants. In 1992, twenty-three indigenous groups were granted qualified sovereignty in their respective indigenous Lands created by an executive decree of President Fernando Collor de Mello. No one, not even the military, is permitted to enter indigenous lands without permission. Yet no one but the military is permitted in the Calha Norte, which overlaps with indigenous lands where periodic small-scale mining takes place. Therefore in remote Amazonian border regions, two groups are legally granted exclusive access and are also legally excluded from the lands in question. In light of the 2011 presidential invitation to open up Brazil's Amazonian reserves as well as grassroots efforts to decriminalize small-scale mining of rare earths, coltan, and gold, long-standing campaigns to rescind the moratorium on mining in indigenous lands were revived by state, corporate, and indigenous actors (Jucá 1996). This means that in the absence of a single definitive and enforceable contemporary legal convention, multiple actors engage in semi-legal prospecting, extraction, and export.

The Moon is governed by two international legal conventions, against which, to date, the United States and Luxembourg have passed legislation to support private property claims. The 1984 Agreement Governing the Activities of States on the Moon and Other Celestial Bodies, also known as the Moon Treaty, bans any ownership of extraterrestrial property by any entity. It requires that an international regime and protocols be established "to govern the exploitation of natural resources of the Moon as such exploitation is about to become feasible" (United Nations 1984, Art. 11, Para. 5). Preceding this is the 1967 United Nations Treaty on Principles Governing the Activities of States in the Exploration and Use of Outer Space, including the Moon and Other Celestial Bodies (OST). The OST frames outer space as the patrimony of all humankind and prohibits assertions of national sovereignty. However, designating the Moon as the patrimony of all humankind effectively reconfigures it as a global commons, which makes it vulnerable to processes of appropriation and enclosure that have characterized the fates of commons elsewhere (Beery 2011). There are two primary interpretations regarding the applicability of the prohibitions against national appropriation to private enterprise. The first argues that limits on national sovereignty do not apply to private enterprise. The second argues that private appropriation must be backed by the state in any case, so private appropriation is state appropriation after all, and therefore prohibited (Carswell 2002). Private sector actors have mounted a regulatory offensive, propagating the fiction that no legal conventions govern the Moon. It is instead a "wild west" (Klotz 2015) that demands US government intervention to protect the rights of US citizens to own and exploit lunar territory against the threat of a "Chinese colony" (Bigelow quoted in Moskowitz 2011). In November 2015,

US president Barack Obama signed the Spurring Private Aerospace Competitiveness and Entrepreneurship Act of 2015, which recognizes the private property rights of US citizens to outer space resources.

In the three cases examined herein, as well as several others beyond the scope of this book, rare earth mining has required a regulatory offensive on the part of pro-mining interests in order to transform laws otherwise established to protect local societies and environments. This is what a frontier narrative does: it conjures a space of regulatory chaos in order to reconfigure laws in a manner that is hospitable to extralocal mining interests. Each of the sites illustrates a different form of this process. In Bayan Obo, an entire census category and zoning code was eliminated in order to erase competing local claims. In Brazil, the question is not whether mining should proceed at all. Rather, the question is which sort of mining regime—extralocal or indigenous—will succeed in having legal changes made according to its terms. With respect to the Moon, the regulatory offensive aims to create the conditions of possibility for private, state-backed enclosure. This would effectively undo robust treaty regimes inhospitable to private sector accumulation.

In all cases, the regulatory offensive is preceded by a discursive offensive, which brings us to the fourth point of comparison. Baotou, São Gabriel da Cachoeira, and the Moon have been discursively cast as "sacrifice zones" whose destruction—while perhaps contested or publicly lamented—has been promoted as necessary for shoring up context-specific notions of military, territorial, or economic security in the name of a greater good. For example, in central government discourse, Bayan Obo is "the rare earth capital of the world," and "a model strategic resource development site" crucial to China's rise (Baogang Xitu 2013). This status is built on tremendous sacrifice. Arsenic, fluoride, thorium, and heavy metals contaminate surrounding soil and water, poisoning livestock and people living nearby and far downstream because of inadequate pollution control measures. For over four decades, this mode of rare earth production suited China's broader development goals and critically co-constituted the global division of toxic labor under neoliberalism—until the twenty-first century. As detailed in chapter 3, China's central government then began to take steps to control production in the face of growing local research and activism that illuminated an epochal threat to one of the key water sources for north and northeastern China. Prior to late 2010, production practices in Baotou and Bayan Obo scarcely caught international attention, with the exception of some investigative journalists (Hilsum 2009) and antirenewable energy lobbyists seeking to expose "clean energy's dirty little secret" (Margonelli 2009). Following the crisis, Anglophone discourse "racially coded" (Chen 2011) toxic mining practices as both a "Chinese" problem as well as a "dirty trick" played by China's central government in order to achieve dominance in rare earth production.

In São Gabriel da Cachoeira, actors within the federal government have been working with corporate mining interests to renege the moratoriums on mining in indigenous lands in order to "develop the world's largest strategic minerals deposit" and "improve Brazil's position in strategic resource markets."[17] Because of the known hazards of rare earth mining, as well as the extensive history of sovereign anxieties with respect to the region, planners in the military and federal government see large-scale mineral exploitation as a way to kill two birds with one stone. After all, infrastructure construction, industrialization, and resource extraction proved effective to industrialize and rationalize China's northwestern frontier. As the reasoning goes in corridors of power, many lives were lost, great sacrifices were made, but China modernized into a superpower, so Brazil should follow the same process. Planners and policymakers in Brazil have hoped to formulate a version of China's northwestern development strategies in order to integrate and capitalize on the resources of the Amazon.[18]

Private space mining entrepreneurs and their legal advocates idealize the Moon as a consequence-free terrain for resource exploitation ostensibly "for the benefit of all humanity," (Moon Express 2013). In their view, space mining must happen in order to rescue the human race, which is otherwise condemned to unavoidable destruction of Earthly resources. In policy, industry, and academic literature, the push to develop extraglobal extraction technology is legitimated according to the apocalyptic imaginary of Earth as planetary sacrifice zone, where neo-Malthusian nightmares prevail as we succumb to scarcity-induced civilizational collapse (Autry 2011; Dolman 2002; Dudley-Flores and Gangale 2013; Guner 2004). Among these actors, the greatest appeal of the Moon is the likelihood of near total automation of mining activities, and the lack of requirements for environmental and social impact assessments. The appeal of a putatively consequence-free terrain for extractivist adventure is charged with neo-Cold War fears over which civilization will be rescued by off-Earth mining: that of China, Russia, or the West?[19] In this view, whichever firm or country first transforms the off-Earth frontier into an extraterrestrial hinterland is a matter of central geopolitical importance. The questions of whether and on what terms outer space should be mined are subsumed by hawkish preoccupations with who will win the race to enclose lunar space and beyond. As detailed in chapter 6, these ideations have spawned a multipronged effort to devalue the Moon, to cast it as ungoverned and inessential to life on Earth.

In all cases, nationalist and geopolitical discourses surrounding rare earth elements argue that destructive mining must be permitted because rare earths are essential to the infrastructure and hardware of global modernity. As shown in subsequent chapters, who is subject to the devastation and who is spared is arbitrated in part by raced and gendered notions of difference in which some landscapes and lives are designated as more sacrificable than others. This defini-tion of the greater good sets up a false notion that terrible health and environ-mental devastation are the unavoidable price to pay for sourcing rare earths. Abundant research on recycling and flex mining shows otherwise (Binnemans et al. 2013; Knapp 2016; Verrax 2015), suggesting that it is not for lack of more just and sustainable alternatives that rare earth sourcing comes at such a high social and environmental cost. Rather, imposing sacrifice through the hazards of mining in specific places serves economic and geopolitical ends. Narrow eco-nomic ends are served by dodging the required investments to clean up the life-cycles of rare earths and rare earth-bearing products. Disciplining or capturing historically elusive territory serves geopolitical ends. These ends constitute much of the so-called greater good served by rare earth prospecting and extraction. The real (in the sense of material goods produced) and the imagined (in the sense of points scored in global geopolitical and economic contests) are inextricably inter-twined in the production of the rare earth frontier.

Finally, in all three sites rare earth exploitation, in itself, is an absurd prop-osition that defies straightforward market logic. Rather, it is from the geopo-litical perspective—both conventional and critical—that the importance of these particular sites to the global rare earth frontier begins to make any sort of sense. Geopolitics is the optic through which states organize territorial politics. The strategic significance of these particular commodities gives "exploration and development . . . a praetorian cast," producing "a frontier of violent accumula-tion working hand in hand with militarism and empire" (Watts 2012, 438). Lo-cal resistance to the privations preceding and resulting from large-scale rare earth mining generates, in military and firm parlance, security problems. At the proj-ect sites, security problems are dealt with by the "securitization" of space at fur-ther expense of local autonomy. In other words, resistance to sacrifice tends to be met with the imposition of still greater regimes sacrifice in order to maintain or advance a territorial order defined by extractive geopolitics.

The myths of the rarity of the elements with which these sites are endowed am-plify their already privileged position in nationalist dreams of territory. These rare earth frontiers are imbued with mythic significance in dominant national imaginaries insofar as they are framed as exotic, undeveloped, or inalienable parts of the national patrimony nevertheless at risk of expropriation by foreign actors. These dynamics which so powerfully define Baotou and São Gabriel da Cachoeira in their respective national polities also illuminate the sorts of spaces that have been sought for reconfiguration into a resource hinterland: specifically marginal, con-tested spaces. While the putative marginalities of these places both precedes and lends itself to the reconfiguration of these spaces into sites of extraction on the global rare earth frontier, such marginality is not given a priori. As subsequent chapters show, these frontiers and their marginalities were produced through hotly contested territorial exercises of imperial, colonial, and state power over time. No place is marginal unless it is forcefully produced as such; marginality implicates the gaze of extralocal power. Their imagined status as marginal does two kinds of work: it legitimates their framing as sacrificable while simultaneously recasting them as vulnerable points that need to be secured for the sake of national sovereignty or geopolitical power.

Indeed, "sacrifice—of people, places and things—is part and parcel to security geopolitics" (Brownlow and Perkins 2014). Security geopolitics are multivalent on the rare earth frontier. First, the space of exploitation itself is geopolitically significant: Baotou and São Gabriel da Cachoeira are in historically contested border regions; the Moon represents "the ultimate military high ground" (Dolman 2002, 151) and the "most compelling" piece of off-Earth real estate that is "under threat" of Chinese colonization (Faust 2011). Second, security geopolitics are reflected in a number of trade, policy or production measures formulated in response to the threat of rare earth scarcity and the presumed military, technological, and economic vulnerability that would result.[20] Third, when historically contested resource extraction is at issue, security geopolitics encompass not just the land in question, but the infrastructure, transport, and processing spaces as well as local labor and reproductive regimes. Each must be controlled against outside threats, real or imagined, whoever the outsider may be. In the case of the Moon, this is illustrated by the agitation of aspiring space miners to capture legal rights of exclusivity to their as-yet hypothetical mining claims. In the case of the Bayan Obo mining district within Baotou municipality, this is manifest in the decades-old prohibition against foreigners in the vicinity of the mine. In São Gabriel da Cachoeira, indigenous peoples remain outsiders on their own lands and within the nation of Brazil so long as outside firms receive mining licenses while indigenous peoples are prohibited from any economically motivated extraction on their own lands.

Because Baotou, São Gabriel da Cachoeira, and the Moon are geostrategically desirable frontiers, resource exploitation is also a means of territorialization. This is evident in the terms in which claims to the resources are framed. Marshaling capital, rewriting laws, and imposing regimes of enclosure-based social control to open up and develop these places are considered necessary in hegemonic political, economic, and technological progress narratives, which I examine in subsequent chapters. These visions, though broad and sometimes incoherent, are specific in their designations of who bears the mandate to execute the geopolitical agenda— at whose expense—within the broader push to conquer the rare earth frontier.

The Research

Rare earths are at the forefront of many global changes. The geography of their extraction both reflects and results in significant global shifts. Understanding these changes requires research that is grounded and global in scope (see figure 2). Rather than attempt to squeeze the local and global complexities of the questions at hand into a set of "impacts" and "outcomes," geography embraces the material and meaningful complexity characterizing a given issue, and examines the unfolding relationship between humans and the environment as a dialectic. This means not only focusing on the mutually influencing interactions among multiple competing actors and interests, but also interrogating the surrounding discourses, or what is taken to be common sense about the issues at hand.

To unpack the strange geography of the global rare earth frontier, I analyzed literatures, archives, expert interviews and ethnographic data across the Anglophone, Sinophone, and Lusophone world, which I gathered in China, Brazil, Germany, Australia, and the United States between 2010 and 2015. The multilingual approach is central to this project, not least because multiple forms of knowledge are evident in the discourses on rare earth elements within the English-speaking world. More still are present in the Portuguese and Chinese spheres. There is a symmetrical inaccessibility to the working rationales across language barriers even as these rationales interpenetrate to shape thought and action across global space. This symmetrical inaccessibility has been deeply reinforced by colonial and Cold War epistemological divisions of the world into mutually unintelligible "areas." The result is that knowledge about rare earths has tended to be formulated reactively during troubled times, and is therefore rife with generalizations amenable to the politically expedient discourses of the given context: among lobbyists and entrepreneurs in Washington, DC, for instance; in the state-run newspapers of Inner Mongolia, or among competing policymakers in Brasília.

Discourses have power. As systems of representation—even when the representations differ wildly from reality—they carry tremendous political power because they shape thought and action in consequential ways (Hall 1992). In Anglophone, Sinophone, and Lusophone discourses, I examined how multiple sets of knowledge from diverse actors and regions entangled with the ongoing production of the global rare earth frontier. As our contemporary global rare earth situation demonstrates, discourse and practice need not be precise or coherent to get results. In fact, fictions abound on the rare earth frontier. Black market and clandestine mining provide a sizable proportion of the oxides in circulation, and each new deposit is fabled to be the largest in the world. Critical natural resources, prospectors' dreams, and nationalist passions generate a politically potent mix of developmentalist mandates that need not ever be attainable in any basic, grounded sense in order to alter the fortunes of empires, to erase a human landscape, or to perpetuate a war, especially when there is money to be made and territory to be claimed. Furthermore, the complex contingencies of local and global factors—which includes environmental constraints, prospector's tales, citizen activism, investment vicissitudes, and changing prices and political sentiments crucial to the global rare earth economy—mean that development and industrialization agendas hardly proceed as planned. It is a mistake to assume that marshaling the capital, technology, labor, and political will to capitalize on the 2010 crisis and actually opening a new site on the rare earth frontier are the same thing, as it is likewise a mistake to assume mandated production improvements, particularly those concerned with technological fixes to environmental and labor hazards (Shaiken 1986) actually manifest in practice.[21] As this research shows, *which* deposits are identified as promising and *where* rare earths are mined has less to do with an allegedly exceptional geological incidence than with a host of other agendas and negotiations, which are simultaneously distinct to their own contexts and related across the multiple sites examined herein. Examining how these distinct histories intersect to form our contemporary rare earth situation brings us closer to a truly global understanding of the issues.

By taking such an approach, I contend that the processes shaping Baotou, São Gabriel da Cachoeira, and the Moon: first, had something to do with each other; second, could be understood in critical comparative perspective; and third, that such a perspective raises important new questions about global arrangements of power and production while clarifying the epistemological foundations of the old arrangements received from the twentieth century. Each of the three sites have been selected to provide a window into the past, present, and prospective instances where rare earth prospecting, mining, and processing served as commodity-based means to the territorial and geopolitical ends of conjuring and taming frontier spaces. To examine the histories and the relations among these sites within the framework of global political economy, I adopted an encompassing comparative approach that considers each site in their transnational historical contexts. Rather than search for equivalent parts in each site or choose one site to uphold as the "norm" against which to measure all other sites,[22] I identified analogous yet distinct processes of territoriality, subject formation, and nation building unfolding in the dialectical production of several key frontier spaces.

My methodological toolkit for researching these processes included analyzing multiple actor-networks, conducting semi-structured serial interviews with a diversity of actors at multiple scales over several years, and completing archival research in China, the United States, Brazil, and Germany. The purpose of my multiscalar inquiries was to understand whether and how rare earth elements intersected with site-specific economic, cultural, political, development, and security issues, as understood by actors in multiple institutions and individuals with multiple stakes in these issues. This meant selecting interviewees who were engaged in rare earth prospecting, production, or trade, as well as selecting those who worked specifically on broader development or security issues in or in relation to the primary sites. In practice, it meant conducting hundreds of meetings in local languages over the years. The purpose of the archival research was to identify historical records of exploration of the sites. In these records, I traced the genealogy of the frontier narrative and placed the role of geological exploration within such histories in order to identify the circumstances under which rare earth elements became strategically valued.

This research was built around immersion in the sites where such frontier spaces are produced. While this includes the primary sites (with the exception of the Moon, of course) it also includes the loci of extralocal power relative to each of the sites. These were located in offices in state and national capitals, in the private cars of powerful officials, in the penthouse bars frequented by billionaire investors, in the heady gatherings of aspiring space miners, in the frontier outposts of police and military, in the modest homes of the laborers literally building these places into being, and in the settlements and shanties of displaced people and clandestine miners. I was able to move across such spaces because of my fluency in multiple languages, facilitated by existing networks built over a decade of life and work in Brazil, China, and the San Francisco Bay Area. But experience and networks are not enough. To execute a project of this scope, the moral and material backing of globally renowned academic institutions and the financial support of the National Science Foundation were crucial.

The sole mandates of my academic and funding institutions were to adhere to ethical codes of conduct and to share my findings with diverse audiences in an ongoing manner to broaden the positive potential impacts of this work. My questions and my data were beholden to no one. These ethical mandates and this freedom of inquiry validated my role as an academic researcher working in the interest of a global public and opened more doors than I could have imagined. So too did the novelty of my multisited approach: in each place—Brazil, China, the United States—my experience in the other two places tended to spark interest among interviewees and in general favorably disposed people from all walks of life to talking with me.

When embarking on extensive ethnographic and archival research, there are a number of sources one can rely on for preparation. Mentors provide invaluable guidance, but so too do one's peers, the growing body of useful publications, and of course, the lessons of one's own experience. But there was one aspect of fieldwork for which I had no way of preparing: the suspicion, encountered at home and abroad, that I was a spy.

In Brazil and China, police and military personnel routinely joked that I was a CIA agent. Such jokes carried an edge: in both countries, terrible atrocities have been visited on innocent people as part of US intelligence operations. This is a history I denounce but whose legacies I cannot escape: periodic revelations highlight the fact that there are plenty of US citizens abroad who simply are not who they claim to be. Such jokes immediately threw differences of power and privilege between myself and my local interlocutors into sharp relief. The mere suggestion that I might be a sensory organ of the US government extending into the intimate spaces of homes and offices on other continents sometimes complicated efforts to build trust, and other times poisoned local rituals of hospitality with fear.

At home in the United States, friends and family unfamiliar with the practice of international ethnographic work would joke that more likely than being a geographer I was probably a spy. Why else would anyone go to remote places, learn the languages, and endure the physical and emotional hardships of life on the frontier? Who but a spook would cultivate the social skills to move from the highest corridors of power to the humblest frontier dwelling? This variety of common sense seemed to maintain that one would embark on grounded ethnographic research on a topic imbued with international strategic significance only if ordered to do so. The extent of this rather paranoid view was driven home at the conclusion of my research, when I received a visit from US intelligence agencies that, as far as I could discern, were attempting to assess whether I had been recruited to spy for China. Such a visit belies the hostile edge between the two most interdependent economies in the world, but also had a chilling effect on my follow-up research plans. How to proceed when social science research is under watch as a potential crime against the state? For someone whose formative years were defined by the triumphalist discourses of US democratic freedom over Soviet totalitarianism, to be suspected simply for pursuing academic research provided me a glimpse of how deeply totalitarian tendencies had metastasized in my home government.

When such suspicions were not offensive they were deeply troubling. They suggest a dismal worldview, one in which international researchers are not ambassadors for peace but potential traitors, where intellectual labor is performed only by tools of a totalizing state apparatus, where shedding light on contemporary global problems is a classified act rather than a civic duty. Such a worldview seems to forget that governments and firms are made of people to whom we are, at a minimum, entitled to pose questions. If we wish to make sense of seemingly intractable global problems, we must not confine ourselves to the classes, cultures, and places in which we feel at home and unchallenged. To wish to demystify an issue that is profoundly changing the way people relate to the environment and each other across the global space, with the humble hope of informing better practices to reduce needless suffering, is reason enough to do research. One need not be a tool to travel afield in search of answers.

To investigate an issue around which so much speculation, hyperbole, and fictions swirled, I conducted repeat visits over several years, triangulated multiple forms of data, and examined how stories cohered, contradicted, and changed over time. This sometimes involved huddling in unheated homes in the sub-zero temperatures of an Inner Mongolian winter, and sometimes involved improvising ways to smooth the wrinkles out of my clothes and the scuffs from my shoes for meetings in the great halls of national ministries. Other times, more than I would care to remember, it involved going through the elaborate drinking rituals essential to building trust in corridors of power and frontier outposts alike. In the same week, I might share meals with mining millionaires, and then bathe in the tap water so polluted by their industries that sores erupted on my skin.

Such variability defined most of the fieldwork process. Accessing archives was no exception. On some occasions, civil servants greeted my requests for information by literally opening their files for me to peruse, grateful for another set of eyes on a seemingly intractable problem. Other times, gathering fragments of historical data required sitting in the empty waiting area of provincial archives for days on end for no other apparent reason than that the archivist was in a foul mood.

I often visited the same place through different means. The first time I might pass through an area in the private jeep of a local power broker, and the second time on a motorcycle guided well away from paved roads to visit clandestine mining sites. Covering a lot of ground also meant being present for whatever unfolded while I was in a particular place. I abided the fear and rage that tore through a community following a land dispute that ended in murder. I stood with a grieving family at the bedside of a cancer victim poisoned by decades of accumulated toxins in the waters and soils of their village. Bearing witness to these unanticipated tragedies illuminated the stakes of the research in ways no amount of discussion could.

Many of these stories, compelling as they are, would reveal too much about specific people in places where talking to outsiders could have negative consequences. Although it was my privilege to hear these stories and to participate briefly in multiple lives, it is not my prerogative to recount their details here. To honor the integrity of those who risked talking with me without burying the significance of what they shared, I use these intimate and sometimes inflammatory accounts as maps to navigate the complex terrains surrounding the rare earths on which we all depend. I used what was shared with me in confidence as a way to decode and curate information that is publicly available and therefore verifiable. Where I cannot relate certain testimonies from the rare earth frontier, I instead illuminate the forces that create such stories to show how they comprise our present moment.

WHAT ARE RARE EARTH ELEMENTS?

These elements perplex us in our searches, baffle us in our speculations, and haunt us in our very dreams. They stretch like an unknown sea before us—mocking, mystifying, and murmuring strange revelations and possibilities.

—Sir William Crookes (February 16, 1887)

Rare earths: neither rare, nor earths.

—BBC World Service (March 23, 2014)

"In a way," writes Abraham (2011), "it begins with semantic confusion" (101). Rare earths are not rare; the name says more about their scientific beginnings than their actual qualities. In 1788 a miner in Ytterby, Sweden, found a strange black rock that was identified in 1794 as a new kind of "earth": an archaic term for acid-soluble elements[1] (Rowlatt 2014). Because it had not been found anywhere else, it was presumed to be scarce. Hence the name, *rare earths*. The implication of rarity mobilizes all sorts of sentiments that have legitimated the ruthless pursuit and capture of these elements over the past century, and perhaps that is why it persists over 125 years after this misnomer was identified among specialist audiences.[2] As illustrated in each of the cases examined in subsequent chapters, this misnomer continues to operate in territorial contests and geopolitical maneuvering. Tremendous sums of capital are mobilized and the sacrifice of certain landscapes and lives is imposed across global space, all in the name of rare earths.

The dark rock unearthed in 1788 was named "gadolinite" after its discoverer Johan Gadolin; it was later found to be a mineral consisting of cerium, lanthanum, yttrium, and iron. When Dmitri Mendeleev, Julius Meyer, and other chemists inspired by the 1860 Karlsruhe conference put together their respective drafts of the periodic table, there was no place for most of the lanthanides, which are the fifteen elements from lanthanum (atomic number fifty-seven) to lutetium (number seventy-one) (Mendelejew 1869; Scerri 2007; Spedding 1961). Yet at the time, a few of the known elements (lanthanum, cerium, terbium, and erbium) suggested the presence of a rare earth family, what would come to be known as the lanthanide series, that "distant island to the south" of the rest of the periodic table (Atkins 1995).

The elements that are included with the lanthanide series in references to rare earths change over time (see figure 4). During the race to build the nuclear bomb, thorium and uranium were referred to as rare earth elements because of their chemical affiliation and frequent geological coincidence. For the same reason, scandium and yttrium are currently counted as rare earths, although they are found elsewhere on the periodic table: twenty-one and thirty-nine, respectively. Niobium, principally mined in Brazil, and tantalum, one of the notorious conflict minerals mined in the eastern Democratic Republic of the Congo, are often grouped with rare earth elements in political and popular discourse. Despite their geological coincidence and similar ductile properties, neither is currently considered a rare earth element beyond occasional instances of political or marketing opportunism. Therefore, at present *rare earths* refers to a group of seventeen chemically similar elements sharing certain exceptional magnetic and conductive properties (Beaudry and Gschneidner 1974; Goldschmidt 1978; Liu 1978). The rare earth group comprises about seventeen percent of all naturally occurring elements (Cardarelli 2008).[3]

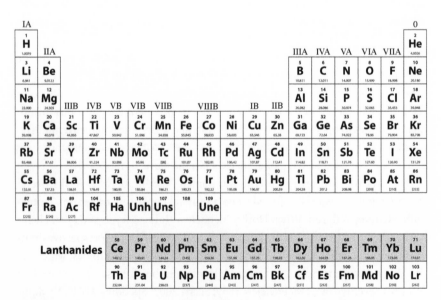

FIGURE 4A. The term "rare earths" refers primarily to the lanthanide series, as depicted by the International Union of Applied and Pure Chemistry.

Source: Image by Molly Roy.

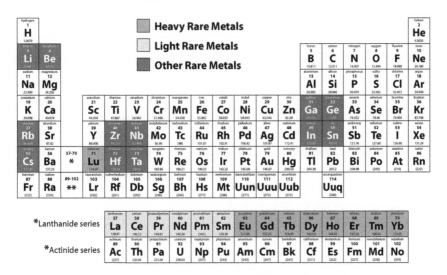

FIGURE 4B. This is a version of the periodic table presented by a mining company with lithium, tungsten, and uranium projects contending to break into the rare earth market in North America. Note the inclusion of "other" elements under the "rare" label. This table labels what are broadly referred to as "technology metals" as "rare." Many contemporary commentators have adopted this practice in order to advance the incorrect thesis that these elements are actually scarce.

Source: Image by Molly Roy.

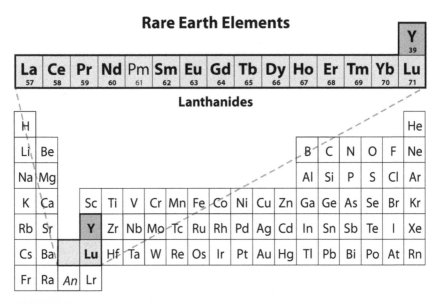

FIGURE 4C. The United States Geological Survey (USGS) uses a version of this table. It differs from the table used by the United States Department of Energy (DOE) by excluding scandium. The DOE, among many others, includes scandium.

Source: Image by Molly Roy.

Although most rare earths are relatively abundant (see table 1 and table 2), they are dispersed throughout Earth's crust, threaded through iron, phosphate, or copper-gold deposits. They are also found in placer and residual deposits formed by the long-term erosion of igneous rocks, which explains why they show up on the black sand beaches of Brazil, India, and elsewhere.

Some rare earths, such as promethium, are not found on Earth outside of nuclear reactors, but are used to produce batteries that power pacemakers and space crafts, as well as to manufacture luminescent paint for watch dials (Krebs 2006). Others, such as thulium, are so scarce that only a few kilograms can be extracted from 500 tonnes of rare earth rich ores (Emsley 2001). Thulium is essential to the production of surgical lasers used to treat neurological and prostate conditions (Duarte 2010) and because it shines blue under ultraviolet light, it is stamped onto Euro banknotes as an anti-counterfeiting measure (Wardle 2009). Then there is scandium, which is so difficult to separate from other rare earths and uranium that annual global trade in the pure metal has yet to exceed 100 kilograms. Scandium is used in the metal halide lamps that illuminate streets, stadiums, and film studios (Krebs 2006) and is part of the secret recipe for high-performance handguns, bicycle frames, and baseball bats (Bjerklie 2006; Wesson 2014; Staff 2009).

TABLE 1 Abundance of rare earth elements in earth's crust, estimates

LANTHANIDES	PARTS PER MILLION (PPM) BY WEIGHT	RANK, IN ORDER OF ABUNDANCE OF ALL KNOWN ELEMENTS
Lutetium	~0.6	61
Thulium	0.45–0.48	58
Terbium	0.94–1.1	57
Holmium	1.2–1.4	56
Europium	1.8–2.1	50
Ytterbium	2.8–3.3	46
Erbium	3–3.8	45
Dysprosium	6–6.2	43
Gadolinium	5.2–7.7	42
Praseodymium	8.7–9.5	40
Samarium	6–7.9	38
Lanthanum	32–34	29
Neodymium	33–38	28
Cerium	60–68	27
Promethium	none[a]	N/A

Note: Table adapted from Zepf (2013).

[a] Perhaps 570 grams of promethium exist in all of Earth's crust (Belli, Bernabei et al. 2007).

TABLE 2   Other elements sometimes classified as rare earths, estimates

NON-LANTHANIDES	PARTS PER MILLION (PPM) BY WEIGHT	RANK, IN ORDER OF ABUNDANCE OF ALL KNOWN ELEMENTS
Tantalum	1.7–2	52
Uranium	1.8–2.7	51
Thorium	6–12	39
Scandium	16–26	36
Niobium	17–20	34

Note: Table adapted from Zepf (2013).

Other rare earth elements are not as scarce but their uses are comparably wide-ranging. Because of their exceptional magnetic and conductive properties, this family of soft, ductile metals is essential for a diverse and expanding array of high-technology applications fundamental to globalized modernity as we know it. Global finance, the Internet, satellite surveillance, oil transport, jet engines, televisions, GPS, and emergency rooms could not function without rare earth elements. They are necessary to produce the navigation components of the most advanced remote warfare technologies, such as drones and smart bombs (Hedrick 2004; Kidman et al. 2012). They are critical components of green technologies, such as wind turbines, solar panels, and hybrid fuel-cell batteries (Armand and Tarascon 2008; Hashimoto et al. 2009; Humphries 2013; Jones 2013). They are essential in the development of nanotechnologies and are used in the production of consumer electronics such as smartphones, hard drives, and flat screen monitors (Krishnamurthy and Gupta 2005).

So thoroughly embedded are rare earths that an analysis of their role in modern life precludes a straightforward commodity-chain or sector-specific analysis, unless, for example, one looks exclusively at a certain kind of magnet (Zepf 2013). There is no singular "rare earth market" to speak of, but rather multiple markets for the seventeen elements (and combinations thereof) with widely divergent availabilities and applications. The production process of a single rare earth resembles a web more than a chain because it graces such an array of goods with its properties. For example, erbium, which turns pink when oxidized, lends its hue to rose-colored glass and porcelain tableware (Hammond 2000) while also acting as an amplifier in fiber optic cables, enabling the construction of global Internet communications networks (Becker, Olsson, and Simpson 1999). This gives rare earths an air of ineffability—they are seemingly everywhere, but in such minute quantities or such sophisticated applications that they are difficult to quantify. This amplifies the effects of supply crises.

The nature of their applications, like their geological occurrence, is both ubiquitous and dispersed. They are most commonly used in alloys, mixed with other elements such as iron or nickel to make them better, stronger, faster, and lighter. Scientific parlance refers to the process of adding rare earths to other elements as "doping," borrowing the slang describing the use of performance-enhancing drugs in competitive sports (Digonnet 2001). In China they are called the "MSG of industry" (Klinger 2011) to capture the sense that, much like how a pinch of MSG enhances one's cooking, just a little bit of rare earth enhances the quality of industrial output. Similarly, German industry refers to them as "spice" metals (Zepf 2013) and the United States Geological Survey (USGS) describes them as "vitamins," which, when added to other elements produce results that neither could achieve alone (Koerth-Baker 2012). In Japan, they are described using the following metaphor: "oil is the blood, steel is the body, and rare earths are the vitamins of a modern economy" (Dent 2012). These metaphors convey a sense of the relatively small quantities generally required to achieve desired effects: most consumer electronics, for example, are composed of only a tiny portion of rare earths. Their dispersal, the difficulties involved in isolating individual elements, and the fact that a few rare earths are actually uncommon excites political economic passions around their scarcity. These passions, somewhat paradoxically, have been most prominent in places where rare earths are plentiful, as in China, Brazil, and the United States.

Although rare earths are now essential to the technological infrastructure of modern life as we know it, for nearly a century after their discovery there was little use for them. During that first long century spanning from 1788 to 1891, scientific progress with rare earth elements was limited: "A great many learned men with famous names busied themselves with rare earth elements and reported interesting work . . . nevertheless, no applications or industrial usage came out of these efforts" (Greinacher 1981, 4). The first successful application addressed a long-standing problem in newly urbanized industrial zones before the advent of urban electricity: how to produce light cheaply and reliably over a large area. This imperative was driven by the industrialist desire to maintain production after dark, especially during long winter nights in northern Europe (Bogard 2013; Ekirch 2005; Koslofsky 2011).

Carl Auer von Welsbach's invention of gas mantles (Eliseeva and Bünzli 2011; Welsbach 1889) at the end of the nineteenth century inaugurated the first phase of industrial usage[4] of mixed or simply separated rare earth elements. Although the gas mantle lantern contained only 1 percent of the rare earth element cerium,[5] the production scale was massive: by the 1930s, over five billion had been sold (Niinistö 1987), providing networks of city lights before the widespread establishment of electrical grids. Welsbach's first invention presaged the second, which addressed a key problem with gas mantles: they were difficult to ignite, and piles of rare earth wastes left over from the production of the incandescent mantles were prone to combustion. By blending these rare earth wastes with 30 percent iron, Welsbach developed the alloy "mischmetall" that sparked when struck. He patented this as the "flint stone," which is still used in all manner of ignition switches, from lanterns to cigarette lighters to automobiles (Krishnamurthy and Gupta 2005).

These initial technological and commercial innovations sparked tremendous interest in the broader applications of rare earths. But it was not until the atomic, television, and computer age that uses beyond the most basic applications would be found for them. Still, the gas mantle and the flint stone were so successful that their invention expanded the rare earth industry dramatically and drove the quest for raw materials to Europe's (post)colonial frontiers in the Americas, India, and China. Until 1895, gadolinite and bastnäsite from Sweden furnished most of the raw materials for rare earth elements and thorium (Greinacher 1981). In 1887, a British mining interest began extracting rare earths from the monazite sands on the beaches of North and South Carolina; the operations were soon taken over by the Welsbach Light Company of New York (Levy 1915).

The German Thorium Syndicate and the Austrian Welsbach Company began exploiting monazite placers in Brazil in 1905 and in India in 1909, which drove US production out of business by 1910, except for a brief interlude during World War I (Mertie 1953). Brazil and India then supplied the global market—consisting of Europe and North America (Russia was self-sufficient)—until 1948. The Indian Atomic Energy Act of 1948 prohibited the export of monazite because radioactive thorium, abundant in the sands, was named a source of atomic energy and therefore a strategic mineral for domestic use only (McMahon 1994). This abrupt interruption in supplies to the United States had a temporary chilling effect on research and industry[6] until domestic production expanded again in late 1952 (Congress 1952).

Part of the lag in identifying applications had to do with early experiments, which generated mistaken perceptions of rare earths, what they are, and how they behave. Frank Spedding, director of the Institute for Atomic Research at Ames Laboratory, wrote:

> Even as late as the early part of this century, one could find statements in textbooks that that the rare earths were all very much alike and resembled lanthanum. As we all know now, this is far from the truth. The differences in the properties of these elements are as great as the differences between the members of other series in the periodic table, such as sodium, potassium, rubidium and cesium, or copper, silver, and gold. The melting points of the rare earth metals vary from around 800 to 1650

degrees Celsius. The vapor pressures of the metals at a given temperature differ by a factor of more than a million from the most volatile
member to the least volatile member. Some of the metals are magnetic, others are not. Lanthanum is a super-conductor. Some, such as
lanthanum and cerium, corrode readily in air, while others corrode in-
appreciably at room temperatures. (1961, 2)

Through most of the twentieth century rare earths were still treated as rare
and research interest was confined to highly specialized audiences, such as readers of *Journal of the Less Common Metals,* inaugurated in 1959. Although small,
this journal drew contributions from materials scientists, chemists, and physicists
experimenting with alloys and compounds during a time of tremendous expansion in communications, military, and aerospace industries. The mid-twentieth
century seemed to be a golden era for experimentation, as there was rising interest in the usefulness of rare earth elements along with political and economic imperatives to exploit potential applications. Yet many basic characteristics of rare
earths still remained unexplored. An excerpt from an article published in 1961
conveys a sense of the times, in which laboratories occasionally caught fire as scientists figured out which elements could and could not be mixed: "Attempts to
make alloys of thorium and ytterbium by arc-melting were unsuccessful; the two
metals appear to be virtually immiscible even at the melting point of thorium. At
this temperature the volatility of ytterbium is serious, resulting in heavy losses of
the metal which, when deposited in the form of a thin film on the inside of equipment proved to be pyrophoric[7] on subsequent exposure to air" (Evans and
Raynor 1961, 179).

Prior to the late 1980s, few rare earth elements were used in electronics; most
of the applications were in the fields of catalysis, glass and ceramics, and metallurgy (Niinistö 1987). It was not until the late 1980s and early 1990s that their
innovative applications in communications and consumer electronics accelerated.
As chapter 3 details, this technological shift occurred contemporaneously with
broader shifts in the global division of labor that were crucial to laying the foundation for China's rare earth monopoly.

In information technology and consumer electronics, the rare earth element
neodymium is especially important. Its exceptional magnetic qualities enabled
the miniaturization of computer hard drives and speakers. Without small hard
drives and tiny speakers, we would not have personal computers, smart phones,
or ear buds, billions of which are in use worldwide (Croat 1997; Guenther 2003).
In fact, no other material comes close to the magnetic power of neodymium, and
it is perhaps the best representation of the most promising and most terrible applications of rare earth elements. Powerful neodymium magnets are essential to

the latest and most efficient renewable energy technologies, including hybrid fuel cell batteries, water and wind turbines, and solar panels (Hatch 2008). A two-megawatt wind turbine contains about 360 kilograms of neodymium and 60 kilograms[8] of dysprosium (Stover 2011), while a three-megawatt turbine contains 1,800 kilograms[9] of these rare earths. Neodymium and neodymium alloys are also fundamental to the hardware of contemporary militarism: cruise missiles, smart bombs, and drones. These armaments also contain praseodymium, terbium, samarium, and dysprosium. Yttrium, europium, and terbium are used in radar, sonar, and radiation detection devices for targeting and detection in urban, maritime, and aerial warfare. Yet these same elements lend their optical properties to medical imaging devices such as X-rays and MRIs.[10] Rare earth elements are the material basis for the hardware of global technological modernity: from the darkest and most dystopic to the greenest and greatest.

The Political Life of Rare Earths

The shifting characterizations and ubiquitous applications of rare earth elements complicate efforts to quantify and trace their value over time. Especially because the contemporary category now excludes certain elements, like thorium and uranium, that were once discussed collectively with rare earths. Such elements were intrinsic to the twentieth century political economy and political ecology of rare earth extraction, fundamentally shaping our contemporary geography of rare earth production. In this way rare earths are qualitatively different from other elements such as gold, silver, or mercury, which were historically accompanied by price tables and moved along a limited set of commodity chains over the centuries. The combined qualities of rare earths as both ubiquitous and dispersed in their geological incidence and applications, as well as their persistent mischaracterization as *rare,* are key to understanding the bizarre potency of their political life and the far-flung geography of their prospecting and production.[11]

The political life of rare earth elements began with the European quest for raw materials in colonial lands around the turn of the twentieth century, when British and German interests prospected in India and the Americas to feed the expanding gas mantle and flint stone industry. Leading up to World War I, not all the rare earths had been properly identified. Likewise, thorium, uranium, tungsten, platinum, and vanadium were grouped with rare earth elements because of their geological coincidence and complementary applications (Martin 1915). During World War I, the pyrophoric properties of rare earths were used in fuses and explosives. Early applications of steel and iron alloys were used in the manufacture of weapons (Martin 1915).

The English physicist Henry Moseley hypothesized that rare earth separation would shed light on nuclear fission and was the first to confirm, in 1914, that the lanthanide series must consist of fifteen members, no more and no less, including promethium, which was not discovered until 1942. Moseley resigned from his research activities in late 1914 to enlist with the Royal Engineers of the British Army. He was shot in the head in 1915 while serving the British Empire in Turkey. It took nearly three decades for the scientific community to pick up his research where World War I had cut it short (Asimov 1982).

In the twentieth century, their political life was shaped by the geopolitical contexts of colonialism, the Cold War, and the rise of neoliberalism. Histories of nuclear weapons development, medical research, and energy politics—to name a few—are inseparable from both applied research in rare earths and shifting political economies that shape funding and research agendas across sectors, among countries, and over time. The nuclear arms race depended on rare earth elements (Chakhmouradian and Wall 2012; Kosynkin et al. 1993); progress in the ongoing fights against cancer are fueled by rare earths (Townley 2013); and the technologies essential for oil prospecting, drilling, transport, and refining all rely on the conductive, magnetic, and enhancement powers of rare earth elements (Sie 1994). The following pages introduce the atomic, environmental, and global trade entanglements of rare earths.

Atomic (Post)colonialities

Rare earths were both inputs and outputs of the nuclear war effort. In 1939, the German scientists Hahn and Strassman discovered the neutron-induced nuclear fission of uranium and identified rare earth elements in fission products (Cardarelli 2008). Although thorium and uranium are no longer grouped with rare earth elements, mid-twentieth century prospecting and procurement procedures sought them collectively, often under the euphemism of "nonferrous" metals. Therefore to make sense of our contemporary rare earth situation, we must consider its origins in the shifting tides of colonialism and nuclear weapons development. The United States and Germany both drew their rare earth and thorium[12] supplies from India and Brazil until World War II broke out in 1939. Germany then dodged British and Allied embargos before ceasing commercial operations with Brazil[13] and India in late 1940. Shortly thereafter, US and British leaders concluded that

> the best future interest of the two countries would be served by a joint effort to seek out and gain control over as much of the world's uranium and thorium deposits as possible; this policy . . . would ensure their governments ready access to major new resources of inestimable

value and would keep these resources out of the hands of their potential enemies. Furthermore, project leaders perceived that, strictly from
the viewpoint of national interest, it would be better for the United States
to conserve its own apparently limited domestic resources and use
whatever raw materials it could acquire from other countries instead.
(Jones 1985, 293)

Executing this agenda required a geological survey of unprecedented scope into
global rare earth, thorium, and uranium resources. Union Carbide, working in
cooperation with the Manhattan Project, assembled a team of approximately 130
geologists, translators, and clerks in New York to search through all available technical literature in any language. In the first six months of 1944, they examined
sixty-five thousand volumes and carried out field expeditions in thirty-seven states
and twenty countries. They determined that the Belgian Congo, Brazil, and India
would provide the most abundant high-quality materials to support the nuclear
arms race, with Canadian and Western US minerals as good alternatives (Jones
1985). The United States could not secure supplies in colonial territories without
the assistance of the British Empire. The British Empire had interest in the global
intelligence capacities of the United States. So the two countries formulated a joint
undertaking to extend their atomic hinterlands to "areas outside of American and
British territory" (Stimson 1944 quoted in Jones 1985, 299). Russia, meanwhile,
extended its own rare earth hinterland into Kyrgyzstan, opening a rare earth-
thorium-uranium mine and processing plant in Ak-Tyuz in 1942 (Djenchuraev
1999) shortly after Stalin learned in April of that year that Allied powers were
developing a nuclear weapon (Kojevnikov 2004).

Getting ahold of minerals from the Belgian Congo was difficult for the United
States. The principal mine of interest, the notorious Shinkolobwe, had flooded
and closed and mine director Edgar Sengier had returned to London. Sengier reportedly understood the potential of harnessing atomic power and the role his
mine could play in such an endeavor. However, he did not want to make any commitments to foreign militaries that he might later have to justify to the Belgian
government (Gowing 1943 quoted in Jones 1985) unless the United States and
Britain could make an offer that "served the interests" of the Belgian government
in exile. In exchange for considerable sums of money, no timetable requirements,
new equipment, and assistance in "procuring labor,"[14] Sengier agreed to reopen
the mine to provide uranium to the United States beginning in mid-1945 (Helmreich 1998).[15] In the meantime, the United States continued to rely on India and
Brazil for thorium and rare earth elements.

The United States was unprepared for the postcolonial disruption to the global
division of extractive labor. When Brazilian production failed to make up the

difference following the 1948 Indian embargo on monazite exports, rare earth and thorium prices rose precipitously between 1948 and 1952 (Mertie 1953). Brazilian monazite production was reportedly exhausted by 1950, but a closer look at the production data suggests that it is more likely that no one was able to fully resume the primarily German-run monazite operations after World War II (USGS 1953). Sensing opportunity, US geologists, prospectors, and mining firms set out in search of lucrative deposits in the American West. In 1949, a uranium prospector discovered the rare earth mine at Mountain Pass, California (Olson et al. 1954), which would dominate global rare earth production from 1960 to 2000. But in the immediate aftermath of World War II, and in a way that is strikingly similar to today, the US government preferred to source the elements from overseas despite known domestic abundance. Congress resolved to slow research and production among rare earth-dependent sectors rather than pursue domestic self-sufficiency (United States House of Representatives 1952). At the expense of domestic firms ramping up production in Idaho, the US Department of State pursued an agenda to liberalize India's monazite exports at the dawn of the Cold War.

The race to build the atomic bomb reconstituted global rare earth politics along the emergent fault lines of the Cold War. India restored its independence in 1947. In the post–World War II, postcolonial contests, nuclear weapons were seen as guarantors of newly won national sovereignty. In 1949, the Soviet Union was nearing its first nuclear bomb test, China was expressing atomic intentions, and the United Nations had called for the elimination of nuclear weapons. Developing nuclear weapons was a top priority for Indian Prime Minister Jawaharlal Nehru's government, along with finding a means to relieve the famine and assuage the violence surrounding the India-Pakistan partition (Chengappa 2000; Lawn and Clarke 2008; Pandey 2001). The Indian Atomic Energy Act of 1948 identified thorium as a source of atomic energy, named it a strategic mineral, and immediately ceased the export of thorium-rich monazite. This embargo seriously disrupted the United States' strategic monazite supply, coinciding with the sharp reorientation of US foreign policy toward containing the spread of Soviet influence and suppressing communist movements in India (Merrill 1990).

India had a famine; the United States had grain. US president Harry Truman reckoned that relieving the misery would undermine the appeal of communism, generate favorable attitudes toward the United States, and open the door to negotiating India's monazite embargo. Although famine relief discussions between Indian Ambassadors to Washington and the US Department of State had been underway since 1947, broad factions in Congress opposed the bill because of Nehru's criticisms of the West, or, as Senator Tom Connelly put it in 1951, his "hatred of every white man" (quoted in McMahon 1994, 93). To the US Department of State's proposal of a gift of food aid as "Indian Food Crisis–Opportunity to

Combat Communist Imperialism," Republican opponents to famine relief responded with a reformulation of the transfer of US grain as a *quid pro quo*: "India needs grain immediately; we have the grain. We need strategic materials from India over a period of years; India has those materials. We should make India a loan which can be repaid in strategic materials" (Congressman John M. Vorys quoted in McMahon 1994, 96).

As this line of thinking gained momentum, the proposed US$190 million gift to India became a loan repayable in strategic materials. India's famine relief became contingent on breaking its embargo against strategic mineral exports. Prime Minister Jawaharlal Nehru refused on the basis that such conditionality would violate India's sovereignty. He later relented with the proviso that India would continue to provide only those strategic materials that could not be used for nuclear weapons development, thus precluding monazite. Hence the plan to "bring India closer to the West" backfired miserably, generating bitterness toward the United States in India, and all manner of US Department of State hand-wringing over the possible expansion of the "communist threat" across South Asia, while leaving the monazite issue unresolved (McMahon 1994). But the failure of this arrangement to procure thorium and rare earths from India created the conditions under which it became politically necessary to identify domestic US sources, and economically possible for industrial mining activities to begin at Mountain Pass, California.

Meanwhile, in Baotou, Inner Mongolia, a comprehensive Sino-Soviet mining, industrialization, and urbanization program was under way to transform the ores at the Bayan Obo Iron-Thorium-Rare Earth mine into steel, machinery, and weapons.[16] Started in 1951, the construction of the Baotou Iron and Steel complex was reportedly the flagship project of a massive aid portfolio of 149 Soviet development projects in China. As discussed in detail in chapter 2, Mao and Stalin intended to convert the windswept steppes of Inner Mongolia into a military-industrial heartland that could provision both Republics in their struggle against capitalism and Western imperialism. But the relationship was tricky. China supplied the Soviet Union with uranium and complied with Soviet military requests to set up communications and military bases throughout northern China. In exchange, the Soviet Union provided training and technology to support China's nuclear weapons program. By the mid-1950s the Chinese counterparts were disappointed at what they viewed as Soviet withholding of nuclear expertise (MNO 1958), and began pursuing their own nuclear agenda outside of the Sino-Soviet Plan (Gobarev 1999). In 1956, the "father of China's rare earth chemistry" Xu Guangxian left his teaching and research position at Peking University to support China's effort to build nuclear weapons. In his memoirs, he explained that his expertise working with rare earths transferred well to his new focus on nuclear fuel extraction, which later inspired his breakthrough in rare earth separation (Jia and Di 2009).[17]

When looking at US, British, Indian, Chinese, and Soviet rare earth extraction and weapons development, it is important to view these Cold War competitions in historical context. Yes, the world was divided into ideologically and culturally oppositional spaces, but these antagonisms emerged from centuries of colonial domination and above all constituted a world-historical moment characterized by the coalescence of modern warfare, atomic aspirations, competing imperialisms, and the exploitation of rare earth elements across global space. It is also worth noting how each colonial and state power chose to construct its own rare earth hinterland, because such patterns indicate the spatiality of power in the allocation of environmental destruction: nineteenth-century German interests went far afield, to the beaches of Brazil and India, to mine monazite sands; Britain opened up interests in the Carolinas and India; New York likewise exploited the Carolinas; the United States and Britain collaborated with Belgian imperialists to exploit the Congo; Russia and China sought these resources in inland West Asia, in lands peopled by "ethnic minorities." Although these mining sites may seem remote from the perspective of Berlin, Moscow, Beijing, New York, or London, they were central to the people who lived there and were not always handed over without a fight (Tipper 1930; Zoellner 2009).

The Tension: Necessity versus Pollution

During the first half of the twentieth century, Euro-American powers drew on colonial networks of exploitation to source the raw materials for industrialization and militarism. At the height of World War II, the United States and Britain framed this global division of toxic labor as essential to preserving the "future interests of civilization," insofar as "the protection of civilization required effective control of said ores" (Spaak 1944 quoted in Jones 1985, 300).

The physical act of mining tends to lay waste to the landscapes and livelihoods preceding this enterprise. Therefore the global mining frontier is characterized by the practice of devaluing certain landscapes and livelihoods relative to their subterranean resources. Once devalued, these places can be made into new frontiers for extraction and investment. The question of which places are devalued is less a matter of objective science, and more a matter of power and vulnerability in determining which landscapes and lives are to be sacrificed in favor of mining. This is especially clear in the case of rare earths, which despite their abundance are mined in a select few places. As the wellspring of national and global industry, rare earth frontiers are etched into the landscape by the neo-imperial edge of militarism, socialist industrialism, and capitalism—which are driven by the relentless quest for territory, minerals, wealth, and the geopolitical power that comes from possessing all three (Graulau 2003; Luxemburg [1913] 2004; Tsing 2005). Yet securing "effective control of said ores" also means contending with the hazards and wastes generated by mining. Every tonne of rare earth produced generates approximately one tonne of radioactive wastewater; seventy-five cubic meters of acid wastewater; 9,600 to 12,000 cubic meters of waste gas containing hydrofluoric acid, sulfur dioxide, and sulfuric acid; and approximately 8.5 kilograms of fluorine (Hurst 2010). A tension exists, therefore, between securing access to vital minerals and isolating the devastation typically resulting from large-scale mining activities.

The geography of the global rare earth frontier is defined by this fundamental tension between the absolute necessity of these elements and the acute environmental and epidemiological costs generated by their extraction.[18] Environmental hazards emerge at four primary stages. The first is the mining process, during which certain rare earths and radioactive elements such as thorium and uranium pose health risks to miners. Then there is the refining process, where toxic chemicals and acids are used to separate elements (Hao and Nakano 2011). Third is the waste management from the primary processing and beneficiation activities, which introduce heavy metals, radioactive salts, and radon gas into the surrounding environment. The fourth stage concerns the disposal of rare earth-containing products, many of which become e-waste, and which has a racialized violence of its own (Gullett et al. 2007; Weber and Reisman 2012). All rare earth elements can cause organ damage if inhaled or ingested; several corrode skin; and five—promethium, gadolinium, terbium, thulium, and holmium—are so toxic that they must be handled with extreme care to avoid radiation poisoning or combustion (Krebs 2006). Further, rare earths tend to coincide with radioactive thorium and uranium, meaning that rare earth mining also creates radioactive waste that must be managed (Bai, Zhang, and Wang 2001). Who is exposed to these hazards, and which landscapes are laid to waste, is seldom a matter of democratic decision making. The spatial allocation of sacrifice illustrates the cartographies of the powerful and the powerless.

Because of their radioactive properties, even the most minimal environmental regulation dramatically increases costs of an already capital-intensive enterprise (Goldenberg 2010; Lazenby 2013). The production site must handle tonnes of radioactive wastewater generated by the separation and refining processes (Ives 2013; Li 1987; Wang 2007) that remain toxic for millennia (Najem and Voyce 1990). Between 1965 and 1980, the deserts of southeastern California generated most of the global supply of rare earth elements, but as discussed in chapter 2, the environmental costs were simply too high: between 1984 and 1998, over sixty radioactive wastewater spills occurred, many of which were unreported. Conservative estimates by the Environmental Protection Agency maintain that over six hundred thousand gallons of radioactive wastewater spilled onto the desert floor, and much more leached into groundwater from unlined holding ponds (Danielski 2009). Because of their toxicity as well as the estimated upfront costs involved in establishing cleaner and safer production systems, "remote" regions are preferred for mining and processing (Bai, Zhang, and Wang 2001; Wang 2007). Nonetheless, mines in relatively sparsely populated, but comparatively better-regulated, southern California and Austria went bankrupt in the late 1990s in the face of cheaper imports from China (Humphries 2013). This ended the brief post–World War II period of Western self-sufficiency by closing down rare earth frontiers in the United States and Austria. Moving the rare earth frontier to China also freed these two places from the environmental burdens of production (A. Ellis 2013).

From the mid-1980s until 2010, mines in China gradually supplied a larger share of the global demand for rare earths. For many in the industry, this shift eased the tension between environmental hazard and industrial necessity by concentrating the most toxic aspects of the commodity chain in places with less power of refusal in the economic globalization game, characterized by "cheap labor" and "lax environmental regulations" (Lin 2011). As discussed in chapter 3, the origins of China's rare earth monopoly are best understood in this context. With the high value of refined rare earths as well as the heavy financial, environmental, and social costs of their extraction and processing, global rare earth production has followed a "race to the bottom" and "environmental outsourcing" trajectory similar to that of other industries over the past four decades of neoliberal globalization (R. E. Ellis 2013; Evans, Goodman, and Landsbury 2002). China's export-oriented growth and the West's deindustrialization are two aspects of the same process characterizing the contemporary global division of toxic labor.

The global division of toxic labor, however, is hardly static. The steps taken by China's central government to rein in production and remediate damaged environments since the early 2000s have renewed prospecting efforts across the globe. With greater understanding of the hazards involved in rare earth mining, the impetus to open up new sources further afield, beyond the reach of enforceable environmental and occupational health and safety regulations, has also increased. The high cost of waste management, the sheer difficulty of carrying out legitimate environmental and social impact assessments, and the political blowback in the case of an accident discourage new production in sites with adequate infrastructure and urbanization to support large-scale vertically integrated industry.

These factors make otherwise impossible places, such as the far northern Amazon and the Moon, particularly appealing to states and private firms eager to leverage unevenness in regulatory monitoring and political accountability on these frontiers. The driving assumption is that extracting from such places, where the wastes are cloistered far away from major population centers, will be more economically sustainable and cost effective in the long run compared to developing advanced recycling facilities to reclaim rare earth elements from waste (Meyer 2011, Jain 2012). However, as will be discussed in chapters 5 and 6, these factors are an insufficient impetus to drive extractive interests to such great lengths. On closer scrutiny, the race to mine impossible places is, in fact, only marginally driven by a perceived need for these elements. Of considerable, if not greater, importance are the other longer-term territorial and geopolitical projects rare earth mining might enable in places that have historically eluded the reach of centralized power. For example, Brazil has abundant rare earth deposits in the already-active Araxá niobium mine, but as chapter 5 details, Brazilian government actors and international firms harnessed the geopolitical moment generated by global concern around China's rare earth monopoly to renew efforts to conquer a contested and historically rebellious region of the Amazon. The Moon, examined in chapter 6, belongs to a class of new resource frontiers that require increasingly specialized knowledge and technology for rare earth exploration: parallels have been made elsewhere between the rediscovery of the deep oceans and exploration of our near solar system in the ongoing race for resources (Macdonald 2007).

These remote sites demonstrate that the geography of the rare earth frontier is driven by environmental tensions. The scope of the rare earth frontier is limited only by technology, political will, and the potential for power and profit, rather than by market logic or even by the confines of land or sea. In any case, the rare earth frontier is no longer limited to Earth. This alone indicates a significant qualitative difference from other strategic commodity frontiers, derived from the peculiar material properties of rare earths: they are sufficiently useful and valuable in small enough quantities that they could, in theory, be economically exploited off-Earth. While exploitation of such far-out sites is hardly as fantastical as one might think, it was not so imminent as to neutralize the political economic effects of China's de facto monopoly in the short term.

Trade, Sacrifice, and Security

The global division of labor for rare earth elements has been characterized by (post)colonial exploitation and environmental injustice. Rare earth production concentrated in China as a result of three converging factors: the central government's emphasis on—and USSR support of—building up its domestic rare earth industry to support technological innovation in the space, defense, and energy sectors; the Reagan-Thatcher era deregulations in the West, which facilitated the export of labor-intensive and environmentally hazardous industry; and China's 1978 economic reforms, which opened the country to foreign investment and provided an industrial platform for the world's heavy enterprises. The monopoly emerged because actors on all sides stood to gain from subcontracting components of rare earth processing to China. Downstream industries in the rest of the world benefitted from the cheap and abundant raw materials coming from China. It was not until exports from China were temporarily halted in 2010 that the rest of the world began to see the global division of toxic labor and China's de facto monopoly as problematic.

Although the production quotas began a decade prior to the 2010 incident, the sudden disruption in rare earth exports to Japan alerted the world to its dependence on China, prompting speculation that China would use its de facto monopoly as leverage in global politics. Despite assurances from then premier Hu Jintao and foreign minister Yang Jiechi that China would never use rare earth elements as a weapon in global affairs (Grasso 2013), the international community was quick to sound the alarm over China's "stranglehold" on the rest of the world (Plumer 2011). Policymakers, lobbyists, military personnel, and pundits warned that China's monopoly could bring global industry to a halt and seriously undermine national security (Galyen 2011; Parthemore 2011). Absent decisive action to diversify supply streams, rare earths could be China's political and economic "ace in the hole" (Hurst 2010) in international negotiations. Indeed, in light of China's production and export restrictions, and without a concerted effort to diversify supply, rare earths could even become the next "element of conflict" in Asia (Ting and Seaman 2013). More concretely, the shortage and attendant price increases brought significant pressure to bear on renewable energy startups, leading US president Barack Obama to blame the renewable energy economy's failure to launch in the United States on the fact that China "broke the rules" (Chapple 2012).[19]

The United States, the European Union, and Japan brought two unsuccessful World Trade Organization (WTO) suits against China's raw materials production quotas in 2009 and 2011. In both cases, the WTO Special Panel ruled that China's export quotas were justified under Article XX of the General Agreement on Tariffs and Trade, which holds that member states may withhold the export of strategic natural resources for environmental or national security reasons. News broke in October 2013 that China had lost a third suit because it failed to provide convincing evidence of environmental harm (WTO 2014).

One local scientist in Bayan Obo commented that those hoping for a WTO victory against China were missing the fact that sustaining the global supply of these resources is less important to China and neighboring countries than stemming the tide of toxic and radioactive waste contaminating extensive areas of

Inner Mongolia, including the Yellow River watershed on which over a hundred million people downstream rely for drinking water, irrigation, fishing, and industry.[20] Some officials considered the merits of selective noncompliance. Interviewees in China's Ministry of Foreign Affairs noted "a precedent of major powers selectively ignoring WTO rules that do not match particular national circumstances."[21] In support of this sentiment, a principle researcher with the Beijing-based China Rare Earth Research Association observed in late 2013 that rare earth elements were not "so hot anymore" because prices had stabilized and China's policy was clear and consistent. As such, it fell to the industrialized countries of the rest of the world to "take responsibility for their own rare earth needs."[22] In some ways, this had already happened: prospectors and speculators have embarked on a "new gold rush" to identify minable deposits in central Asia, the South Pacific, Greenland, southern Africa, and of course the Amazon and the Moon (Farias 2013b; Jeffries 2014; Lazaro 2014). A week after the leaked decision from the third WTO ruling, China's Ministry of Commerce announced rare earth production quotas for 2014, but liberalized rare earth exports in January 2015 following a significant consolidation of domestic industry.

The political lives of rare earths are many and complex, but the complexity boils down to questions of power. Because rare earths confer tremendous power to those who acquire them, power is exercised in the capacity to make hegemonic claims to subsoils containing rare earths. Power is manifest in the ability to subject some and exempt others from the toxic and radioactive byproducts of mining and processing.[23] Therefore, these elements are contentious at every scale, from the local to the (extra)global.

Although China looms large in contemporary rare earth politics, the global rare earth economy is not so simple that China could simply "hold the world hostage," as some have alleged. The 2010 crisis, examined in depth in chapter 4, certainly highlighted global dependence on China's rare earth industry. What was entirely ignored yet remains equally important is China's dependence on the rest of the world to purchase rare earths and value-added components. This dependence is evident in the ongoing overcapacity problems plaguing China's many downstream industries. The contemporary challenge in China is not only to source rare earth ores in a way that does not destroy domestic environments, but also to sustain enough global demand for value-added products to sustain domestic industry. Although the global production asymmetry is important, global rare earth markets are much more entangled and much less unidirectional than typically portrayed.

The many sorts of connections and interests comprising our contemporary global rare earth situation are only revealed if we carry out a multisited analysis such as the one presented in these pages. By examining three key sites comparatively and in world-historical perspective, we are able to root out the concrete and multi-scaled practices that are crucial to thinking globally. Looking at the connections among these sites—as well as our connections to them, however inadvertent—illuminates the global arrangements of power and sacrifice on which our contemporary economy runs and to which we are often rendered blind. Power is crucial to shaping the geography of the global rare earth frontier, but geology matters, too.

Geological Incidence, Formation, and the Contemporary Geography of the Global Rare Earth Frontier

Despite what these contentious politics might suggest, the global rare earth frontier is vast, dynamic, and plentiful. There are currently over eight hundred identified land-based deposits of sufficient concentration to be feasibly mined (USGS 2013), bringing the total known land-based deposits to over 110 million tonnes (USGS 2011). In addition, recent exploration of the Pacific has identified deposits potentially totaling over one thousand times as much (Kato et al 2011; Pritchard 2013).

Rare earths have the maddening characteristic of being relatively common yet incredibly difficult to exploit and amass. Many of them occur between twenty and thirty parts per million (ppm) in the earth's crust (compared to lead at fourteen ppm), but in deposits that are difficult to mine (Long and Van Gosen 2010). This is because of the peculiarities of their formation and their geological coincidence with hazardous minerals.

Because there are seventeen separate elements, the possible combinations and compositions of rare earth deposits vary such that no two deposits are alike; likewise, no two extraction or separation processes are identical (Li, Zuo, and Meng 2004). Furthermore, each element requires a distinct chemically intensive process to separate it from its medium. This means that wherever a minable concentration[24] is discovered, the ores must undergo extensive tests to develop optimal extraction and processing techniques. Unlike other metals, which can be refined through a few smelting and separation steps, rare earth ores must undergo several dozen chemical processes, such as acid baths and controlled heating, to separate them from surrounding minerals and remove impurities, including radioactive thorium and uranium (Li, Zuo, and Meng 2004; White and Kimble 1979).[25]

Rare earths are geological peculiarities. The elements are born in the hearts of exploding supernova and eventually consolidated into deposits through intricate geological processes in Earth's mantle. Deposits form in alkaline magmas, which are distinguished from the more common tholeiitic magmas by their chemical composition; alkaline magmas are higher in iron and magnesium, which keeps them more stable than tholeiitic magmas as they cool. This stability is important for the separation and formation of deposits of rare earths and related elements such as thorium and uranium.[26] As alkaline magmas cool and iron begins to solidify, a process called "fractional crystallization" begins, wherein certain minerals solidify as the temperature drops below their melting point, changing the composition of the remaining magma.

Imagine that as hot magmas cool, they sweat solidified elements. Once these elements solidify into discreet crystals, they no longer constitute the chemical makeup of the liquid magma. The elements that do not solidify into crystals during this initial phase of cooling are called "incompatible elements," and it is from this soup of incompatibles that rare earth deposits eventually form—if the conditions are right. For the formation of rare earth deposits, the critical difference between alkaline and tholeiitic magmas is that in the latter, the fractional crystallization process radically alters the chemistry of the magma and destabilizes the medium. But in alkaline magmas, the high iron and magnesium content facilitates the formation of relatively stable lattice structures that cradle the incompatible elements which slowly solidify into concentrated deposits of rare earth elements, niobium, uranium, and thorium over repeated cycles of gradual heating and cooling. A dramatic temperature change in either direction—such as that caused by a volcanic eruption or tectonic upheaval—and rare earths fail to coalesce in useful concentrations.

This means that the places most amenable to the formation of alkaline magmas are the seismically quieter zones, tucked away from the edges of tectonic plates in the anorogenic plains and plateaus. Although "anorogenic" means free from mountain-making disturbances, it does not mean stillness. If the spasms at the edges of tectonic plates are too great for the formation of rare earth elements, then the slow undulations toward the centers of tectonic plates are generally too stable to provide gradual heating and cooling cycles. It is in the "Goldilocks zone" between mountain-building activity and the sleepy center of the tectonic plate that the deposits of Brazil's São Gabriel da Cachoeira and Inner Mongolia's Baotou are situated, along accretionary orogenic belts between major intracontinental cratons (Lujan and Armbruster 2011; Nutman, Windley, and Xiao 2007).

A "craton" is an old, stable part of the lithosphere that has survived repeated cycles of rifting, merging, and mountain building. Craton refers to vast pieces of Earth's crust: the shields and platforms that form the continents. The deposits in northwestern Brazil and China formed after the subduction of the Mongolian and Guianan oceanic plates beneath the northern cratons of their respective continents (Chao et al. 1997; Voiçu, Bardoux, and Stevenson 2001), which formed the landmasses we now recognize as North Central Asia and South America. It was in the following extended period of relative geological calm within the orogenic belts that the rare earth-bearing bastnäsite were formed in gradual magmatic upwellings protruding from the mantle into the crust (Bai et al. 1996; Giovannini 2013).

The bastnäsite mining era began with Mountain Pass, California, and Bayan Obo in the 1950s. The sites that had earlier supplied Welsbach's gas mantles and the nuclear arms race featured monazite sands which are comparatively more abundant, but not as highly concentrated as rare earth bearing bastnäsite. Monazites also have their origins in alkaline magmas. Although alkaline magmas are less common than tholeiitic magmas, they occur in all tectonic plates. Many igneous and metamorphic rocks produce rare earth-bearing minerals such as monazite and xenotime which, when weathered, produce the monazite-bearing placers found in the rivers of Idaho and the black sand beaches of Brazil, India, and the Carolinas. To extract monazite sands requires shallow surface mining or riverbed dredging, as rare earth elements tend to be present between the surface and a depth of a few meters. The rare earths encased in bastnäsite require blasting, grinding, and several more sophisticated separation processes.

On the Moon, geological[27] conditions amenable to the formation of KREEP[28] deposits occurred under very different circumstances. As noted in the introduction, the prevailing hypothesis is that the Moon formed after a Mars-sized object smashed into Earth and broke off debris that eventually consolidated into our Moon. The power of the collision liquefied much of the debris. The lower density of the Moon left an ocean of this magma trapped between its mantle and its crust, which cooled very gradually. The conditions of lower gravity slowed the temperature flux, resulting in the formation of reportedly higher-concentrated deposits of rare earth elements than typically found on Earth (Heiken, Vaniman, and French 1991). On the Moon, the surface features of the KREEP zone are characterized by highlands and depressions, referred to as seas, which suggests that cooling processes in earlier years of the Moon's formation were shaped by impact events that altered the form and chemistry of the Lunar subsurface magma ocean (McSween, Harry, and Huss 2010). In Inner Mongolia, the surface features of the orogenic belt are characterized by high mountain ranges emerging out of wide desert and grasslands leading to the Mongolian Steppe in the north. A similar geomorphology characterizes northern Brazil, where reserves are found in the northern Amazon, in the foothills of the Neblina range leading to the plateaus of the Guiana shield to the northeast. These geologic formations have been used as national barriers by past empires, imperial powers, and contemporary states. This is emphatically not to suggest environmental determinism, but rather to point out that these geographical features have historically been used in the territorial exercise of power in a way that is consequential to contemporary rare earth politics.

That rare earths tend to be prospected for in border regions is crucial insofar as they are further imbued with a nationalist economic and geopolitical significance. Therefore, like petroleum, they tend to become the subject of territorial disputes where borders are not unanimously agreed on (Nevins 2004; Triggs and Bialek 2002). As illustrated by the cases of Baotou and São Gabriel da Cachoeira, the uneven geological distribution of rare earth-bearing bastnäsites and their incidence in historically contested areas compels state-making and border-marking activities by national actors intent on securing reserves. Based on this, it could be compelling to conclude that rare earth elements have a preternaturally high incidence in border regions, but this would be a false conclusion drawn from the fact that prospecting for economic minerals tends to be carried out in places often characterized as "marginal," such as in border regions.[29] In fact, a closer look at figure 5 shows a curious absence of rare earth deposits in continental Europe. This is not because there are no rare earths in Europe—quite the contrary. Sweden hosted the world's first rare earth mines. Rather, it is because mining rare earths in most of Europe had never been considered politically feasible enough to put such deposits on the map.

Geological surveys have historically played an important role in militarist adventures and nation building. The rare earth deposits in Bayan Obo, São Gabriel da Cachoeira, and Mountain Pass were found when prospectors were looking for other things: iron, gold, and uranium, respectively. In both China and Brazil, initial surveys were undertaken at a time when their central governments as well as foreign powers were seeking to rationalize national territory and take stock of domestic mineral wealth in order to fuel national development schemes. The geological survey of the Moon began as a minor outcome of Cold War superpower politics, but has come to provide the basis for renewed military ambition and a privatized space race. In his history of the role of geological rationality in settler colonialism, Braun (2000) argues for a consideration of "the consequences of this 'geologizing' of the space of the nation-state for forms of economic and political rationality, including efforts by the state to compel individual and corporate actors to 'do the right thing' in relation to a territory that now had an important sense of verticality" (14). As subsequent chapters show, this "right thing" is to exploit resources in the name of national development, geopolitical power, and the greater good. Under such a framing, mining is recast as the lodestar of progress. The capacity of states to implement or enforce large-scale extractive regimes becomes a measure of national sovereignty (as in Brazil) or of civilizational advancement (as in China, and with respect to outer space). Once mining is understood in this way, leaving an extractivist agenda unexecuted becomes a dereliction of moral obligation felt—among mining proponents—as impotence.

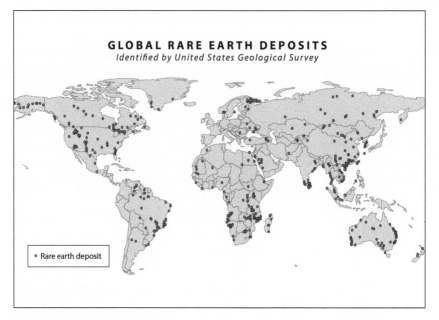

FIGURE 5. Map of global rare earth deposits identified by the United States Geological Survey, 2015.

Source: Image by Molly Roy.

In this way, the geography of rare earth extraction is inseparable from geographies of power and vulnerability. On the one hand, it is far easier to prospect for rare earths in "remote" regions populated by already-marginalized communities where legal regimes are either ambiguous or lacking teeth. On the other hand, the imperative to capture strategically valued resources serves as a convenient pretext for a host of territorializing agendas that allow the state to bring historically independent spaces within the purview of political economic control.

As the three primary cases examined in this book show, the formation of localized hegemony[30] necessary to carry forth rare earth extraction, processing, and production becomes part of the pedagogy of the state. In all cases examined in this book, rare earth resources are framed in terms of a collectively held entitlement despite the limited distribution of benefits, despite struggles over the terms of extraction and distribution, and despite the expropriation and dispossession that precedes production. Rare earths, like other strategically valued elements, stimulate the invocation of ethno-national "community" (Watts 1999) in order to mobilize the necessary social, political, and financial capital to extract them. Such was the case between Japan and China with respect to maritime boundaries. It is also true of Brazil, Colombia, and Venezuela, which periodically militarize their strategic Amazonian reserves (Maize 2012). Although assertions of national sovereignty are forbidden in outer space (UN 1967), the Moon is legally defined as a global commons, and therefore subject to familiar processes of use-based enclosure (Beery 2011). This was dramatically demonstrated in late 2015 when President Obama signed legislation recognizing the private property claims of US citizens to outer space resources. As discussed in chapter 6, the agitation over whom shall be entitled to exploit lunar resources is fueling a debate in Anglophone and Sinophone discourse over the relevance of existing legal conventions safeguarding outer space from enclosure or colonization. In all cases examined herein, collective rhetoric concerning rare earth production is invoked to legitimate multiple forms of dispossession.

Although practical applications for the rare earth discoveries in China and Brazil were not immediately apparent, their discoverers reportedly asserted their strategic value. Ding Daoheng, the Chinese geologist credited with the 1933 discovery of rare earth elements at Bayan Obo reportedly said, "these deposits will become an important treasury of China," nearly thirty years before rare earth processing began (Shi 2012). Similarly, in Brazil, the military geology expedition that publicly announced the discovery of the Seis Lagos deposits in 1972 declared that the yet-to-be exploited minerals located on indigenous lands "are the patrimony of Brazil, for the good of the Brazilian people" (Oliveira 2013). Analysis of the first lunar rock samples in 1969 ignited cornucopian fantasies (Lewis 1996) that are now at the heart of the contemporary resource-driven space race: National Aeronautics and Space Administration, private space companies, and the China National Space Administration have invoked the moon as "an offshore island," "rich with strategic natural resources" that are "desperately needed on Earth" (CM022 2013; Jain 2012). Indeed, the specific industrial uses of geological knowledge need not be readily apparent in order to direct territorial ambitions. Discovery is enough. Geologizing history and space is an important part of any territorial project; the newfound legibility afforded by geological knowledge transforms the complexity of existing landscapes, lives, and legal conventions into a stratified schemata of mineral wealth waiting to be put to productive use.

Conclusion

This chapter has shown that the term "rare earths" is political shorthand for a rotating cast of strategic minerals based around the lanthanide series at the bottom of the periodic table of elements. The term is extended to incorporate other elements as technology and politics change. Although the name is hardly an accurate descriptor for this suite of soft, ductile metals, it continues to function as a politically expedient term in the ongoing quest to acquire these resources, whether the intention is to avoid or impose the hazards associated with their extraction in specific places.

The political life of rare earth elements extends well over a century preceding the 2010 crisis. The geography of the rare earth frontier follows the reach of colonial, Cold War, and neoliberal power over the long twentieth century. Indeed, the history of rare earth production and innovation has been shaped by successive and overlapping epochs of violence and domination. Although violence has been inherent to our practices of rare earth prospecting, production, and consumption, it is too simplistic to describe rare earths as "conflict minerals." Unlike diamonds or coltan, mining revenues do not directly fuel armed conflict. Furthermore, rare earth elements enable technologies that bring great help: renewable energy generation, life-saving medical treatments, and the physical infrastructure of global communications networks. But rare earth elements are also critical to technologies expressly designed to do great harm: drones, smart bombs, and infrastructure for mass surveillance. In this way, the diversity of their applications mirrors the best and worst of the world we are building.

2

PLACING CHINA IN THE WORLD HISTORY OF DISCOVERY, PRODUCTION, AND USE

The Bayan Obo mine in Inner Mongolia, China, was once a sacred Mongolian site and an area in which competing empires fought for control. How did this place become the single greatest source of rare earth elements in the world?

Even after several years of diversification in the global supply chain, the Bayan Obo mine remains the single greatest source of rare earth elements in the world, producing approximately 50 percent of the global supply. As recently as 2013, it was thought to be the largest rare earth deposit on Earth (Ling et al. 2013). As this chapter shows, the discovery of Bayan Obo's mineral wealth was an outcome of competing territorial, industrial, and cultural campaigns to transform the frontier into a larger resource hinterland for China, the former USSR, and other imperial powers. These campaigns laid the foundation for the Bayan Obo mine in Baotou, Inner Mongolia Autonomous Region (IMAR), shown in figures 6 and 7, to later become the "Rare Earth Capital of the World," as it is called in contemporary literature and domestic parlance in China.

China's de facto monopoly is mythologized in contemporary rare earth discourse: "China as the world's richest rare earth nation" or "the only nation possessing all seventeen types of rare earth elements" (Wang 2010, 15) was a common trope in popular commentary in China and across the globe in the years immediately following the 2010 crisis. This trope revived the theories of Malthus (1798) and Hotelling (1931), which stated that as population grows and demand

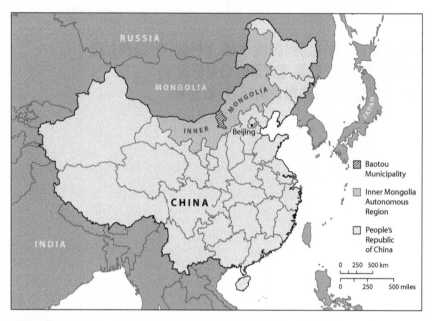

FIGURE 6. Baotou municipality is located in Inner Mongolia Autonomous Region, People's Republic of China. It borders Mongolia to the north and the Yellow River to the south.

Source: Image by Molly Roy.

increases, nonrenewable resources will be depleted, leading to higher prices, conflict, and suffering.

These myths of China's legendary rare earth endowment, buttressed with neo-Malthusian logic, have generated three contradictory outcomes. First, the myth of China as the world's richest rare earth nation has been deployed in domestic discourse to justify recentralization and greater state control over production (Cheng and Che 2010). This has had both positive and negative consequences. On the one hand, the state has placed greater importance on environmental conservation and regulation. On the other hand, state-run media has used the rare earth scarcity myth to reframe legitimate concerns over production practices into a nationalist paranoia about how the rest of the world is conspiring to pillage China's strategic mineral wealth (Wang 2010). Second, outside of China, the myth of rare earth scarcity justified prospecting in places previously considered off-limits, such as the Brazilian Amazon, Greenland, and the Moon. Third, China's purported geological privilege has been useful internationally to justify the World Trade Organization (WTO) suits brought by the United States, Japan, and the European Union against China's production and export quotas, based on the

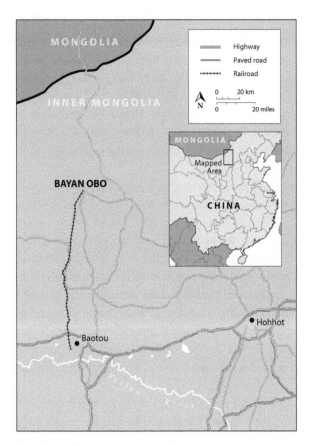

FIGURE 7. The Bayan Obo mine is located in the northernmost portion of Baotou Municipality. One rail and one road connect it with Baotou City, located in southern Baotou Municipality along the Yellow River. Due to the scarcity of water in the north, the heavy industry, urban, military, and supporting agricultural bases are located in Baotou City, in the south. Major highway and rail lines connect Baotou City to the south, east, and west.

Source: Image by Molly Roy.

argument that because the rest of the world depended on China's reserves, the quotas gave China an unfair advantage. Paradoxically, the successful suit against China to remove its production and export quotas in 2014 had the effect of bringing prices of some elements closer to precrisis levels, thereby eliminating the incentive for others outside of China to invest in the technological development necessary to mine rare earths in a more sustainable manner.

Although the Malthus and Hotelling theories possess a compelling logic, logic is not the same as truth. As with most nonrenewable resources, the global rare earth supply is not fixed. It is true that at any given period there is only a certain quantity of known reserves; however, the historical tendency shows that as resources are depleted, new deposits are found. Peak resource paranoia aside, we are so far from exhausting global rare earth reserves that the possibility does not merit serious consideration. Indeed, since the 2010 crisis, estimates of China's reserves as a percentage of the global total have been decreasing annually, from 50 percent in 2009 to 23 percent in 2012 (Chen 2011; United States Geological Survey 2014; Zhao quoted in Hilsum 2009; see also Information Office of the State Council 2012).

This percentage does not account for the discovery of the largest known global deposit at the bottom of the South Pacific or the undetermined but reportedly immense finds in North Korea (Bruce 2012; Evans-Pritchard 2013; Kato et al. 2011; Schearf 2014). As the post-2010 wave in global prospecting and research illustrated, scarcity-induced price increases incentivized exploration ever further afield, as well as investment in the technologies required to exploit resources in remote places. Indeed, China's Bayan Obo deposits were discovered under precisely these conditions several decades ago when competing imperial powers were seeking iron to build military-industrial complexes and uranium for their nuclear efforts. Therefore, China's contemporary rare earth dominance is only marginally explained by geological circumstance. Most countries have enough reserves to be self-sufficient (Baltz 2013).[1]

Chapter 1 detailed the pre- and postwar production history of rare earth elements—a history that spanned the globe. This chapter centers that global history within China, whose mineral resources, struggles over their definitions, and their role in state building played a critical role in the global politics of the early twentieth century. In the mid-twentieth century, Mao Zedong and Joseph Stalin's agenda to convert the windswept steppes of Inner Mongolia into a military-industrial heartland that could provision both Republics in the struggle against western capitalism and Japanese imperialism laid the foundation for China's long-term rare earth development.

This history underscores the point that China's rare earth monopoly arose from particular political and economic circumstances—some only marginally related to rare earth elements—rather than fabled resource abundance. The concentration of global production in China followed a contingent and uneven process over the course of several decades of cultural, political, and economic transformations across the globe. These transformations included post–World War II decolonization, Cold War politics, the global liberalization of capital, and the emergence of a shifting global division of toxic labor. This chapter and the following one seek to overcome the tendency in Sinophone, Anglophone, and Lusophone literature to attribute China's rare earth monopoly to geological determinism and the foresight of the Chinese state, on the one hand, or the vagaries of global (Western) capitalism, on the other. Rather, these chapters tell a very different story, presented in two parts: chapter 2 considers the territorial contests that were instrumental to carving IMAR, Baotou, and Bayan Obo out of the expanse of the Mongolian Steppe, whereas chapter 3 examines subsequent global shifts in political economy characterized by Deng Xiaoping's reforms and the Reagan/Thatcher counterrevolution.

Too often, the rise of Bayan Obo is explained in terms of Western industry migrating to China. Such analyses follow closely on the "impact model of globalization" (Hart 2002), wherein change unilaterally emanates from "the West" to remake "the Rest" in its own image (Hall 1992). There is some truth to such models, but also a great deal of historical illiteracy. To understand how and why Bayan Obo emerged as the rare earth capital of the world requires an in-depth examination of the world-historical processes from which our contemporary situation emerged. Without the foundations laid by decades of prospecting, frontier taming, and hinterland development activities, Baotou and Bayan Obo would not have been in the position to assume primacy in rare earth mining and processing when the industry left the west a few decades ago.

Most contemporary stories of China begin in 1949 and position the central government as the primary driver of history. However significant state planning is to contemporary industrial development, it is important to remember that China's de facto monopoly was not predestined. For the first half of the twentieth century, Bayan Obo, and indeed much of what is today IMAR was claimed by several competing powers: Japan, Chinese Nationalists, Communists, Mongolians, and Soviets, while more remote powers in the Euro-American world coveted IMAR's undiscovered mineral wealth.

To illustrate the uncertainty of China's consolidation of power over Bayan Obo, and to show the monumental scale of the task of industrializing Bayan Obo and Baotou, this chapter examines how multiple territorial ambitions—even those which were ultimately unfulfilled—were crucial to laying the foundation for Bayan Obo to emerge as the hometown of rare earths. Exploring this history provides two key insights. First, the establishment of Bayan Obo as the primary source of rare earths consumed worldwide was far from inevitable. The second insight, which holds important lessons for the future, is that the world's only long-running, diversified, and vertically integrated rare earth production site to date was built through deliberate, sustained, and internationally collaborative efforts.

There are several important lessons to be drawn from the historical record as we consider our possible pathways forward. The complexity of the history presented here shows that the development and management of rare earth resources has always been a complex and uncertain endeavor. Our current state of affairs was built on a combination of myths, violently exercised territorial ambitions, and breathtaking scientific and technological change. Terrible and wonderful things both emerged from these historical processes. We choose which we carry forward. To make an informed decision, we must look squarely at the past.

The foundation for China's rare earth monopoly was laid through three key processes, each discussed in turn. I first discuss how Baotou was the object of the territorial ambitions of several powers in the first half of the twentieth century. Geological knowledge production served as a tool of competing empires, and later, after the founding of the People's Republic of China (PRC) in 1949, as a basis for nation building and internal colonization. Next, I demonstrate that, contrary to possessing a grand strategy to use its rare earth monopoly to hold the world hostage (as many Anglophone and Lusophone commentators have alleged), the development of China's rare earth capital emerged from a broader policy to tame a border and create a red hinterland out of a historically stateless steppe using violence, military-industrial development, and cultural campaigns targeted toward different ethnic and gender groups. "Red hinterland" is a term that captures the production of inland resource bases as part of the socialist industrialization project between China and the USSR in the 1950s, as both powers sought to gain control over the historically dynamic territorial politics in what is today contemporary IMAR. I then examine the significance of atomic aspirations to consolidating the transformation of the inland frontier into a military-industrial hinterland, even amid national and international turmoil. These atomic aspirations and attendant visions of a socialist nation leading the global revolution laid the foundation for China's contemporary rare earth dominance. Over time, a changing cast of actors ascribed different meanings to rare earths as knowledge and politics changed. Each played a critical role in China's emergence as the rare earth capital of the world in ways that could hardly be foreseen.

Inland Frontier: Geology, Imperialism, and Nationalism

> And China began its survey in the early years of the twentieth century as it endeavoured to build up military and economic strength to combat Western and Japanese imperialist threats. It was the powerful countries that interested themselves in matters geological; and so it has continued to the present day (with power now projected under the oceans and toward other regions of the solar system).
>
> —Oldroyd (1996, 123)

Global rare earth extraction in the first half of the twentieth century was defined by colonial power, driven by the resource demands of war and industrial modernization, and accelerated by the fervor of the atomic age. In this time of shifting empires, revolutions, and world wars, the sacrifice of certain peoples and places was taken for granted in the service of grand ambitions. The complexity of this situation in China deserves special treatment, as competing imperialist and nationalist interests surveyed China's terrain with the intention of rationalizing a mysterious empire. Fundamental to this was the expanding scope of global geological prospecting alongside the reconfiguration of geological knowledge as a tool of state power in the global struggle between communism and capitalism (Underwood and Guth 1998; United States Department of the Army 1952). The deposits at Bayan Obo were discovered in this context.

Geology evolved as a science of territoriality (Braun 2000; Winchester 2009). It is a way to rationalize terra incognita in the service of state, empire, and nation building. The evolution of geological science in China is inseparable from imperial designs on China's territory and resources, beginning in the late Qing dynasty during the latter decades of the nineteenth century (Shen 2014; Wu 2010). In the 1880s, colonial actors in the German Foreign Ministry looked to China to expand their reach with the objective of eclipsing the more extensive British and French empires. The strategy for achieving this was overwhelmingly material: diplomatic transmissions from both the Chinese and the German sides were dominated with concerns over mining technology transfer, land use, and mining rights (Wu 2010). During this period, Western missionaries sought to capitalize on Chinese interest in the industrialization of the Euro-American world. Their strategy was to educate their target populations in science and technology in order to legitimate ongoing efforts at religious conversion. Hence the quest for souls on the southern Mongolian steppe directed competing imperial interests through the late nineteenth and early twentieth century, which geological interests followed.[2]

Geological survey teams from Germany, Japan, the Soviet Union, and China prospected in the steppe and desert of what was to become, in 1947, the Inner Mongolia Autonomous Region. Each team came to survey the region with the intention of bounding its geological wealth into a larger resource hinterland, whether for imperial Europe, Nazi Germany, imperial Japan, or the USSR. It is worth noting that the industrial orientation of geological survey activities, the cartographic portrayal of mineral wealth, and the construction of the infrastructure required to extract it were viewed as symbols of progress and modernity for imperialist, nationalist, capitalist, and communist interests alike (Davis 1926). Struggles around these developments were not necessarily struggles against industrialization or mining per se, but rather struggles over their meaning and purpose. Where imperial powers saw the extension of their surveyors and technologies to China as tangible evidence of the reach of their empires, Chinese nationalists viewed foreign-constructed mines, railways, roads, and ports as symbols of imperialist humiliation, which had to be reclaimed and improved for the development of the Chinese nation (Chen and Yao 1954).

China's Geological Society was the first scientific institution established in modern China in 1922 under the Republican government, with the express purpose of understanding the nation. As Shen (2014, 13) observes: "any viable understanding of the nation had to suit the twin criteria of protecting Chinese existence and promoting geological activity, and often the boundaries of one effort would shift to accommodate the other." Early geological research activity in China was characterized by international collaboration and open exchange of information.[3] The first meetings of the geological society featured international speakers, which cultivated Beijing's reputation as "the center of scientific life in Asia" (Shen 2014, 102). There was considerable inequality between Chinese and foreign researchers, however. The former were cash-strapped and relied on state directives and commissions from mining companies to keep China's Geological Society afloat. The latter were convinced that only an established colonial power could tackle the vast unknown of China's geology. To these observers, Japan appeared to be particularly well positioned, with control over key infrastructure extending inland from northeastern and southeastern China (Margerie in Wu 2014, 5n).

This infrastructure was built to consolidate imperial Japanese control over northern China, where Japan organized local puppet governments, engaged in prospecting activities, and took over heavy industry and munitions factories. In the 1930s, Japan occupied almost one-third of China, primarily the coastal and northern regions where the majority of China's government, research, and industry were located. The joint Communist-Guomindang (KMT)[4] forces organized a mass retreat to the interior while the militaries attempted to block the Japanese army's westward advance. Chinese geologists and other technically trained personnel focused on opening up new resource bases in central and western China to provide the raw materials needed in the war of resistance against Japan (Wu 2010). The force of China's resistance surprised imperial Japan and Western observers; instead of conquering the whole of China in three months' time as predicted (Utley 1937), China's defensive compelled Japan to pour ever more resources into their conquest of the mainland until their surrender in 1945.

A key part of China's resistance involved building an international presence. From the late 1920s through the 1930s, the KMT sought to reunify China, integrate its economy with the world economy, and engage as equals in international relations. During this time, Germany exerted arguably the greatest influence among the KMT's governing elite. Chiang Kai-shek viewed Prussian fascism as a model of rapid national development to emulate in order to mobilize and discipline the populace into breathing "New Life"[5] into the nation (Kirby 1984). To revive the German economy struggling in the aftermath of World War I and the global slowdown of the Great Depression, the Nazi leadership looked to China as both a resource hinterland and an immense potential market for German industry. The two countries brokered a set of barter agreements in which China would exchange raw materials for German military equipment,[6] railroad materials, and industrial equipment. The German government also sponsored Chinese students to receive training in Germany—when they returned to China, many staffed agencies overseeing the country's industrial and military modernization (Kirby 1984).

China exported tungsten, antimony, tin, and copper, which were crucial for Germany's post–World War I rearmament. Tungsten is an important element of war because it is extremely hard, so it is used to make projectiles. It also has the lowest coefficient of thermal expansion of any pure metal, making it an important predecessor to rare earth superalloys in the construction of airplane engines, tanks, rockets, and other steel alloys (Li 1955). Antimony was used to build ignition switches, produce flame retardants, and harden the lead used in bullets (Butterman 2004). Both preceded rare earth elements in the development of modern industry and warfare.[7] However, they were heavy and cracked unpredictably, so scientists sought replacements. It is important to note that while the 2010 crisis stimulated the search for rare earth replacements, rare earths themselves were once replacements for other materials.

Germany provided the majority of Nationalist China's foreign credit, so the KMT worked to increase China's mineral output as much as possible in order to generate foreign exchange (Kirby 1984). This meant consolidating KMT control over a fragmented territory and acquiring the geological knowledge necessary to exploit it. Geological prospecting is one means to settle territory, but building sovereignty through geological survey is a risky affair when multiple powers are involved. The KMT enlisted German, Swiss, and Danish experts to explore and map the subsoils of Inner Mongolia and Xinjiang. The international teams of geologists, archeologists (including John Gunnar Andersson), and geographers (including prominent Swiss geographer Sven Hedin) formed the Northwestern Scientific Expedition Team (*xibei kexue kaocha tuan*), which identified mineral, fossil, and archeological treasures in this inland frontier (Deng 2007; Hedin et al. 1944).

In April 1927, this team of forty left Beijing by train and traveled to Baotou—then a border outpost before the "uninhabited" steppe and desert—where they provisioned themselves for the long prospecting journey by mule and camel from Baotou to Alashan tribe in Ejina Banner (Xing and Lin 1992). In July of that year, when visiting a sacred mountain in southern Mongolia, geologist Ding Daoheng discovered the resources at Bayan Obo (Ding 1933). Although he is upheld as a national hero for identifying what was, for a long time, thought to be the world's largest rare earth deposit, he was part of a group within the expedition that was entirely focused on identifying iron sources to provision the German, Soviet, and nascent KMT industries. The presence of rare earths at Bayan Obo was not demonstrated until ten years later by the chemist He Zuolin[8] (Zhang 1995). Ding Daoheng's legacy as the discoverer of Bayan Obo's rare earth elements was established only at the end of the twentieth century.

Since Ding Daoheng remained active in the expedition for the entirety of its five-year duration—discovering dinosaur fossils and leading a KMT political expedition through Xinjiang—he did not publish his results until 1933 (Luo 2007). The following year he went to Berlin to complete his PhD, a fact that is often excised from popular retellings of his illustrious career, which also tend to omit international collaboration or the fact that the purpose of the prospecting was to identify mineral resources for multiple states. He returned in 1937 and took up a teaching post in Wuhan during the Western retreat (Luo 2007). After the founding of the PRC in 1949, he served on various high-level committees in central and western China, but never returned to Bayan Obo. The narratives emerging in the early twenty-first century lionize him as a local hero. The *Annals of the District of Bayan Obo* reports that he wished deeply to return to the northern borderland so loaded with significance in his life and for the nation. When he received the "joyous news" of the government-planned "geological assault" on Bayan Obo in 1950, he reportedly looked northward, wept tears of excitement, and fervently uttered his best wishes for the exploitation of Bayan Obo to begin as quickly as possible (Zhao 2010, 386).

Technically, these resources were not China's to sell until after the founding of the PRC, and even then, jurisdiction over IMAR remained ambiguous. Despite retrospective accounts in China and elsewhere that treat the consolidation of Chinese control over Bayan Obo as inevitable, any number of territorial power arrangements were possible. At the time of the Northwest Scientific Expedition, Inner Mongolia was officially under the control of the Nationalist government of Chiang Kai-shek, headquartered in Nanjing over one thousand kilometers away. The native Mongolians organized themselves into unions (*meng*), which were further divided into banners (*qi*), the latter of which collectively owned the soils and subsoils of Inner Mongolia. Although they maintained a certain degree of autonomy under the KMT despite the best efforts of the latter, there was a growing Pan-Mongolian independence movement. This was further provoked by the Chinese policy, initiated in the Qing and continued under the KMT, of encouraging Han migration to the region (Foreign Affairs Bureau 1916; Guomindang 1941). For centuries the sandy soils of Inner Mongolia had mostly discouraged large-scale migration by the agrarian Han Chinese. The soils can support fodder for no-madic pastoralists, but generally not for settled agriculture. Despite this, the Qing dynasty (1644–1911) began a resettlement campaign for Chinese peasants who had lost their land to war or natural disasters elsewhere in China. Although the Qing dynasty collapsed in 1911, before agrarian colonization extended into Mongolia, the Nationalist government adapted this strategy. It was renamed the "land reclamation program" (*fangken*) and was intended to alleviate land crises elsewhere while also taming the northern frontier by peopling it with Han Chinese. Over the first half of the twentieth century, this and related policies stimu-lated the migration of an estimated 4.5 million people, primarily Han Chinese, to the region that is now IMAR (Cai, Hai, and Sudehualige 2007).

Although Mongolia was officially under control of the KMT, southern Mon-golian leadership strove for autonomy. Locals feared that their lands would be overwhelmed by the growing Han Chinese population and wanted independence from China. Mongolian Prince Demchegdongrov[9] held a conference on South-ern Mongolian self-rule in 1933, the year after the completion of the Northwest Scientific Expedition. The Imperial Japanese Army exploited this desire for inde-pendence by establishing a puppet government and organizing armies under Mongolian princes to fight on their side against China. Prince Demchegdongrov announced his break with the Chinese Nationalists in February 1936 and gathered his military at Bailingmiao located within present-day Baotou municipality, forty kilometers southeast of the Bayan Obo mining district. At the time, Bailingmiao and Bayan Obo were in two neighboring provinces, Chahar and Suiyuan, that have since been subsumed into Baotou municipality under Inner Mongolia's contemporary administrative units (Tighe 2005). On November 14, 1936, the Japanese-backed Mongolian army invaded Suiyuan in order to set up a puppet government and exploit the region's mineral wealth.

The Japanese-backed Mongolian Army was devastated by the ferocity of the KMT resistance led by General Fu Zuoyi.[10] These battles over the future rare earth capital of the world marked the first time that the international press reported that Chinese forces successfully fought off Japanese-led forces, though Anglophone media was quick to attribute the success to military aid from Siberia, Czechoslo-vakia, and Indo-China (Jowett 2005; Syndicated Press 1937). Determined to capture Suiyuan and its mineral wealth under the guise of incorporating the ter-ritory into an independent Mongolian state[11] (Associated Press 1937b), the Japa-nese and Mongolian forces led small-scale skirmishes over the next eight months (Jowett 2005). Bayan Obo remained under KMT control,[12] but the historical co-incidence of Ding Daoheng's discoveries and the threat of Japanese occupation are frequently cited in local histories in order to portray Bayan Obo as a national treasure nearly lost.

The Inner Mongolian Construction Committee (*Neimenggu laoqu jianshe cu-jinhui*) under the Communist Party has worked hard to maintain a revised but living memory of local conflicts. Mongolian desires for independence and their collaboration with Japan have been purged from Chinese accounts of the War of Resistance against Japan in this area. In contemporary versions, competing Mongolian, European, and Soviet claims to the region are erased, and the Nationalist leadership is described as seeing the errors of their ways and joining the Communists (Bai 1999).[13] What is emphasized in the stories, local monuments, and official history around Inner Mongolia—particularly in Hohhot, Baotou, and Bayan Obo—is the unity of the ethnically diverse Chinese nation in the name of revolutionary Communism (Li 2005). As one official in Hohhot wrote: "In Inner Mongolia's glorious revolutionary tradition of opposing imperialism, opposing feudalism and opposing bourgeois capitalism in a revolutionary struggle of over a hundred years, countless Inner Mongolian revolutionaries of Mongolian, Han, and other ethnicities shed their blood and laid down their precious lives" (Chao 2000, 1).

The Mongolian heartland was seen as a frontier to be conquered by Japanese, Soviet, Communist, and Nationalist forces, with various Euro-American interests acting through each of them in different ways. For the first half of the twentieth century, ownership over the region—and therefore the legality of resource extraction—was far from settled. In the Sino-Soviet Treaty of 1945,[14] Stalin agreed that the Soviet Union would enter into the war against Japan on the condition that China recognize the de facto independence of Outer Mongolia (Rupen 1955).[15] The Nationalist diplomats opposed this, but eventually capitulated when Stalin offered Southern Mongolia to China against the wishes of his Mongolian allies. This is one example in which the frontiers argued over by distant powers are seldom as empty or lawless as imagined in the corridors of power.

This was the first formal division of Mongolia. After Japan's surrender in 1945, Mao Zedong leveraged the rift between the Mongolians and the Nationalists by offering autonomous governance to Southern Mongolia, renamed IMAR, if they joined the Communists in fighting off the Nationalists. After two betrayals of Mongolian hopes for self-rule—once from Japan and again from the KMT-Soviet alliance—and stirred by the hope that Southern Mongolian autonomy might lead to Pan-Mongolian unification, the Inner Mongolian leadership accepted the offer (Liu 2006). These were the conditions under which IMAR was established as an autonomous region in law but as a resource hinterland in practice. An intensive Communist Party education campaign followed, carried out by an ethnic Mongolian cadre elite committed to building a socialist Inner Mongolia (Wu 1999). Those who continued to fight for true independence or reunification with Mongolia were branded Japanese collaborators or "counterrevolutionaries" and later purged (Brown 2007; Bulag 1998). According to some estimates, 10 percent of the region's 1.5 million ethnic Mongolians were executed between 1945 and 1949 (Oyunbilig 1997). Although the establishment of the territory is framed as a glorious sacrifice in the name of Chinese Communism, the establishment of IMAR was necropolitical; that is, built on the politics of death in order to erase competing claims to the region's land and mineral resources.

In two of the primary cases examined in this book, geological prospecting and campaigns of mass death unfolded under the aegis of resource-driven development and nation building on the frontier.[16] When discussing this history with industry and policy interlocutors across my research sites, I consistently encountered the objection that geology, as an objective science, had nothing to do with political matters such as state-directed campaigns of violence aiming to rid the frontier of those with competing territorial claims. Such sentiments fail to distinguish the collection of facts—which are indeed, in themselves, points of objective data incapable of both help and harm—from the conditions under which such facts were collected and the ends toward which they were deployed. However removed we may perceive ourselves to be from the necropolitical campaigns that cleared the landscape for the benefit of rare earth industries and our easy access to rare earth-enabled technologies, it is absurd to propose that the collection of geological data, their recognition as valid, and their valorization as strategic can be considered in isolation of their social and cultural context. This is especially important in the case of Baotou and Bayan Obo, where violence along ethnic and ideological lines fundamentally shaped the subsequent distribution of benefits and hazards in a manner that continues to inform contemporary policy and politics in the rare earth capital of the world.

The question of who should benefit from exploiting the region's wealth hinged on the question of who conquered this frontier. On the eve of the 1949 revolution, the US Department of State was in negotiations with the KMT to collaborate in the geological "exploration of China for minerals of importance in the atomic energy programs of the two governments" ("Negotiations" 1948, 1018). The Atomic Energy Commission and affiliated private firms sought to secure low-cost monazite sands outside of India, while the Nationalists hoped that guaranteeing high volume sales to the United States would help generate foreign exchange that could then be used to purchase equipment to develop its own nuclear program ("Negotiations" 1948, 748). In exchange, the US Department of State arranged for Chinese scientists to receive scientific training in the United States. This agreement, which was all but approved by late November 1948, never reached fruition as the People's Liberation Army (PLA) defeated the KMT south of Baotou, drove them out of the Mongolian frontier, forced their surrender in the Northeast, and retook Beijing. Shortly thereafter in 1949, the KMT government fled to

Taiwan with the Sino-American survey documents for Chinese uranium and other minerals, where they would be kept safe from the "unauthorized" hands of Chinese Communists ("Negotiations" 1948, 751). But the geologists, by and large, stayed on the mainland. They maintained that "governments might come and go, but geological knowledge would always benefit the nation, so the development of a geological enterprise was inherently patriotic" (Shen 2014, 186).

The members of China's Geological Society adapted to the prevailing ideology immediately following the founding of the People's Republic of China in 1949. They denounced Nationalist-era practices as "capitalist" and criticized their international collaborators as conducting shallow research to serve narrow business interests and Western imperialism. They characterized US and British activities in the Middle East as capitalist geology, lacking both objectivity and long-term viability. They argued that Socialist geology, by contrast, had a boundless and bright future supporting the development of society and improving the standard of living for all workers (Cheng 1950). To correct the error of their past ways, the Society advised geologists to "learn to grasp the Marxist-Leninist standpoint, perspective, and method" (Shen 2014, 187).

Although much has been made of the setbacks dealt to the development of modern Chinese science by Maoism and the Cultural Revolution,[17] Oldroyd (1996) observed that the East-West divergence in geological theory during the Cold War was because much geological data was being gathered for military purposes, and was therefore kept secret. In contrast to the early twentieth century, scientists across the globe could not access each other's data. Geopolitical tensions directly hindered the advancement of scientific knowledge. In this case, it was not possible to reach a scientific consensus on the formation of the Earth when scientists in one region could not collaborate with scientists in another. Since China's 1978 reforms and the end of the Cold War, the ideological orientation of China's geological sciences has undergone further revision. Contemporary geologists now criticize the Marxist-Leninist approach as hindering the development of new theories necessary for bringing the discipline closer to the "objective reality of the Earth" (Li 1996).

Nonetheless, geological knowledge continues to be highly politicized and controlled—perhaps because foreign interest in China's mineral wealth persists in living memory. In China, the WTO cases were compared to Soviet, Japanese, and Euro-American imperial interests in the region during the first half of the twentieth century (Wang 2010). As a result, geological practitioners in the government, academies, and state-owned enterprises maintained an air of priesthood: most of the pertinent knowledge one might need was available in writing if the researcher knew precisely where to look, which conferences to attend, and which questions to ask, but there were numerous self-styled gatekeepers who insisted that what lies in Bayan Obo was a secret known only to the initiated. The operating theology was nationalism positioned against perceived threats of international resource appropriation.

But the professional pressures to publish in internationally recognized journals means that the debate over the formation of Bayan Obo has migrated from the cloisters of government agencies to international forums. This is somewhat distinct from the question of politics shaping the formation of geological knowledge about the region, but it is important to note that the project of geological knowledge formation about Bayan Obo is not yet complete. Papers continue to be published in Chinese and English on the "Bayan Obo Controversy," which refers exclusively to the geological debate surrounding its origins and formation. Briefly, the controversy concerns how the deposits at Bayan Obo—which are rich not only in rare earth elements, iron, and niobium, but also gold, uranium, and thorium—were formed (Bai et al. 1996; Wu 2007). One camp maintains that they were formed through sedimentary and later low-grade metamorphic processes, citing evidence of the presence of fossils and high amounts of silica, which support theories of an ancient inland sea (Meng 1982). Another argues that, like the general process of formation described in the introduction and chapter 1, they are the result of a carbonatitic intrusion typical of infrequent, low-grade seismic activity (Fan et al. 2006; Le Bas 2006; Le Bas et al. 1992). A third camp has attempted to unify the theories by distinguishing their underlying models and placing them both in a geochronology, concluding that the abundance of minerals suggests that sedimentary, metamorphic, and carbonatitic processes occurred over several heating and cooling periods (Smith, Campbell, and Kynicky 2014, Ling et al. 2013; Ren, Zhan, and Zhang 1994; Wu 2007).

The scientific stakes of the Bayan Obo controversy concern not only the foundational theories of the geological life cycle of Inland East Asia, but also the range of possible conditions under which rare earths are formed. The political stakes concern the validity of the claim that Bayan Obo is in fact an exceptional geological formation endowing China with a unique quantity of rare earth elements (Smith et al. 2014). If Bayan Obo is found to be a truly exceptional formation, then popular fictions concerning China's unique geological endowment gain legitimacy and both sides of the central postcrisis debate gain credence. On the one hand, central government initiatives to slow production, restructure industry, and control the production of geological knowledge about Bayan Obo would then be reasonable measures to preserve an exceptional deposit. International claims that such measures disadvantage the rest of the world, where large rare earth deposits are supposedly lacking, would likewise gain credence. However, if the Bayan Obo deposits were found to be unexceptional, most likely through ongoing comparisons with other rare earth deposits elsewhere, this would validate criticisms that both sides engaged in thinly disguised geopolitical exercises with little basis in geological fact, while also underscoring the fact that other countries do have the means to assume greater responsibility for fulfilling their own rare earth needs. This would in turn cast a new light on Deng Xiaoping's often misquoted statement: "The Middle East has oil; China has rare earths." Although provocative, it is too often invoked in Chinese and international discourses to naturalize the dominance of these regions in the production of strategic resources—when in fact the status of the Middle East as an oil exporter is an outcome of historical and political processes rather than geological fact: the top three largest known oil reserves are in the Americas,[18] and estimates of China's portion of known global rare earth reserves continue to fall.

Deng's comments were made in 1992, decades after oil deposits in the Americas had displaced Middle Eastern sources as the largest known reserves. Viewed in this context, it is hardly a statement of geological determinism. Rather, it is a savvy geopolitical observation on China's position in the global division of toxic labor made at a critical historical moment. The year 1992 marked the conclusion of the first Iraq war and the first year that China's trade and GDP returned to pre-1989 levels, driven in no small part by the proliferation of subcontracting networks into China's hinterland as international firms sought to cut production costs by moving the dirtiest and most labor intensive portions of their production chains overseas (Muldavin 1993, 2003). In China's orthodox political discourse of the time, the Persian Gulf War was understood as the enforcement of a resource-based status quo under the banner of capitalist democracy.[19] In other words, the war was viewed as an exercise to preserve the global division of extractive labor. Deng Xiaoping was commenting not just on the significance of natural resources, but also on the global geography of their production in the first major military exercise following the Cold War. The quote was widely circulated in domestic and international discourse in the years following 2010, as though the crisis represented some intended endpoint of long-term Chinese strategy. Such assertions are ahistorical.

While the production of geological knowledge was central to territorializing Inner Mongolia over the first half of the twentieth century, geological fact has not been the primary driver in contemporary politics surrounding China's rare earth monopoly. Rather, it is the fictions that have driven domestic resource nationalism and revived protectionist tendencies in China, Europe, and the United States. Even as geological fact percolates into popular and official discourse—most commonly with the prim acknowledgement that rare earths are actually not that rare—it has not addressed the often unspoken fears expressed in the logic underlying the Malthusian and Hotelling theories of resource scarcity. Thus, the power of those fictitious logics continues to fuel the fear that the wrong party—whomever that may be—will seize control of these strategically vital elements. This is why actual and structural scarcity are so seldom distinguished, and why rare earth elements are classified as critical and spoken of as treasures, despite their relative ubiquity (Associated Press 2013; United States Congress 2011; Wang 2010).

It was these epochal struggles over the Mongolian frontier that provided the motivations and the means to identify, characterize, and produce Bayan Obo's resources as a national treasure. Geology has served as both a practice of territoriality and an exercise of power in multiple contests for the land and resources of Inner Mongolia. Far from being an isolationist measure by a single-minded party state, the process by which Inner Mongolia, Baotou, and Bayan Obo were explored, captured, defended, and rationalized was in fact a transnational military and ideological struggle among powers great and small, all of whom saw in Bayan Obo the raw materials needed to reconfigure a historically contested frontier region into a strategic hinterland from which the enterprises of war and industry might be nourished. Once the sacred mountain was expropriated from Mongolian pastoralists, whether the resources at Bayan Obo should be exploited was never in question. Then as now, the struggles concern the meaning, control, and purpose of mining Bayan Obo.

Death, Life, and Development in the Red Hinterland

If the material bases of modern statecraft are war and industry,[20] then the material bases of war and industry are land, labor, and minerals. To transform the minerals within the subsoils of frontier regions into the substance of war and industry, labor must be mobilized on the land to build grand visions into reality. The process through which people reorganize space to sustain themselves and their ambitions is called territorialization. New territorial orders are built with the pieces of the old. In other words, a new order cannot be imposed without expropriating that which came before, yet the imposition of one order does not mean total obliteration of what precedes it (Deleuze and Guattari 1987). What emerges then are "territorial assemblages": spaces characterized by multiple and often antagonistic social orders struggling over the meaning and control of specific landscapes and livelihoods.[21] When we look at contemporary Bayan Obo as a territorial project, we see the unsettled land use regimes under which contemporary extraction has been carried out as an outcome of historical and ongoing state efforts to consolidate an industrial mining regime—and a particular form of social organization—to suit its evolving purposes. There are many forms this may take in struggles for control over land and resources on the rare earth frontier. In China, this has involved violence, military-industrial development, and social engineering campaigns directed at specific ethnic and gender groups.

The remainder of this chapter examines how the state selectively used necro- and biopolitics to build a red hinterland. Substituting propaganda for guns as a matter of both compromise and strategy, the state switched from the necropoli- tics of ethnic and ideological "purges" of the Revolutionary days to the biopoli- tics of creating desirable frontier subjects in order to achieve developmentalist and geopolitical ends on the IMAR frontier. Necropolitics describes the use of mass death and state-sanctioned killings to achieve political and economic ends (Mbembe and Meintjes 2003). This was one key strategy used to tame the his- torically autonomous southern Mongolian region. Building up a red hinterland in this space required a different approach. Biopolitics, by contrast, are concerned with managing human life toward political and economic ends (Foucault 2010). In other words, people—with a particular set of cultural values—are essential to de- veloping a robust regional economy. Both necropolitics and biopolitics are manifest in the revolutionary, cultural, and development campaigns that laid the foundation for Baotou and Bayan Obo to emerge as the rare earth capital of the world.

Territorializing IMAR has been a geopolitical project in the conventional and critical senses. Conquering this rare earth frontier was crucial to consolidating national territory, reordering landscapes and lives in profound and irreversible ways. Since 1947, campaigns to transform Baotou and Bayan Obo from Mongo- lian frontiers into an industrial mining hinterland have proceeded according to raced and gendered social mobilization programs. These campaigns channeled labor and resources to the region while constructing a particular ethnonational- ist culture around the mineral resources in Bayan Obo. This involved not only physically relocating people, but also mandating miscegenation, punishing dis- senters, and elevating mining-driven industrialization above all else. Mao con- tinued the migration and agrarian resettlement campaign initiated under the Qing dynasty and furthered by the Nationalists. Former nationalist general Fu Zuoyi, who was key in resisting Mongolian-Japanese attempts to conquer Bayan Obo and Baotou, oversaw the project to populate the border region with Han military and civilian personnel (Tighe 2005). The Chinese Communist Party launched a massive "Ethnic Unity" (*Minzu Tuanjie*) propaganda and education campaign to integrate Mongolians and Han Chinese in the Inner Mongolia Autonomous Region, thereby circumscribing any promised autonomy for southern Mongo- lians (Various 1948–58). This hinged on an explicitly raced and gendered divi- sion of labor orchestrated to advance urban and industrial development on the Inner Mongolian frontier. The objective was to conquer the frontier by building it into the national hinterland through complementary industrial and reproduc- tive measures (Wang 2000). When we examine these campaigns, we are able to understand how Bayan Obo in particular emerged as the single greatest source of rare earth elements worldwide. We are also able to better grasp the politics of sacrifice on which China's de facto monopoly is built.

The new Communist leadership was intent on taming the northern frontier and set to work marshaling its people and resources for the project of building "New China." Three core issues defined this agenda. The first were border threats: there remained a restive Mongolian independence movement, while the Soviet Union was determined to incorporate these same regions into its own industrial hinterland (Liu 2006; Nachukdorji 1955; Tighe 2005). The second issue stemmed from the first: the question of urban infrastructure construction and transportation. Third was the convergence of housing, land, and employment crises elsewhere in the country, particularly in postwar Manchuria, in a region where the Soviet Union held investments and privileges on par with those enjoyed by former colonial powers—particularly Britain and Japan. Mao's security strategy relied first and foremost on mobilizing people through migration and social engineering projects meant to Sinicize the frontier[22] and solve all three problems at once. However unwitting, this strategy laid the industrial, social, and geopolitical foundation for China's de facto global monopoly to later emerge.

Both China and the USSR had territorial ambitions for Baotou that figured into the earliest official relations between the two republics. On December 16, 1949, ten weeks after the founding of the People's Republic of China, the newly established Ministry of Heavy Industry (*Zhong Gongye Bu*) held its first meeting in Beijing to plan a new steel and metallurgical center in China's interior. That same month, Chairman Mao Zedong left China for the first time on a diplomatic visit to the Soviet Union. Shortly thereafter, in January 1950, Premier Zhou Enlai traveled to Moscow with a group of officials to negotiate a strategic partnership and development assistance portfolio with the Soviet Union (Su 2004). In those cold winter months, both sides hammered out the details of the Sino-Soviet Treaty of Friendship, Alliance, and Mutual Assistance in Moscow, which contained overlapping and conflicting visions for the IMAR frontier. In exchange for development aid and the promise of comprehensive military, industrial, and scientific knowledge-transfer, Mao's delegation granted the Soviet Union special privileges in Xinjiang, Manchuria, and parts of Inner Mongolia.

These territorial privileges represented a continuation of Soviet geostrategic practices since the 1930s, which had been formalized in concessions granted by Chiang Kai-shek in the Sino-Soviet Treaty of 1945 (Kraus 2010). The regions—and their mineral endowments—attributed to the "Soviet sphere of influence" were vast. They included Manchuria, Xinjiang, and what is now eastern and western Inner Mongolia (Li 1998). But it is a mistake to assume continuous rule in practice simply because it was agreed on in Moscow and Beijing. These two metropoles were thousands of kilometers away from these immense spaces peopled by nomadic civilizations and divided by mountains and deserts. Although imperial Japan had made efforts at regional integration in Manchuria and parts of Inner Mongolia by constructing roads and railroads, there was very little infrastructure integrating these regions with Beijing, and even less with Moscow. At least twenty-two different ethnic groups inhabited this region from Manchuria to Xinjiang, many of whom maintained strong cultural memory or an ongoing practice of independent self-rule (Liu 2006).[23] Under the Soviet and the Han Chinese, there were periodic independence movements from East Turkestan, in what is today western Xinjiang, to the eastern reaches of what is today the Inner Mongolia Autonomous Region, a region spanning some four thousand kilometers (Bulag 2004; Rossabi 2004). Across many of these areas, Communist Sino-Soviet rule was little more than a marginal concept with little bearing on everyday life.

State presence had a fluid history in these regions, varying between occasional military occupations, religious fiefdoms, and the periodic rise of city-states and oasis towns (Barfield 1989; Lattimore 1962). Peopled by ethnic others atop immense mineral bounty just beyond the reach of centralized national power, the regions were problematized by Mao and Stalin as lawless and underutilized.

This was precisely Beijing's "frontier problem," as defined in the introduction. The Communist leadership aimed to solve this problem through militarization, industrialization, and resource extraction. One way to make a regime real is to quite literally build it into a concrete, territorial fact. China's formidable contemporary rare earth enterprise emerged from this context of domestic and international efforts to build a diversified industrial base on the Mongolian frontier. However, lacking the necessary infrastructure and technology, Beijing granted resource and military concessions to the Soviet Union to develop oil and other mining interests in Xinjiang and parts of Inner Mongolia. There are two divergent interpretations of this world historical transformation of the Mongolian frontier in both English and Chinese literature. The first is that Stalin forced his Chinese counterparts to accept unfavorable terms in the territorial concessions, in essence creating semi-colonies on Chinese territory that Mao later found repugnant (Khrushchev 1971; Zubok 2001). The second view holds that Mao intended to build an industrial corridor to connect the Soviet and Chinese economies and saw sites of ongoing Soviet presence as potential industrial hubs to jumpstart China's economic recovery and stabilization (Kraus 2010, 135; Westad 2003).

Based on the portrayal of these concessions in Chinese newspapers of the time, it appears that central government leadership sought to build Baotou into an industrialized mining hinterland using the means offered by the Soviet Union. In exchange for technology, expertise, and aid, Beijing offered the Soviet Union qualified access to the mineral wealth and territorial privileges in Xinjiang and Inner

Mongolia's vast spaces (Various 1948–58). Regional state-run newspapers exhorted their readers to strive in every way to deepen the friendship between China and the Soviet Union. The fruits of this relationship were displayed in photographs of military technology, survey reports, and industrial construction featuring Chinese and Soviet personnel ("The People's Construction" 1949; Various 1948–58). Indeed, in assessing "the present character and probable future courses" of Sino-Soviet relations, the US Intelligence Advisory Committee[24] surmised that China "provides the USSR with a defense in depth, constitutes a valuable potential source of manpower and other resources, and is an important political and psychological asset" (Central Intelligence Agency [CIA] 1952, 4). Although Soviet political and economic influence along China's northern border was extensive, a 1952 CIA report found that "Chinese political and territorial interests have apparently not been sacrificed in the interest of Soviet expansion," and further predicted "an increase in Chinese Communist administrative control" in the region (CIA 1952, 4). However, were the frontier project so firmly and explicitly within Chinese control, it is unlikely that territorial tensions between the two powers would have culminated in the Sino-Soviet split in 1959–60.

The ongoing confusion appears to stem from a tendency to accept orthodox narratives of the founding of the PRC as a stark break with the colonial and feudal past instead of as a dynamic set of historical continuities, where Soviet privilege in Chinese territory was especially salient. Postcolonial China scholarship[25] has largely accepted the epistemic rupture insisted on in the orthodox founding narrative of the People's Republic of China as historical ontology. This is illustrated by, among other things, the manner in which the Soviet presence in China tends to be treated distinctly from Japanese and European colonialism. As a result, this scholarship has missed the fact that the enduring Soviet presence in northwest China was a direct contradiction to the territorial ideals of new China. The ongoing censorship of archival materials generated by Soviet activities in China from over half a century ago attests to the enduring sensitivity of the territorial narrative.

Yet it is unlikely that the Soviet investment in developing industry, infrastructure, and expertise in Baotou and Bayan Obo would have proceeded without the presumption that these regions would serve as the resource base for a world communist revolution controlled from Moscow—especially given that Mongolia was a Soviet satellite at the time. Provisioning this revolution required transforming the mineral resources in Bayan Obo into the tools of war and industry. This could not be done without a multiscalar effort, from international treaties to enlisting the populace in the cause of building a red hinterland.

Tremendous labor power was required to transform Baotou and Bayan Obo from historically autonomous frontier regions belonging to nomadic polities into a red hinterland for Moscow and Beijing. Taming the frontier involved necropolitics, in which some ten percent of the region's ethnic Mongolians were executed, and uncounted others killed in the immediate post–World War II struggles for control over the region. However, it is exceedingly difficult to build an industrial base using mass death. Life and labor were needed. Securing long-term extraction and industrial modernization depended not only on having plentiful labor, but also on establishing a stable, hegemonic claim to the land above these rich geological deposits. Therefore, the state coordinated labor migration in such a way as to encourage permanent settlement and multiethnic integration. For Maoist planners, migration was the solution to the housing, food, and work shortages in eastern China, which would also securitize the frontier by peopling it with Han Chinese. As a result, between 1953 and 1982 the Han population grew by 79.9 percent in Inner Mongolia (Chou 1982 and State Statistical Bureau 1986 quoted in Li 1989, 503–4).

Conditions in war-torn postrevolutionary Baotou and Bayan Obo were hostile compared to those in more established cities (Bi 2007). Winters were bitter, travel was arduous, and the work was dangerous. The first generations that migrated to IMAR via Xikou (Wang 2012) or along the Japanese-built Tianjin-Baotou railroad are regarded as pioneers who toiled hard and "ate bitterness" (*chile henduo ku*) in order to build modern China (Chao 2000). State propaganda promoted migration to IMAR as part of national construction and reunification, promising migrants not only the chance to settle down and live the proletarian dream of new China, but also to participate in the glory of restoring millennia-old peaceful and prosperous relations between Han and Mongolian peoples. These campaigns combined a curious mix of socialist modernist sensibilities and sentimental invocations of the second century BCE tradition of sending imperial Han concubines to marry Mongolian tribesmen.[26]

Mao's famous maxim that "women hold up half the sky" was motivated by the need to unleash the labor power of the majority sex in order to realize the herculean task of national construction. The egalitarian recruitment plans were driven by an explicitly reproductivist agenda to territorialize the lands beyond the Great Wall. By exhorting Han women to love "ethnic others," the state intended to quite literally incorporate them into the nation. In addition to these reproductive politics, the PLA enlisted women soldiers to "liberate" the northern frontier regions (Gao 2007).[27] Others were recruited as teachers, doctors, laborers, and farmers.

Furthermore, there was a common concern that a male-dominated Han population transplanted to the frontier would be unsustainable. As a solution, the PLA recruited Han women from eastern China and sent them to the Northwestern frontier to marry soldiers in couplings arranged by the party: "They were told that the guiding principle was 'the party assigns and the woman agrees'" (Gao 2007,

198). Women's inclusion in the public sphere was a revolutionary development, but it did not signal liberation from the gendered division of labor: "Although the Communist Party required the employment of more women in government offices and encouraged women to participate in society, the family was the focus of a woman's life, and the ideology of domesticity predominated" on the frontier (Gao 2007, 199).

In Baotou, high-ranking soldiers looked forward to being "rewarded" with a "pretty young woman" from Chengdu or the Northeast[28] in exchange for their service building and securing the IMAR frontier.[29] Men came first to build the state-owned Baotou Iron and Steel company (*Baogang*) and to open the Bayan Obo mine, but "those early leaders knew that having a lot of young men working in the mines and smelting plants would quickly create the conditions for chaos. A lot of men in one place—such a social arrangement is unsustainable. So they recruited women from Sichuan and the Northeast to come live here, and built textile *danwei* where they could live, work, and start families."[30]

Organizing and mobilizing women in Baotou began almost immediately after the founding of the PRC (see figures 8 and 10). The Baotou Municipal Women's

FIGURE 8. Photo from an *Inner Mongolia Daily* front page article on women's committee progress, June 12, 1952. The photo shows a Han women's committee representative posing with new ethnic Mongolian delegates from a pastoral region.

Source: Inner Mongolian Autonomous Region Archive, Hohhot, Inner Mongolia.

FIGURE 9.   An editorial comic on the back page of the *Inner Mongolia Daily*, March 5, 1952. The comic portrays a male industrial worker thwarting Western imperialists with his machine. The caption reads: "Speed up production, support the frontline."

Source: Inner Mongolian Autonomous Region Archive, Hohhot, Inner Mongolia.

Committee held its inaugural meeting in early December 1949 under the leadership of the Communist Party, less than three months after the liberation of Suiyuan[31] in September 1949. At this meeting, the women's committee coordinated with the outgoing Nationalist military government planning committee to develop a work plan and organize working groups to build up Baotou. Their pri-

FIGURE 10.   A team composed of young Han women shown surveying and cataloging Inner Mongolia's natural resources. Photos from a special to the *Inner Mongolia Daily*, titled "Young Women Build Socialism!," June 12, 1953.

Source: Inner Mongolian Autonomous Region Archive, Hohhot, Inner Mongolia.

mary tasks were to: "deepen the women's masses at every level, understand their conditions, study thoughts and feelings, propagandize the party's policies [and] build a solid base of mass feeling" (Zuo 2000, 244). This involved visiting every village and town in the municipality and establishing local women's committee offices. The Baotou women's committee focused on integrating women under its jurisdiction into the national military-industrial project. The committee was an important force for integration and subject formation. It identified and celebrated Han and Mongolian peoples' heroines and organized study groups to disseminate examples of virtuous, revolutionary women who helped liberate China, defend Suiyuan, or resist imperialism. Those who opposed the women's movement were branded counterrevolutionaries for failing to follow Mao Zedong's thought and for failing to respect the party's leadership.

In 1957, the Third National Women's Committee Congress resolved to mobilize more women to work in strategic national industries. This was significant for Baotou, where five integrated industries had been established in the preceding five years to build up China's military-industrial complex: Baotou Iron and Steel, which oversaw mining and beneficiation of the Bayan Obo output; Baotou Smelting Plant, which separated metals from the ores; Inner Mongolia First Aviation Factory, which built airplanes for China's nascent air force; Inner Mongolia Northern Heavy Industry Factory, which built tanks and trucks; and China 202 Nuclear Fuels Production Facility, which built critical components for China's atomic and hydrogen bombs. Women's labor and the participation of ethnic Mongolians were essential to building and sustaining these industries.

Mao's comment that women hold up half the sky was meant to liberate them from the "whole feudal-patriarchal system and ideology" (Mao 1967, 44). One conspicuous practice of this former system was the practice of offering high-status Han women to outlying princes in order to cement interstate relations and integrate outlying peoples into the Chinese nation through transmission of the Han "kin substance" (Bulag 2002). The practice was vilified even before Mao (Eoyang 1982). But with the intensification of migration it was revived and revised to emphasize revolutionary necessity, and romanticized as an ideal way to integrate IMAR with China.

Women were thus liberated from imperial reproductive servitude, but their reproductive capacity was put into the service of furthering the state's reach into the northern frontier. "Mongolian-Han blood mixing" continues to be fetishized in local and popular discourse.[32] But this excitement is differentially raced and gendered. Observing the attitudes in one IMAR village, veteran ethnographer D. M. Williams (2002, 97) noted that "nobody seems to mind if Mongol men take a Han wife, but residents frown upon 'losing their women' to the Han, even if the couple remains in the village." Conversely, Han male contacts casually expressed the view that a Han man marrying a Mongolian woman was positive because he would get a spirited and healthy wife, while a Han woman marrying a Mongolian was potentially endangered by his wild masculinity. But in most cases, such commentators were quick to point out that a Han woman would have a positive influence on her Mongolian husband, and was thereby serving the nation, just as the ancient imperial brides had.

Such raced and gendered tension cuts across questions of values, territory, and memory, which the state attempts to recast as positive and conciliatory in official publications. One telling example is an attempt to overwrite a particularly painful incident in which PLA soldiers distributed poisoned moon cakes to Mongolian nomads on Mid-Autumn Day during a "Han-Mongolian Social Gathering Festival." One official work published by the Bayan Obo Historical Literature Editing Committee and distributed to all local government offices concedes that "owing to a historical incident," Mid-Autumn Day is "the only unhappy date between the Chinese and Mongolian peoples." However, the "Social Gathering Festival" is still observed because Mongolians reportedly assisted a Central Government geological survey team in Bayan Obo in during the same period in 1950 (Zhao 2010).

The *Annals of Bayan Obo* extoll the virtues and wisdom of exploiting the local mineral resources. The official narrative praises visionary national investment and party leadership for making mining paramount in Bayan Obo, supposedly bringing international fame, development, prosperity, and happiness to all local peoples. Yet interviews with Mongolians in and around Bayan Obo conveyed the opposite. The name "Bayan Obo" referred to a sacred mountain, which resembled a great stone yurt.[33] Mining desecrated the sacred peak by literally emptying it out and inverting the mountain into a great hole in the ground: "this is what the Han do to us; they take everything that is ours, even the mountains," said one elder Mongolian resident of Bayan Obo.[34] The feeling of ethnocultural injury was present even among "Sinicized" Mongolians interviewed who were decorated members of the Communist Party. While passing a plaque marking the northernmost point of the Great Wall outside of Bayan Obo, my escort, who was a Communist Party youth volunteer, explained: 'the Han are crazy about marking everything in IMAR. They do that to make the statement: "look everyone, this land is ours. You Mongolians trespassed."'[35] Codifying locals as trespassers helps to legitimate the destruction of their lands, whether through mining, urbanization, or other military-industrial campaigns.

The complex historical relations between Han settlers and native Mongolians constitute the cultural landscape on which the military-industrial base in Baotou and Bayan Obo was built, thus informing the distribution of risk and vulnerability to this day. Over several decades, all aspects of life in Baotou, Bayan Obo, and vicinity were reoriented toward building the rare earth-fed military industrial base.

Atomic Aspirations in Cold War Baotou

These integrated industrial, military, and reproductive projects sustained the atomic aspirations of Cold War Baotou through national and international turmoil. During the Cold War, China's proximity to the "hot" sites of Hiroshima and Nagasaki made revolutionary modernization inseparable from militarization. China's efforts to build the bomb were an anticolonial reaction to both the Japanese imperial occupation of China and the US bombing of Japan. In the shadow of the atomic bombings of those two cities, conquering the frontier to build an integrated military-industrial hinterland took on an urgency that can be difficult for Euro-American observers to appreciate. Building the bomb was a means for the PRC and USSR to build an empire, not just to consolidate power on the Mongolian steppe, but also to project power globally in support of world communist revolution (see figure 9). This fundamentally shaped the industrial geography and cultural landscape of Baotou and Bayan Obo, out of which China's rare earth dominance emerged.

Mining, migration, and transforming the frontier into a red hinterland required extensive cultural and military campaigns, which were organized according to raced and gendered regimes of labor and sacrifice. Fulfilling China's atomic aspirations was no different. In the Museum of Inner Mongolia in Hohhot, the history of ethnic displacement during the development of the nuclear weapons program is memorialized as a patriotic cause that united all ethnicities. Figure 11 shows a life-sized diorama of a Mongolian family leaving their ancestral land on

FIGURE 11.   A life-size diorama at the Inner Mongolia Autonomous Region Museum in Hohhot, Inner Mongolia, depicting migrating Mongolian families leaving their ancestral lands to support the glorious nuclear achievements of the motherland.

Source: Photo by author, July 2011.

foot to clear the way for the PRC to conduct its nuclear weapons tests. While Mongolian independence activists cite these weapons tests as evidence of Chinese destruction and contamination of Mongolian lands (Oyunbilig 1997), the description at the museum explains that Mongolians made this choice for the glory of the People's Republic of China: "To support the Motherland's national defense and aerospace industry, from April 1958, the Ejina banner let go of tens of thousands of square kilometers of land. The banner government moved to a place 140 kilometers away from the original site. . . . The entire population of the migration covered one third of the population of the banner, more than half of the total livestock, and more than 3,000 mu of arable land" (Klinger 2013a).

On the wall beside the display is a quote from Marshal Nie Rongzhen, architect of China's nuclear strategy and head of the nuclear commission that oversaw the development of the bomb. It reads: "All ethnic peoples made a mighty sacrifice in the service of the national defense of the Motherland. We must work for them in return" (Klinger 2011, 4). This mighty sacrifice demanded different things from different people. While many Han migrants sacrificed their bodies and their lives in the hard labor of building up a Han-dominated military-industrial complex, many Mongolian natives were forced to resettle. Women of both ethnicities were recruited and coerced into reproductive labor, while men of both ethnicities were compelled to join military and construction campaigns. Resistance was counterrevolutionary and punished by imprisonment, hard labor, and execution.

However, as discussed in the previous section, developing the capacity to mine, process, and produce rare earth-doped goods first required a massive reorganization of society on the order of building entire cities and extensive infrastructure networks. Geological prospecting, the creation of a securitized mining and weapons-manufacturing landscape, and the training of thousands of personnel in the context of the Cold War produced the cultural landscape recognizable in IMAR today and laid the foundation for establishing the "rare earth capital of the world" in China. Much of this history is obscured by the propensity in Anglophone analysis to treat the founding of the PRC in 1949 and Deng Xiaoping's reforms in 1978 as historical breaks beyond which the past could not reach, as though our contemporary rare earth situation arose uniquely from the twenty-first century. One aspect of this misreading is the tendency in Anglophone literature to characterize the Cold War as "cold." The atomic bomb radically transformed the political calculus of the post–World War II world, but it was difficult for leading analysts in the United States to discern these rationalities in Mao-era governance: "Mao's mind may still envision catastrophe, like victory, as a cumulative effect. He may, therefore, fail to appreciate the dangers of atomic war, the war of massive and instantaneous destruction . . . his values are in a different balance from those of the Western world" (Katzenbach and Hanrahan 1955, 338–39).

This is, of course, inaccurate. But it illustrates US establishment thinking about China during the Cold War, wherein those vested with the responsibility to produce knowledge about China did so from afar with few checks in place to correct gross errors in analysis. The correctives that did emerge came from Westerners living and working in China who possessed language competency and ethnographic knowledge, but whose findings tended to be seized or suppressed during the years of the "Red Scare" in the West (Hinton 1966). With a few exceptions,[36] Anglophone accounts of heavy industry development in mid-twentieth century China tend to emphasize the weaknesses of the communist foe by citing its dependence on the USSR. For example, CIA intelligence report from 1952 found that "except for captured equipment, the Chinese Communist forces are wholly dependent on the USSR for heavy items of military equipment, and the large scale of Soviet logistics support has presumably further increased Moscow's influence with the Chinese military. The Chinese Communist Air Force is largely a Soviet creation and is wholly dependent on the USSR for equipment and supply" (CIA 1952, 3).

But there is a parallel story of carefully planned, generously funded, and judiciously managed iron and steel-based industrial production in the Sino-Soviet programs that focused on building an innovation-oriented industrial base in northern China to supply both republics with the hardware of industrial modernity and war. What Anglophone reports missed was that much of Soviet military aid was projected to come from the shared red hinterland of Bayan Obo and Baotou, which would provision the military industrial needs of both countries from a site considered to be remote and secure. The Baotou Project was the most prominent among these joint projects, giving it pride of place as the most successful and comprehensive Sino-Soviet military-industrial project. It was treated as a model for establishing a modern industrial base to produce some of the latest and greatest technology to fuel urbanization, infrastructure, and defense programs.

While detonating atomic bombs on the other side of the world might facilitate the conception of the Cold War as "cold" in the West, it is important to appreciate how differently the US bombing of Hiroshima and Nagasaki were perceived by Japan's neighbors. The Sino-Soviet response was to create a credible deterrent to the threat of a US nuclear attack by building an atomic bomb in China. Set in the Mongolian frontier, the two powers began building up China's nuclear capacity shortly after 1949. China's rare earth and nuclear industry have symbiotic beginnings.

The modernization of war and industry as we have known it in the twentieth and twenty-first centuries was realized to a significant degree through the discovery of new applications of rare earth elements, which, like uranium and thorium, belong to the nonferrous metals family. In the early 1950s, researchers across Eurasia and the Americas were developing rare earth alloys to use in the steel production process, thus transforming the skeletal system of modernity from heavy, rust-prone, and brittle to stronger, lighter, and more durable (Morena 1956, Kent 1953). Rare earths also helped make weapons of war more precise, long-range, and devastating (Bungardt 1959, Hickman 1955). Beginning in 1950 Soviet researchers experimented with nickel-based rare earth super alloys in order to move away from the high-temperature instability of iron-tungsten alloys used during World War II. Rare earths are the key to developing materials that remain stable in temperatures as high as 1,500 degrees Celsius, the sorts of temperatures needed for rockets and airplane engines. Soviet experts shared their discoveries with Chinese researchers in the early 1950s, and by 1956, trial super alloys were being developed in Baotou as a necessary step in China's quest to develop its own aircraft and ballistic missiles (Jiang 2013). In the first and second five-year plans of the

People's Republic of China, developing these technologies was of utmost importance, not just because they signaled the establishment of a modern industrial society, but also because they were viewed as essential tools for bringing about a world communist revolution.

The first and second five-year plans designated Baotou as a national industrial foundation (*gongye jidi*) (Geng 2007) and emphasized the importance of rare earth metallurgy in national industrialization:

> Industrial development calls for increased production of non-ferrous metals. Since it is still a weak link in our heavy industry, the expansion of non-ferrous metals industry is one of the important tasks of industrial construction under the First Five-Year Plan. . . . We must improve and expand experimental and research work to raise the quality of [metallurgical] products to higher standards. We must increase the output of raw materials needed for production of iron and steel, fluxes and refractory metals. (Congress 1955, 58–59)

Exactly *when* the search for uranium began in China under the PRC is the subject of debate. US intelligence reports from the time estimate that the search for uranium began in 1950, under the auspices of the newly founded Sino-Soviet Non-Ferrous and Rare Metals Corporation joint venture. Following conventional usage, Western observers assumed that "non-ferrous" included uranium and thorium, and so dated the official beginning of prospecting efforts in 1950 (Committee 1960).[37] However, Lewis and Xue (1988) cite Chinese documents[38] and interviews with former nuclear program personnel who insist that uranium and thorium were not part of the nonferrous and rare earth prospecting activities in northern China in the early 1950s: "According to one official Chinese account, in 1954 the corporation had put into production the largest ore-dressing plant for non-ferrous metals (but not uranium) in Northwest China, and thousands of technicians had been trained with young Chinese in charge of most operations. . . . The Chinese, in working alongside Soviet specialists in the Non-Ferrous and Rare Metals Corporation, undoubtedly did gain experience that was useful to their overall nuclear program" (Lewis and Xue 1988, 76–77).

The official narrative maintains that Chinese geologists had not developed sufficient know-how or technologies to begin uranium prospecting until 1956 (Lewis and Xue 1988). While this is possible, it leaves unanswered the question of what sort of prospecting activities were taking place while Soviet, Polish, and Hungarian experts were training Chinese geologists in the early days of the liberation (Lewis and Xue 1988, 76) or on what basis the United States and Germany were negotiating with the KMT government in the 1930s and 1940s. It is difficult to reconcile the accelerated production of nuclear weapons manufacturing facilities—most of which were approved for construction between 1951 and 1956—with the claim that uranium prospecting did not begin until 1956.

While the authors deserve credit for producing the most authoritative Anglophone account of China's nuclear development to date, I question their conclusion that repeated denial that the Non-Ferrous and Rare Metals Corporation was prospecting for uranium and thorium is sufficient to undermine the findings of other reports published in China and elsewhere.[39] In my interview experience from 2011 to 2013, it appeared to be standard practice to deny the existence or knowledge of radioactive materials regardless of known facts: only one official working in a public health capacity acknowledged the presence of thorium in the Bayan Obo mine output despite the fact that the rich thorium content of the mine is routinely referenced in scientific and official literature. The operating rationale seemed to be that radioactive materials are a sensitive topic and it was best not to talk about them. Academic specialists, by contrast, exhibited no such compunction.

In sum, the exact year that geological prospecting for uranium began in earnest remains disputed. The stakes of this dispute concern how much credit the USSR, United States, or other European countries should receive for developing China's nuclear weapons program, which later precipitated the development of China's rare earths separations program. If it occurred after 1956, then China had already taken over full control of the nonferrous metals companies as well as other key Sino-Soviet joint ventures. If it occurred before 1956, then it would be necessary to acknowledge that uranium prospecting in China had been led, at least in part, by outside powers (Chi 1990).

Nestled safely between "brotherly states," Baotou was a key site in the production of postliberation China's military-industrial complex. In 1956, the Second Ministry of Machine Building selected Baotou as the location for a large plant to produce uranium tetraflouride, nuclear fuel rods, and materials for hydrogen bombs. With the assistance of Soviet advisors, Plant 202 was built just north of Baotou city and integrated with other industrial facilities built up around the Bayan Obo mine. China's primary producers of tanks and other armored vehicles were also located in Baotou. These facilities were built with the original thinking that Baotou's position between China and the Soviet Union would provide strategic access to the resources at Bayan Obo to provision both powers and be well insulated from potential imperialist aggressors from the West. When China and Russia were on the brink of war in the 1960s, these weapons manufacturing facilities suddenly became very vulnerable. There were periodic shut-downs and evacuation exercises to protect those possessing the nuclear know-how to produce uranium-235 in the event of a Soviet attack on Baotou (Bachman 2007).

One of the notable things about China's nuclear program is the fact that it developed even in a period of intense turmoil: the Great Leap Forward (1958–61), which precipitated one of the greatest human disasters in history; the Sino-Soviet split (1960–89), which resulted in the abrupt loss of equipment and expertise as Soviet advisors withdrew and cancelled their outstanding equipment deliveries; and the Sino-Soviet border disputes (1961–69), which threatened China's grip on the northern frontier and exposed strategic industries to possible Soviet takeover. Despite this upheaval, nuclear construction—as well as the aforementioned social mobilization campaigns—continued to transform the Mongolian frontier into China's red hinterland.

The Sino-Soviet split and border disputes in the 1960s turned on both powers' tenuous claims to regions peopled by diverse autonomous groups, and highlighted the fact that China's northern border was largely unguarded (Li 1999). In the "Yita" incident of 1962, Soviet agents allegedly urged tens of thousands of local Mongolians and Uighurs in northern Xinjiang to flee to the Soviet Union. Chinese analysts suspected Soviet instigation of numerous other "separatist" incidents along the Mongolia-IMAR border during that time (Fravel 2008). China's response was to depopulate majority-minority regions in border areas deemed "sensitive" and resettle inhabitants in more "interior" regions, effectively emptying the frontier of those who would have competing claims to statehood. Some were conscripted for armed service or manual labor in other northwest development projects (Lewis and Xue 1988). Histories of forced resettlement have been revised to demonstrate a spirit of volunteerism and patriotism on the part of non-Han peoples (see figure 8).

The Sino-Soviet split created a state of emergency as Soviet experts withdrew from Baotou, taking much of their equipment with them. But this state of emergency catalyzed important research and technological developments for China's rare earth and nuclear industries. Innovation in one industry fed innovations in the other. Premier Zhou Enlai prioritized the production of nickel-rare earth alloys, which were essential to make the equipment used to produce uranium tetra-flouride at Baotou's 202 Nuclear Fuel Component Plant. As Lewis and Xue (1988, 98) note: "To meet the emergency, Premier Zhou Enlai authorized the Baotou Plant to draw 40 tons of these alloys from the central state reserves. The ministry asked a factory, a welding research institute, and an iron and steel academy to cooperate in fabricating the missing apparatus, and the result was the manufacture of China's first heavy equipment made with metal alloys."

This "heavy equipment" was put to work at 202, which produced China's first atomic bomb. After the first nuclear weapon detonation on October 16, 1964, Mao declared that China would not use nuclear weapons in a first strike and would designate them "for defense and for protecting the Chinese people from US threats to launch a nuclear war." With China's nuclear weapons manufacturing facilities up and running, senior researchers and technicians branched out to other political and industrial enterprises, such as planning, education, and research.

The nuclear program provided the physical, intellectual, and political infrastructure to further reorganize society in service of military-industrial expansion. Without this, the emergence of Baotou as the rare earth capital of the world would be unimaginable. The symbiotic knowledge economies of the nuclear program planted the seed for China to assume prominence not just with rare earth extraction but, importantly, rare earth separation. Xu Guangxian, considered the father of China's rare earth industry, had attended Washington University in St. Louis, Missouri, to conduct graduate work in chemistry in 1946. He finished his PhD at Columbia University and returned to China just as the Korean War broke out in 1951; he went to work on China's nuclear program in 1956. There, Xu participated in the race to separate uranium isotopes for China's atomic program. After China's first successful nuclear weapon detonation in 1964, he joined the chemistry faculty at Beijing University where he resumed work on rare earth chemistry. During the Cultural Revolution in 1969, he and his wife's US education implicated them as foreign spies. After being detained in his university dormitory for six months, they were placed in a rehabilitative agricultural labor camp in eastern China until 1971 (Stone 2009).

On his return to Beijing in 1971, Xu was given an urgent military mission to devise a way to separate praseodymium and neodymium (Deng 2009). He applied his previous research in extracting uranium isotopes to rare earth extraction and succeeded after four years (Jia and Di 2009). Separating uranium isotopes is complicated enough, but the proximity in atomic structures and chemical properties among certain rare earth elements posed considerable challenges that had not yet been resolved in any scientific community, anywhere. Recalling the four years of concerted effort required to develop the separation theory and method, he explained in an interview with the *Bulletin of the China Academy of Sciences*: "In Latin, praseodymium means 'the green twin'; neodymium 'the new twin.' They were the most inseparable twins at that time" (Xu quoted in Xin 2009, 100).

The now elderly scientist relates stories of working long days, following established but time-intensive methods of "shaking funnels"—that is, using centrifuges—and recording extraction data. Then one day he was fed up with established methods. He started to think back on his graduate work in quantum chemistry and the glory days of successful uranium and plutonium extraction in China's nuclear program. He came up with the "cascade theory of countercurrent extraction,"[40] which achieved rare earth oxides of 99.999 percent purity (Xin 2009). In 1974, he traveled to the Number Three Baotou Rare Earth Factory (*Baotou Xitu Sanchang*) to develop the techniques needed to implement this theory on an industrial scale. When he announced these findings at the first national rare earths meeting in August 1975, it signaled a fundamental shift in the global division of labor. Until this point, China had exported raw materials and imported separated and refined rare earths from the United States, France, and Japan, who previously controlled rare earth extraction processes (Deng 2009). With this development, it became possible for Western industries to subcontract specific steps in the rare earth refining process to China. The transformation did not happen overnight, but it was the beginning of China's technological superiority in the rare earth sector.

In addition to the political and economic considerations, this theory and method is hailed as a milestone in rare earth separation because it can be applied to ores of different composition on an industrial scale. Much of the research since this discovery has focused on expanding or applying the theory to separating other rare earth elements (Yan et al. 2006). This finding stimulated research throughout China, as reflected in the expansion of the Baotou Research Institute on Rare Earths. Established in 1963 under the former Ministry of Metallurgy Industry (Yang 2013), in its early days the institute was primarily focused on developing alloys for use in regional weapons and heavy industry manufacturing. Xu's innovations stimulated a research and industry renaissance, as government support for identifying new processing methods and technological applications expanded. The center now includes twenty-five research groups totaling approximately seven hundred employees. It also fosters international research collaborations built on the ties established by researchers who completed graduate work overseas, mainly in the United States, France, Germany, and Japan. Collaborative international research and exchange among the world's rare earth scientists has continued, even amid the diplomatic tensions that periodically flare among these states.

Conclusion

This chapter has shown that that the foundation for China's rare earth monopoly was laid through several distinct yet broadly related processes. Baotou was the site of territorial ambitions for several powers in the first half of the twentieth century. Multiple empires sought to conquer the Mongolian frontier through violence, the production of geological knowledge, and international negotiation.

Contrary to the claims of many Western commentators, China did not possess a grand strategy to use its rare earth monopoly to hold the world hostage. Rather, the development of China's rare earth base emerged from the drive to conquer a frontier, securitize a border and build a red hinterland on the Mongolian steppe between the brotherly states of the PRC and the USSR. At the outset, the aim was to provision both republics in their pursuit of world communist revolution. In the aftermath of decades of war and attempts by multiple competing powers to eliminate ethnic Mongolian claims to the region, Mao's government relied on Soviet assistance, domestic migration, and mandated miscegenation to transform the mineral resources in Bayan Obo into the substance of war and industry.

Rare earths were, for much of the time, one part of an immensely ambitious regional development plan. They were strategically valued and therefore became a common link among multiple heavy industries built up around the resources at Bayan Obo. This culminated in a robust, diversified, and in some sectors, a vertically integrated rare earth industry in the red hinterland.

A red hinterland is a comprehensive attempt at regional development, thus conceived in order to sustain a communist society rather than a singularly focused corporate agenda. This historical context is perhaps the least appreciated yet most crucial factor that explains China's rare earth dominance and perceived geopolitical advantage. This is also the most conspicuous difference between the long-term industrial success of Bayan Obo as a site of large-scale industrial rare earth production and the many rare earth mines around the world that have floundered or failed to launch at all. History suggests that the most successful production strategy is one in which rare earth elements are part of a diversified local and regional economy. *Laissez-faire* political economy has thus far failed to sustain strategic resource production for more than a few years, or in very rare cases, a few decades. But acknowledging this must not be confused with a naïve celebration of Maoist industrialism. How we might take the best lessons from across history to build a more sustainable future is taken up in the conclusion to this book.

The origins of China's rare earth monopoly lie in larger-scale and many seemingly marginally related processes. Reproductivist campaigns directed at different ethnic and gender groups, and the race to build the bomb were two such processes. Atomic aspirations drove the establishment of a military-industrial base on the inland frontier, sustained by an intense cultural commitment even amid national and international turmoil. The nuclear program fit within broader industrial and migration campaigns to stimulate rare earth research and produce today's cultural landscape. These various ideological and geopolitical quests among competing powers laid the foundation for establishing the rare earth capital of the world in China. This came at a tremendous cost, but neither the sacrifices nor the ambitions necessarily offer a replicable model for rare earth dominance, nor are the sacrifices confined to history.

"WELCOME TO THE HOMETOWN OF RARE EARTHS"

1980–2010

From 1980 to 2010, China's rare earths strategy shifted from export dominance to resource conservation. What prompted these changes?

Chapter 2 placed China in the world history of rare earth discovery, production, and use by considering the territorial contests that were instrumental to carving Inner Mongolia Autonomous Region (IMAR), Baotou, and Bayan Obo out of the expanse of the Mongolian Steppe. This chapter moves from the era of late colonialism, World War II and the early Cold War examined in the last chapter to the changing geography of the global rare earth frontier brought about by the dawn of neoliberalism and the turn of the millennium. This historical analysis continues the task of addressing the apparent paradox of China's near total dominance of global rare earth production despite the relative ubiquity of these elements in Earth's crust.

The rise of China's rare earth monopoly was a gradual process. Bayan Obo and Baotou were regionally important for rare earth elements and related industries from the 1960s to the 1980s. During this time, it was one of a sparse constellation of regionally important sites across the globe. In a way, each hemisphere had its own rare earth hinterland, built according to prevailing political economic ideologies in each context.

The contemporary global organization of production emerged from the historical conjuncture between mid-twentieth century regional developments centered on Baotou and late-twentieth century shifts in global political economy.

Beginning in the 1980s, ideological transformations in both contexts produced a dynamic that remade the global division of labor and precipitated the rise of neoliberalism. In this context, rare earth beneficiation and value-added processing began transferring from the West to China, culminating in the concentration of global rare earth production in Baotou and Bayan Obo by the turn of the millennium. Although the historic deregulations of the Reagan/Thatcher era in the late 1970s and early 1980s are generally treated as separate from Deng Xiaoping's 1978 reforms, both were fundamental to reshaping global economic geography in general during the final two decades of the twentieth century. This broader political economic context is essential to understanding the origins of China's rare earth monopoly.

As we follow the historical evolution of the global rare earth frontier in specific places, we see the larger political and economic currents that drive it, and we see that it cuts literally and figuratively across human bodies. Sending the hazards of rare earth production "away" from one population brings them "home" to someone else. Then, as now, the geography of rare earth production is driven by a fundamental tension between the desire to control their production and the need to confine the hazards in places deemed sacrificable. As this chapter shows, that tension is never resolved. Rather, it is temporarily settled through periodic fixes that depend as much on complex historical contingencies as on grassroots movements for environmental justice.

The cumulative environmental and epidemiological effects of the concentration of rare earth production in Bayan Obo and Baotou reached such extreme levels that in the first decade of the twenty-first century they were designated a threat to China's national security. This recognition from the central government—in response to local activism in Baotou and supported by decades of domestic scientific research—provoked serious reevaluations in domestic policy concerning the role of rare earth elements in China's development.

Although many of the postcrisis analyses of the concentration of rare earth production in China allege conspiracy or foul play on the part of the central government, a minority characterizes China's rare earth monopoly as an unintended consequence of the vagaries of Western capitalism. There are a few grains of truth to be found within these narratives, but both are too simplistic. Ascribing total causal force either to a masterminding Chinese monolith or to the invisible hand of laissez-faire capitalism conjures a world that does not exist. In such a fictitious world, nebulous forces on one side or the other push events toward predetermined outcomes, to which there was never any alternative and for which no one is ultimately responsible.

To erase the cooperative logics that created the present problems is to undermine our capacity to even conceptualize meaningful solutions. Against the backdrop of the struggles around Baotou and Bayan Obo's military-industrial mining base, this chapter connects the global shift to neoliberalism with the consolidation of global rare earth production in China, the human-environmental disasters that resulted, and sweeping domestic changes provoked by citizen mobilization against this state of affairs.

Global Shifts: Deregulation, Reform, and the Rise of China's Rare Earth Monopoly

> This issue has the potential to bring down America and greatly undermine its well-being and security, as well as the rest of the world. American prosperity and stability, and the world's strategic balance, may well hang in the balance on the REE [rare earth element] issue. If China consolidates its position and maintains a long-term monopoly of REEs, then it will ensure that most high-end and valuable products with REEs will be manufactured eventually in China and, thus, much of the world's wealth will shift to China and be utilized there. China, then, will attain the level of the new—and, possibly, sole—superpower in the world in the coming decades.
>
> —Dobransky (2013)

This statement, written in 2013, describes in the future tense a process that had, in fact, already happened. Shortly after the 2010 crisis, China's monopoly was cast as either an attack on the US defense industry or an outcome of China's long-term strategy, devised over decades to put the West in a "stranglehold" and demonstrate its economic prowess (Anthony 2010; Czarnecki 2010; Dobransky 2013; Martyn 2012). By late 2011, more historically literate interpretations of the rise of China's rare earth monopoly began emerging. These latter commentators observed that the transfer of rare earth production to China was at least partially caused by policy changes in the West (Baltz 2013; Maginnis 2011). Yet many of these observations tended to miss the mark by pointing to US environmental regulations as the reason why Western firms could no longer compete with China (Jasper 2011; Rowley 2013). In a classic example of neoliberal logic, the problems generated by neoliberal practices on a global scale were to be solved by the enforcement of neoliberal rules: if China simply "played by the rules" set by the World Trade Organization (WTO), this logic contends, the crisis never would have happened (Bradsher 2012; Office of the United States Trade Representative 2014a). While regulatory differences are certainly a key part of puzzle, more significant is the capacity of firms to exploit regulatory differences across hinterland spaces and across national boundaries, as illustrated by the Australia-Malaysia case discussed in the introduction, and the US-China case discussed in this chapter. This capacity grew precipitously in the last quarter of the twentieth century, with the rise of globalized neoliberalism.[1]

Before proceeding, I want to point out a broader tendency in the scholarship on the global changes of the past four decades that have muddled understandings of China's rare earth monopoly: the tendency to view China as an exception to processes of neoliberal globalization while simultaneously viewing neoliberalism as inexorable. What this presumption obscures is the key role played by China's people and environments in the global rise of neoliberal hegemony, as well as the necessary and conspicuous role of states in producing global neoliberalism. This is important to get right: global rare earth production became concentrated in China contemporaneously with the rise of globalized neoliberalism. Thus, our characterizations of the latter shape our understanding of the former.

As economic geographer David Harvey (2005) notes, neoliberalism rose to prominence relatively quickly beginning in the late 1970s from "several epicenters" across China, the United States, and United Kingdom. Harvey's characterization may be read to mean that Deng Xiaoping's reforms were proto-neoliberal policies, an interpretation that reads causality backward in time. It is more precise to say that contemporary global neoliberalism as we know it is impossible to understand without taking into account Deng's reforms and their consequences. More precisely still, the position of China's labor and resources in global political economy is constitutive of global neoliberalism in very concrete ways. Deng's reforms built on Mao's economic geography to enable the rise of far-flung subcontracting networks and the introduction of the largest pool of unorganized labor into the global workforce. Both attracted international firms and enabled them to escape organized labor at home. Without this, neoliberalism as we know it would be unrecognizable. It is nonsensical for a space to be constitutive of global neoliberalism while somehow remaining outside it, even if multiple forms of economic relations in addition to neoliberal practices characterize that particular space.[2] Therefore, China's domestic political economy contains characteristics recognizable as a *variety* of neoliberalism. If we accept this proposition, as many do (Kipnis 2007; Nonini 2008; Ong 2007; Rofel 2007; Wu 2010), then we must reexamine a basic tenet of many theorizations of neoliberalism, such as the withdrawal of the state (Harvey 2005; Strange 1996). The clear role of an authoritarian state in China's political economy allows us to better recognize the role of the state in neoliberalism more generally.[3] This, in turn, should quell the urge to explain China's rare earth monopoly as a result of *either* the state *or* the market, on the one hand, or processes driven by the East *or* the West, on the other.

Although the policy practices enforced by Deng, Thatcher, Reagan, and subsequent administrations differ with regard to the discursive role of the state, their common features indicate that neoliberalism is more precisely characterized by an antidemocratic reconfiguration of state functions to support transnational interests that profit from generalized precarity. This realignment of state functions is most vividly illustrated by the redistribution of public assets into private hands and the reorientation of police and military actions toward suppressing claims against the mass privations generated by neoliberal policies. Moving away from a position of "China as exception" to one that recognizes China's mutual constitution with contemporary global neoliberalism allows us to see more clearly pivotal role of the state in neoliberalism.

Neoliberalism requires, from start to finish, people determined to facilitate the colonization of public institutions by the private sector and who are in a position to marshal the power of the state to enforce ongoing expropriation.[4] In other words, the supposed "inevitability" of neoliberalism is the result of varying, yet intentional, antidemocratic state practices. This is, of course, a selective, uneven, and contested process that assumes different forms in different sectors.

Most scholarship has readily identified the central role of the state in producing China's market economy, but the dramatic sweep of these policies in the West has been characterized in opposite terms, as a retreat of the state. This despite the central role of Western governments in inaugurating and enforcing the necessary conditions to establish a global system of production and consumption defined, above all, by social and regulatory unevenness leveraged by footloose capital.[5] This paradoxical reading of global neoliberalism relies on the presence of the state to make sense of it in the context of China, while insisting that it is the absence of the state that characterizes neoliberalism in other contexts. The result is a tendency to frame neoliberalism as "self-actualizing" rather than actively produced, and to see it as an unstoppable force penetrating all things[6] rather than as an incomplete project that requires periodic authoritarian measures to violently impose it.

A unifying feature of neoliberal practices across diverse contexts is the forced privatization of public goods alongside the forced socialization of risk, pollution, and vulnerability as two complementary strategies that serve to enrich the wealthiest class at the expense of all others. Beginning in the late 1970s, such dynamics across national and international space—driven by state and private actors—were instrumental to the reorganization of the global geography of rare earth production. China's rare earth monopoly was definitively established at the turn of the millennium, yet the foundation for this development rests on the world-historical convergence of two distinct transformations unfolding in different hemispheres: Deng's 1978 economic reforms and Reagan and Thatcher's deregulations of capital, all of which gathered momentum in the 1980s. It is important to view these two developments as convergences that produced global neoliberalism as we know it, rather than in terms of a "Chinese exception" to "global" neoliberalism. As the early postcrisis interpretations show, if we approach the problem of China's rare earth monopoly with the view that China exists in a state of exception to the global neoliberal norm—rather than thoroughly embedded in and mutually constituted with contemporary globalization—we come up short.

Deng Xiaoping assumed leadership of China after Mao Zedong's death in 1976. In a series of market-oriented reforms initiated in late 1978, Deng ordered the dismantling of rural communes and initiated the "opening up" of the country to foreign investment. In his early international visits to Southeast Asia, Europe, and the United States to promote this policy change, Deng emphasized that China's policy priorities were economic and technological development. Although policies were articulated in Communist rhetoric, they were used to legitimate the reinstatement of the landlord class and the redomestication of women by decollectivizing commonly held assets and deeding them to male heads of households (Hinton 1990). Without rights to land or community services, the four hundred million peasants "liberated" from their rural livelihoods became the "cheap labor" that attracted foreign direct investment (FDI) and subcontracting to China (Erickson 2008; Muldavin 2000). This contemporary instance of primitive accumulation was crucial to the development of global neoliberalism.

Beneath the many important differences among the ideological frameworks within which Reagan, Thatcher, and Deng's policies unfolded, there was also a conspicuous—if unspoken—consensus concerning the acceptability of social brutality and irremediable environmental degradation as an intrinsic part of the modern production process. This is not unique to the contemporary era. Colonial relations rested on just such an assumption, explicitly rationalized with racist ideologies concerning the expendability of non-European "others." Perhaps it is a holdover of these ideologies that racially code China as a space of exception in the making of the contemporary neoliberal world order. This too would help explain the postcrisis portrayals of China as an inscrutable pollution haven. If we examine the spatial politics, we see that the dynamic reconfiguration of the global rare earth frontier is driven by the shared—again, if unspoken—ethos among policy elites that certain people and places must be sacrificed in order to maximize economic gains.

This shared ethos is lost in analyses that insist on reifying difference and mutual unintelligibility among China, the West, and the multiple political economies with which neoliberalism coexists. Take, for example, the question of primitive accumulation. There is a healthy body of work arguing that primitive accumulation refers exclusively to the first mass expropriation of English peasants from their land during the British Industrial Revolution, which gave rise to the working class.[7] However, Sinophone debates on primitive accumulation rest on a different translation and interpretation of Marx's concept that decouples primitive accumulation from the rise of capitalism: "'In reality, Marx never saw primitive accumulation as an early period of capitalism. He never even used capitalist primitive accumulation . . .' Under the rubric of primitive accumulation, 'all sorts of brutal behavior of a non-capitalist nature' was attributed to capitalism, just as in present day China, much 'pernicious behavior' of a non-market nature is being understood as rooted in the market itself" (Qin 1997, 4–5 quoted in Day 2013, 63).

In China, the transfer of rural and household capital to urban and industrial centers during the Mao era was described as "primitive socialist accumulation" and was seen as a positive development in service of socialist industrialization. In contemporary official narratives in IMAR, the mass expropriation of Mongolian pastoralists is likewise described as a positive transformation in the rise of Baotou and Bayan Obo. Prominent Chinese political economist Wen Tiejun (2004b) argues that industrialization—whether socialist or capitalist—fundamentally requires primitive accumulation in order to emerge in a specific place. This standpoint maintains that China's marketization was not the result of entirely exogenous factors, in the sense of global capitalism "penetrating" China. Rather, endogenous transformations[8]—not only Deng's policies but also the primitive socialist accumulation necessary for 1950s industrialization—created the conditions of possibility for capitalist relations to emerge within, as well as extend into, China (Wen 2004a). None of this is to say that primitive accumulation is a universal fact, but that it is a common feature of centralized industrial production regimes, whether communist or capitalist.

Translating this view to the political economy of rare earths allows us to see the origins of China's monopoly with greater clarity. The historical development of China's rare earth industry required a comprehensive suite of state policies and market stimuli in order to assume global prominence. In other words, the Mao-era frontier industrialization strategies involving mass expropriation and migration, as well as the political economic liberalizations initiated by Deng Xiaoping, were essential to positioning Baotou and Bayan Obo as the rare earth capital of the world—with sufficient cheap labor and state subsidies to undersell firms established elsewhere. But this is only half of the story.

The other half of the story lies in the deregulation of capital in the West initiated by the complementary policies of Margaret Thatcher and Ronald Reagan. Among their many policies, most consequential for the rise of China's rare earth industry and its decline in the West was the deregulation of corporate and financial capital in the 1970s and 1980s. This reduction or elimination of government power in a range of industries was enacted under the banner of creating more competition and innovation. In practice, it meant the reduction of corporate taxes and a significant reduction in territorial constraints on corporate and financial operations. The elimination of most corporate standards other than profit maximization unleashed a global corporate rush to maximize profit by lowering labor and environmental compliance costs in a renewed "race to the bottom" (Berle and Means 1932). This entailed states jockeying to present themselves as the most alluring frontier for transnational capital by underbidding each other to offer fewer taxes, environmental regulations, and safety protections in order to attract investment in the hopes of generating employment. These state practices, and their immediate consequences for working peoples and environments, continue to be bitterly opposed across the globe.[9]

Harvey notes that "the spectacular emergence of China as a global economic power after 1980 was in part the unintended consequence of the neoliberal turn in the advanced capitalist world" (2005, 121). However, it is more cogent to say that globalization as we know it would be unthinkable without China's 1978 reforms, which recast the third largest country in the world as a pollution haven and introduced four hundred million newly disempowered laborers into the global market. After all, how impactful could the neoliberal turn in the West be without the deregulation of the East? These shifts allowed firms to escape the disciplining of organized labor and environmental regulation in certain Western countries and then undersell firms that were slow to follow suit (Muldavin 2003).

Hence, China's policies, laborers, and environments played a key role in the global political economic restructuring that resulted from globally institutionalized neoliberal prescriptions. These prescriptions have been harnessed by footloose capital, on the one hand, and channeled by China's state-supported development strategies, on the other. In other words, key actors in China's state apparatus were "willing partners with global capital in a restructuring process long advocated by Western economists" and global economic institutions (Muldavin 2003, 9). Under such circumstances, any firm unwilling to leverage the global race to the bottom became a target for corporate takeover or faced bankruptcy. This pressure was vividly illustrated in the textile and manufacturing sectors, in which US manufacturing migrated first to Mexico and then, in the face of gains by organized labor, relocated to China. Such industrial mobility was possible because business operations could be disassembled and moved in relatively short order. The situation in the mining sector unfolded differently, due to the fixed nature of the assets, the sheer size of the capital goods, and the logistical and technological impossibility of performing immediate separation processes far off-site.

Despite these obstacles, the geography of the global rare earth industry followed the trajectory of other labor-intensive industries with the rise of global neoliberalism. The rare earth industry, too, participated in the construction of a new international division of labor insofar as extractive industries embarked on a race to the bottom, circling the globe in search of new frontiers where the ideal combination of lax environmental regulations and stringently business-friendly practices could bring the fastest rate of accumulation. This is not to suggest, however, that rare earth firms relocated wholesale to China. While the fixed constraints on mining activities kept US- and Europe-based operations running well into the late 1990s, the material peculiarities of rare earths allowed for the subcontracting of certain portions of the beneficiation process to China beginning in the 1980s. With the aid of state and private actors on both sides of the Pacific, industrial capacity was transferred piecemeal to China as Western firms sought to cut costs associated with the most hazardous aspects of rare earth refining. One particularly vivid illustration is the key role played by Edward Nixon in transferring various aspects of the beneficiation process from the Mountain Pass mine in Southern California to China, beginning in the 1980s.

Edward Nixon played an interesting role in the development of US environmental policy as well as the relocation of the US rare earth industry to China. A graduate of Duke University's geological sciences program, he reportedly persuaded his brother, President Richard Nixon, to establish the Environmental Protection Agency (EPA). He was active in the private sector in the late 1960s and early 1970s, occupying leadership roles in consulting firms that claimed to provide environment-related services—though information about their precise operations is difficult to find. For example, he was the president of Oceanographic Mutual Funds until the SEC opened a "gross misconduct" investigation against the company in 1971. He left because he reportedly "had nothing to do with it" (Turner 1973). He then worked as the vice president of Ecoforum, Inc., a Seattle-based firm "engaged in furnishing environmental information primarily to industrial corporations" (Lemon 1972). The heir to the Giannini fortune in Los Angeles had set up the company as an environmental consultancy. The idea was to provide "help to industrial concerns that had pollution problems." Reportedly, the company was never successful, Nixon was never paid, and he never approached the government on behalf of environmental or industrial concerns (Turner 1973). There are several other nominally environmental or resource-based firms that Nixon was involved in, but for which he was reportedly never paid, and for which he reportedly never performed any government outreach. The record of the younger Nixon's activities has been carefully buried in English-language archives, but sales receipts and other archival materials held in China help uncover some of this history.

In 1982, the well-established Mountain Pass mine in Southern California had just completed a US$15 million separation plant to allow for a 35 percent production increase. At the time, the mine was responsible for 70 percent of the annual global production of rare earth oxides (Goldman 2014). As the EPA began more regularly monitoring environmental practices in US mines in the late 1970s and early 1980s (EPA 2014), this Unocal-operated mine was subjected to increased scrutiny. In the early 1980s, Nixon reportedly approached mining executives with a proposal to subcontract some of the beneficiation processes to China in order to reduce environmental liabilities in California and to save costs. It was more cost effective to ship tons of minimally processed ore concentrates to China rather than ensure environmentally sound practices on US soil. For the brief period between 2012 and 2015 that production at Mountain Pass resumed, this continued to be the case, as export data at the Port of Oakland indicates (PMSA-WIL 2013). But the subcontracting went beyond this, driven by the imperative to move ever-greater portions of production offshore in order to reduce costs and increase profits. Over the course of the 1980s, Nixon's firm was one of several that facilitated the transfer of magnet production to China, setting up joint ventures with Chinese partners in order to produce cheaper rare earth magnets. These relatively early postreform joint ventures also provided technology transfer from international firms to Chinese firms. Technology transfer—combined with the mid-century industrial and scientific foundation laid by Sino-Soviet investment and ongoing state support—contributed to the steady growth of China's rare earth industry. Edward Nixon subsequently founded Great Circle Resources, Inc., to help downstream industries in the West purchase refined rare earth oxides from China. Thus, Nixon facilitated the transfer of production and technology to China and facilitated the resale of cheaper commodities to US firms.

None of this purports to hold a single person responsible for the decline of rare earth mining in the United States or its rise in China, but rather to point out the fluidity between state and private sector actors on both sides of the Pacific in leveraging the policy changes in both countries. Nixon, who was one of many, found a way to leverage the deregulation of capital on both sides of the process by helping a US firm subcontract the dirtiest aspects of production to China to cut costs and by facilitating the import of cheaper Chinese magnets to the United States. Within the context of the deregulation unfolding in both hemispheres, the profit-maximizing behaviors undertaken by executives of the Mountain Pass mine, facilitated by key politically connected actors, contributed to the demise of rare earth mining in the West and the rise of China's rare earth monopoly.

However, the transfer of rare earth production from the United States to China was not simply a story of *de*regulation, but also one of *re*regulation. Although Deng's reforms welcomed FDI and created a new workforce vulnerable to the whims of global capital, postreform China was by no means a free-for-all for international capital. Incoming foreign investment in China is channeled into three categories: encouraged, restricted, and prohibited. As a strategic national resource embedded in the defense sector, the exploration, mining, and processing of rare earth falls under the prohibited category. Investment in smelting and sep-aration is restricted to equity or contractual joint ventures whereas downstream industries—such as the production of dyes, fiber optic cables, and lithium-ion batteries—are encouraged to set up shop in China (Ministry of Commerce 2012). This strategy encourages the transfer of production capacity and technology to China, while also protecting intelligence on the country's geological endowments, for reasons noted in the previous chapter. Technology transfer is, and has been since the reforms, a cornerstone of the joint venture model. Still, China is far from a "wild west" of intellectual property rights theft, as many have alleged. The US China Business Council explains: "Technology is typically licensed to a China-based entity in which the foreign company has an ownership stake. In many cases the foreign company owns 100 percent of the entity in China; in some cases, the foreign company must form a joint venture with a Chinese partner. In exchange, the company determines the value of the technology to be transferred and negoti-ates a payment—the technology is rarely 'given' for 'free'" (US China Business Council 2011, 20).

The cost of purchasing exported goods produced with transferred technology is routinely taken into consideration during joint-venture negotiations. In 1985, China's Ministry of Finance, the General Administration of Customs, and State Administration of Taxation began offering rebates to Chinese companies and joint ventures for rare earth exports (Chen 2010), which attracted FDI and incentiv-ized the development of China's rare earth industry through export-oriented growth. At the same time, China's state-owned banks granted subsidized loans to joint ventures and state-owned enterprises (SOEs) to promote full employment as a means of maintaining social stability. Throughout the 1990s, those loans helped expand the mining sector while state investment in rare earth applications research stimulated production and drove down global prices. Meanwhile in the United States, troubles were mounting at Mountain Pass.

Beginning in the 1980s, the Unocal-operated Mountain Pass mine began piping wastewater to evaporation ponds fourteen miles away at the Ivanpah dry lake, which was just east of US Interstate 15 near Nevada in the Mojave National Preserve. The partially buried pipeline was structurally unsound and tended to rupture during routine cleaning operations to remove the buildup of mineral deposits, spraying the soil and surrounding vegetation with mineral slurry containing toxic concentrations of lead, uranium, barium, thorium, and radium 100–200 times above background levels (Associated Press 1998). An investigation by the EPA later found that sixty spills had occurred between 1984 and 1998, several of which were unreported. Between 1994 and 1997, seven spills reportedly releasing 350,000 gallons occurred along a stretch of pipeline near the eastern entrance of Mojave National Preserve, near a school and California Transportation

(Caltrans) employee housing (Cone 1997). Federal authorities calculated that roughly 600,000 gallons of radioactive wastewater spilled into the desert. In 1998, the mine halted processing operations when the San Bernardino district attorney issued a lawsuit and cleanup order (Margonelli 2009). According to public records, the company paid US$1.4 million in fines and settlements (Danielski 2009), but interlocutors close to the issue stated that the company paid much more.

Faced with fines for environmental violations and price pressure from cheaper commodities coming from China, the Mountain Pass mine closed in 2000 (Baltz 2013; Bourzac 2011; Coppel 2011; Galyen 2011; Goldman 2014; Zepf 2013; Zielinski 2010). This consolidated 97 percent of rare earth production in China, with the remaining global supply furnished by sales of stockpiles in the United States and Russia. By most accounts, the story stops there, concluding that the perfect storm of environmental regulation and cheap Chinese commodities condemned Western industry to bankruptcy. Such a discourse attributes causal force to broader market and regulatory mechanisms, and acquits specific actors of their key roles in shaping the contemporary geography of the global rare earth frontier. Given the former technological leadership of the United States in rare earth processing, the fixed nature of mining and beneficiation assets, and the supreme difficulty of working with these elements, it is absurd to explain the transfer of an entire industry by market mechanisms alone. Rather, it required multiple exercises of politically connected economic power, as illustrated by the case of Edward Nixon. Another relative of a powerful DC leader played a key role in transferring the last of US rare earth magnet production to China, in a series of events that have become known as the Magnequench Saga, chronicled by St. Clair (2003).

Key actors in China were willing partners to profit-maximizing firms seeking to capitalize on the country's lax labor and environmental regulations, and certain Western counterparts knowingly transferred technology and industry to China despite national security concerns. Magnequench was an Indianapolis-based company specialized in producing neodymium magnets necessary for information and aviation technology as well as the guidance systems of cruise missiles. It began as a subsidiary of General Motors (GM) that used Pentagon grants to develop permanent magnet materials, which it began manufacturing in 1987 at a facility in Anderson, Indiana. During the downsizing of the 1990s, GM sold Magnequench to the Sextant Group, an investment firm owned by Archibald Cox, Jr., son of the Watergate prosecutor who then became CEO. Sextant's primary clients were the San Huan New Materials and the China National Non-Ferrous Metals Import and Export Corporation, which continue to be two major players in China's rare earth trade. In 1998, Cox closed one plant and shipped its assembly line to China. GA Powders, a subsidiary of

Magnequench that produced the rare earth oxide powders used in the production of magnets, met a similar fate. The company originated in a Department of Energy Research Group at the Idaho National Engineering and Environmental Laboratory. After Lockheed Martin took over the national laboratory, it sold GA Powders to Magnequench in 1998. In June 2000, Magnequench management closed the Idaho production facilities and moved them to Tianjin, China. In 2003, Cox closed the final Magnequench facility in the United States, which was the only remaining domestic source of the magnets used in cruise missiles (St. Clair 2003).

Bipartisan members of Congress demanded that the Congressional Committee on Foreign Investment in the United States (CFIUS) review the takeover. They asked President George W. Bush to intervene on the basis of the Exxon-Florio Amendment to the 1988 Defense Appropriation Act, which requires presidential review of foreign investments and acquisitions that pose a threat to US national security. No such action was taken. This prompted a Government Accountability Office review of CFIUS, which found that most companies did not file until after the acquisitions were complete. When they were required to answer for national security concerns, companies withdrew their applications and CFIUS did not follow up. Members of the US-China Economic Security Review Commission attributed CFIUS inaction to the fact that it is led by the Treasury, which has conflicting interests with respect to China. Because the Treasury funds the US budget deficit, it has relied on China to purchase US debt; this made the department reluctant to alienate Chinese counterparts by halting the takeover bid of a US company (Greising 2005). Although it was clear to Pentagon and Cabinet national security advisors that China's firms had targeted Magnequench in order to develop long-range cruise missiles, the sale went ahead.

Neither the Nixon nor the Cox vignettes are intended to suggest conspiracy. Rather, it is important to recognize how Nixon and Cox were following particular economic rationales that, whether intentional or not, resulted in facilitating the rise of China's rare earth monopoly. Key individuals, market processes, or policy (in)actions cannot independently transform the geography of an entire resource frontier. It is the result of a combination of the three. Edward Nixon's legacy is that of a forward-thinking environmentalist. By all accounts, he was a patriotic capitalist looking to minimize environmental problems of rare earth mining on US soil while maximizing profits from rearranging production.[10] Cox's illustrious career at the helm of major investment firms (including Barclay's as it took on the assets of the ruined Lehman Brothers following the 2008 crisis) indicates that he was a savvy investor representing the interests of his clients who, in the case of Magnequench and GA Powders, happened to be Chinese state-owned enterprises. It is possible to remove the conventional US-China geopolitics from the analysis and to explain the actions of both sets of actors solely according to profit maximization. Despite their patriotic credentials, they are striking embodiments of the Jeffersonian observation that: "Merchants have no country. The mere spot they stand on does not constitute so strong an attachment as that from which they draw their gains" (Jefferson 1814). It is, in fact, openly acknowledged in the US foreign policy community that "many politicians and corporate executives have transnational interests and cannot argue for more aggressive policies without risking possible severe consequences" (Dobransky 2013).

This is less a story about China threatening US national security than the symbiotic relationships between varieties of neoliberalism: two rapacious political economic ideologies put into practice. In the West, corporate competition measured by greater profitability became the highest ideal of the marketplace (Foucault 2010; Harvey 2005), while democratic governance elided into defending and enforcing this state of affairs. In China, the idea that economic growth was to be achieved at all human and environmental costs was overseen by a modernizing state (Lin 2009; Qiu 2007; So 2007). China's rare earth monopoly, and Western dependence, was one result of this symbiosis between varieties of neoliberalism. The true costs of these transformations are borne by the landscapes and lives in Bayan Obo and Baotou.

FIGURE 12.   Satellite image of industrial geography of Baotou City and vicinity.

Source: Image by Molly Roy.

Radioactive Rivers, Cancer Villages, and Long Tooth Disease

> With the development of the rare earth industry and the rare earth smelting pro-
> duction processes, the "three wastes" of radioactive material, thorium, and rare
> earth associated environmental radiation pollution have become increasingly se-
> rious. China's rare earth enterprises are compelled to take this seriously.
>
> —Wang (2007, 2)

In the process of extracting rare earth elements from the Earth, many other things
are brought up with the ores. Hazardous materials such as arsenic, heavy metals,
and radioactive materials are unleashed from their subterranean confines and
released into the air, soil, surface water, and groundwater. The resulting radioactive rivers, cancer villages, acute chronic arsenic toxicity, and long tooth disease
constitute an environmental and epidemiological crisis so grave and expansive
that addressing it is now viewed as a matter of national security and territorial
integrity.[11]

There is no shortage of documentation on China's intensifying environmental
challenges (Smil 1993; Wu 2009; Yang 2005). That the environmental conditions
of rare earth production emerged in Western consciousness following the 2010
crisis is not the result of deliberate secrecy on the part of China. To put it bluntly,
no one bothered to write about these critical issues until the end of the first
decade of the twenty-first century, much less examine the detailed Sinophone research literature. This lacunae with respect to rare earths has had the effect of
framing environmental and epidemiological harms as both inherent to rare earth
production and a distinctly "Chinese" characteristic. This presents rare earths as
an unavoidably dirty industry and racializes pollution as "Chinese," thereby ob-
scuring the contingency of such hazards and the broader global investment in
these destructive practices. This section clarifies the precise nature and origin of
these hazards as they emerge in rare earth mining in Bayan Obo and processing
in Baotou. The formation of radioactive rivers and cancer villages as well as the
scourge of arsenic toxicity and long tooth disease directly informed changes in
China's rare earth policy and practice.

The hazards of rare earth mining and processing begin with geology (Zhou
et al. 2002). Recalling the discussion in chapter 1, the conditions under which rare
earth deposits are formed are also amenable to the formation of naturally occurring radioactive mineral deposits, which accounts for their frequent coincidence.
In the 2003 People's Republic of China Radioactive Pollution Prevention Law, rare
earth mines were given specific mention as sites containing high levels of radioactive ore concentration (Hu 2003). But neither the stones nor the sands are

hazardous in themselves. The hazards emerge when the elements are removed from their earthly confines; dispersed through crushing, transport, and benefi- ciation processes; and concentrated in local and downstream water sources. In her extensive study on the environmental dynamics of local rare earth produc- tion, Bai Lina (2004) identified ten points at which radioactive and toxic materi- als are released into the environment.

These findings indicate two key points. First, Chinese scientists have been care- fully studying the hazards for years. Second, viewing table 3 in relation to the

TABLE 3 Distribution of Thorium in Bayan Obo iron, steel, and rare earth production, 2002

PROCESS	RAW MATERIALS AND PRODUCTS METRIC TONS PER YEAR	THORIUM (%)	TOTAL THORIUM METRIC TONS PER YEAR
Bayan Obo ore	88,436,000	0.0364	3219.07
Rare earth concentrates	993,000	0.191	189.61
Iron ore	33,366,000	0.013	433.76
Tailings	54,077,000	0.048	2596.7
Rare earth iron ore slag	3,690	0.095	0.35
Sinter	74,918,000	0.0057	426.49
Sintering smoke and dust	1,078,000	0.0064	6.92
Steel forge	49,672,000	Undetected	Ignored
Smelting slag	20,862,000	0.02	418.42
Smelting dust	111,000	0.004	0.44
Smelting gaseous emissions	746,000	0.01	7.46
Rare earth rich slag	39,470	0.0555	2.19
Rare earth concentrates REO content 50%	7,160	0.191	1.36
Primary rare earth alloy	9,120	0.18	1.64
Baogang smoke and dust	28,000	0.015	0.42
Rare earth alloy slag	121,000	0.0152	1.84
Locally consumed rare earth concentrates	686,000	0.191	130.95
Rare earth processing acid concentrates	598,000	0.184	109.83
Rare earth processing soda concentrates	88,000	0.24	21.12
Exported rare earth concentrates	300,000	0.191	57.3

Source: Bai (2004).

regional industrial geography shown in figure 7 shows that the hazards emerge along the 241-kilometer stretch between the Bayan Obo mine to the Baotou processing facilities, shown in figure 12. To simplify, there are four main stages where environmental hazards emerge. The first is the mining process, which generates dusts laden with heavy metals and radioactive materials. Then there is the refining process, where sulfuric or hydrochloric acids are used to separate elements from their parent rock (Hao and Nakano 2011). The third is the waste management from the primary processing and beneficiation activities, which generate slag containing high levels of radioactivity (Wang et al. 2009).[12] The fourth concerns the disposal of rare earth-containing products (Gullett et al. 2007; Weber and Reisman 2012). All rare earth elements cause organ damage if inhaled or ingested; several corrode skin; and five elements—promethium, gadolinium, terbium, thulium, and holmium—are so toxic that they must be handled with extreme care to avoid radiation poisoning or combustion (Krebs 2006). Because rare earths tend to coincide with radioactive thorium and uranium, mining is also a radioactive waste management situation (Bai, Licheng, and Lingxiu 2001), which compounds the difficulty and expense.

The ores extracted from Bayan Obo are processed into concentrates using high temperature roasting with sulfuric acid. This energy-intensive process "cracks" rare earths out of their parent rock through repeated cycles of acid baths, smelting, rinsing, and cooling. Every ton of rare earth concentrate produced generates approximately one ton of radioactive wastewater; seventy-five cubic meters of acid wastewater; 9,600 to 12,000 cubic meters of waste gas containing radon, hydrofluoric acid, sulfur dioxide, and sulfuric acid; and approximately 8.5 kilograms of fluorine (Hurst 2010).

Because of the chemical similarities between rare earths, uranium, and thorium, separation is extremely difficult and requires roasting at temperatures above 300 degrees Celsius. The high temperatures convert thorium to a mobile and water-insoluble form, thorium pyrophosphate, which accumulates in the mine tailings and is difficult to recover or reuse. To be used for nuclear fuel, thorium must be purified and converted to thorium nitrate, which is difficult to do with thorium pyrophosphate. Nevertheless, because thorium has the highest melting point of all oxides at 3,300 degrees Celsius, it is required only for a small set of highly specialized industrial applications (Cardarelli 2008).[13] This is significant for two reasons: first, rare earth processing concentrates thorium in tailings in forms that are especially mobile and extremely difficult to reuse, and second, given the limited current applications for thorium,[14] there are few incentives to invest in the development of more efficient techniques to recapture radioactive waste material. The high cost of thorium storage further discourages initiatives to reprocess mine tailings in order to separate out the thorium pyrophosphate, since doing so would create another expensive waste management problem distinct from the tailings pond (Xin 2006).

Separating thorium and uranium from the tailings does not eliminate the radioactive threat. As Marie Curie discovered, as much as 85 percent of the radioactivity remains in the host material after the element is removed (Edwards 1992). How does this work? Unlike nonradioactive elements, the atoms of radioactive elements are unstable. This means that the atoms explode, releasing highly charged alpha and beta particles. It is helpful to think of these particles as shrapnel from microscopic explosions that rip through microscopic material such as cells, nuclei, and DNA. This shrapnel randomly breaks or burns chemical bonds, and this is what causes many of the harmful effects of radiation exposure.

The "explosions" described here are actually radioactive decay. As elements decay, they transform into other elements: thorium disintegrates into uranium, which disintegrates into protactinium, which disintegrates into radium, which disintegrates into radon gas and polonium. Radon gas atoms disintegrate into "radon daughters," which include another half dozen solid radioactive materials that stick to surfaces such as dust particles and are easily inhaled. The end result is lead.[15] If inhaled, these particles stick to the airways of the lung and increase the risk of lung cancer (EPA 1990; Liu 1996). A few micrograms of radium in the body will cause the bones to go soft, teeth to fall out, gums to bleed, and cancers of the bone and soft tissues to develop.[16] The tailings are hazardous because the bulk of the radioactivity is left behind in the slurry, which continues to generate radon gas (Edwards 1992). Finely pulverized powder circulates in the air and water, introduced into the wider environment by wind, rain, leaching, and industrial accidents. It is through these molecular, unintended, and poorly controlled processes that the rare earth frontier literally cuts through the human body at the cellular and atomic level.

Radioactive Rivers and Cancer Villages

Heavy industry requires water. The industrial geography of Baotou, shown in figure 12, is fundamentally shaped by this imperative. The Bayan Obo mine is located eighty kilometers from the border with Mongolia, and 241 kilometers north of Baotou's iron and rare earth processing industry. The processing industry is built around the state-owned enterprise Baotou Iron and Steel, or Baogang for short. Baogang set up its operations near the abundant waters of the Yellow River, which bends northward through Ordos and Baotou across the Hetao Plain. Most of the heavy industry and industrial waste is concentrated in Baotou, far from the site of extraction (see figure 7). Nevertheless, the cancer mortality rate in the Bayan Obo mining district rose from 107.93 per 100,000 in 1989–90 (three times the national average and five times the average for western China), to 155.7 per 100,000 in 1997 (Chen, Siwei, and Xiaonong 2010; Liu 1996). In the mining district, the three leading causes of death are cancer; unspecified poisoning and accidents; and infant mortality (Zhang et al. 2001).

The Baogang tailings dam has been growing since the late 1950s to the point where it has become the "world's largest rare earth lake," containing two hundred million tonnes of radioactive slurry. This thirteen square kilometer lake is located ten kilometers north of the Yellow River and twelve kilometers west of the Baotou city center. Since 2005, the rare earth working group of the State Planning Commission has issued warnings about the structural vulnerabilities of the tailings pond in light of seismic activity and climate variability (Xin 2006). An earthquake or unseasonably heavy rains could cause the tailings lake to burst, the consequences of which, officials warn, would dwarf even the most alarming river pollution incidents to date, such as the Songhua River incident in 2005. The situation has received domestic media attention over the past fifteen years, and has been the subject of local, regional, and national research initiatives since at least 1975, several of which are available to the public in China. While the gravity of the situation may not be translated for non-Sinophone audiences, the issues are nonetheless well known in Chinese academic, official, and popular discourse. As one reporter put it: "It is not only about the risk of dam failure. Around the dam, sheep have suffered from long tooth disease, villagers are suffering from cancer, and this once fertile vegetable garden has become a place where seeds do not grow and the water cannot be drunk" (Li 2010).

The Baogang Environmental Monitoring Station released reports showing that the area of the most severe surface contamination[17] in 1998, 2002, and 2006 increased threefold over the eight-year period, from 4.92 square kilometers (about a kilometer out from the edge of the tailings dam) in 1998 to fifteen square kilometers in 2006 (Wang et al. 2009). This is an indicator, first, of the strict delimitation of the areas of investigation under Baogang jurisdiction, and second, of the intensity of contamination generated by wind transport. Of course, the circulation of radioactive residues is not confined to this limited area. Because radon gas is a decay product of uranium and thorium, the tailings lake releases it continuously. It is more dense than air, so it travels close to the ground and can cover one thousand kilometers in a couple of days on a steady breeze (Edwards 1992).

The tailings pond does not have any sort of lining to prevent seepage (Wang, Xie, and Chen 2007). It sits at 1,045 meters above sea level, while the agricultural villages between the tailings pond and the Yellow River are at 700–1,000 meters above sea level on sandy soils. Villagers in this area describe the tailings pond as a "hanging lake" (*xuan hu*) over their heads. Drawn by gravity and larger regional drainage dynamics, contaminated water travels through the sandy subsoil down to the Yellow River at a rate of 300 meters per year. Atop the sandy subsoil is Baotou's declining "vegetable base," which historically supplied the city with its produce. The problems with the tailings dam have been documented since the 1970s, when farmers in the surrounding villages noticed decreasing yields in the vegetable plots. A 1994 survey found elevated levels of radioactive contamination in vegetables produced downstream (Zhao et al. 1994), and a 2002 study of village well water released by the environmental monitoring station found that radioactive salts exceeded the safety threshold by a factor of ten (Xin 2006).

Chinese journalists have coined the phrase "cancer village" to capture the manner in which pollution from rare earth beneficiation has defined death for residents surrounding the tailings ponds. In one small village of seventy-five households, locals reported six cancer deaths a year, and frequent strokes among adults (Liu 2013). In a neighboring village, residents coined the phrase "one in seven" to draw attention the rate of middle-aged cancer deaths (Wang 2006). This is compared to the national average of two cancer deaths per one thousand rural inhabitants (Guo et al. 2012).

In early 2004, villagers initiated a dialogue with Baogang and the Baotou city government to demand compensation and resettlement. Local officials responded by digging a deep well to provide potable water, but villagers opted not to use it after finding that boiling the water generated a gooey white residue (Ren 2013). Furthermore, the geological profile of the area is such that high levels of arsenic and fluoride are concentrated in hydrothermal deposits, so while deep wells may escape the surface seepage, they expose users to different hazards. Unsatisfied with local responses, villagers successfully demanded environmental monitoring reports and used them to petition higher levels of government in Hohhot and Beijing. They have had some success reframing the plight of their villages as a threat to regional food security and a major regional water supply because the pollution contaminating their land is also contaminating the Yellow River, which flows through five provinces after passing by Baotou and supplies water to more than one hundred million people. Their efforts have helped shape national and international consciousness about the human cost of China's rare earth monopoly (Hilsum 2009; Jeffries 2014). In compensation, Baogang agreed to give five villages surrounding the tailings dam ¥5 million (US$814,000) per year. However, divided equally among the twenty-five thousand inhabitants, this comes to about ¥200 (US$32) per person, which is a tiny fraction of the amount needed to relocate from the polluted village. "It's not even enough to buy water," noted one citizen. One newspaper article quoted a villager as saying, "Soon our lives will be over, what good is this money? We just want to move away. Not moving means waiting for death" (Hui 2013).

Since beginning their campaign ten years ago, villagers have successfully compelled the state to action. As of 2015, the tailings pond has been shrinking. Where the waste is going has not yet been disclosed. Villagers have also gained better access to healthcare at the specialized osteology hospitals in Baotou city, but doctors reportedly stop short of naming the causes of ailments. One villager interviewed in 2013 reported: "The doctor told me that the bones of people from my village are different from the bones of people who live in other places. Our bones are weak, they grow strangely, and they break."[18] Although some doctors may be hesitant to share the realities of the ailments with their patients, local citizens are well aware of the situation: "In the dry season we breathe the dust. In the wet season it comes into our water. We eat the pollution in our food. The livestock eats it in the grass, and we eat the livestock. It is very dangerous. Everyone knows the problem, but it is too big to manage."[19]

Chronic Arsenic Poisoning, Skeletal Fluorosis, and Long Tooth Disease

Because of its deadly, uncontrollable character, discourses on radioactive waste contamination tend to crowd out the maladies caused by other toxins proliferated by rare earth mining and processing. Outside of Baotou, where the majority of the processing takes place, arsenic and fluoride are the primary contaminants (Liu et al. 2005; Xia and Liu 2004; Yu et al. 2005; Zhao et al. 2013). Their proliferation in the environment surrounding Bayan Obo and Baotou has profoundly altered the human landscape in this region formerly populated by Mongolian nomadic pastoralists. It is, heartbreakingly enough, often possible to distinguish true natives from migrants by the skin lesions caused by arsenic poisoning, and by the malformed bones and decaying teeth which are symptoms of chronic fluorosis. It is estimated that 40 percent of the peri-urban and rural inhabitants of the Hetao Plain, or roughly three hundred thousand people, are suffering from arsenical dermatosis (Mao et al. 2010). The social and economic calamities caused by arsenic in drinking water are considered comparable to secondhand smoke and indoor radon gas (Chowdhury et al. 2006). Chronic arsenic toxicity is linked to "cardiovascular, hepatic, renal, gastrointestinal, neurological, reproductive problems, and malignancies" (Mao et al. 2010) and has been demonstrated to hinder the cognitive and intellectual development of children in Baotou (Li et al. 2003).

Arsenic and fluoride are naturally occurring elements that do not become hazardous to humans until they are liberated from their earthly confines and concentrated in soil and drinking water. Arsenic and fluoride enter the human body by ingestion or respiration. Windborne residues from mining activities build up on the surface of the soil and are absorbed by food crops and grazing livestock. As rainwater carries the elements into the soil, these elements can build up in the water of shallow wells. But digging deeper wells is no escape: as they are bored deeper to escape surface pollution, hydrogeochemical deposits rich in arsenic and fluoride are tapped, sometimes at 200 or 400 hundred meters in depth (Wen et al. 2013). Thus, escaping one problem creates another. As with the scholarship on radioactive hazards around the tailings pond, there has been extensive Sinophone research on these issues dating back several decades (Luo 1993; Wang, Kawahara, and Guo 1999).

Fluorosis has serious effects on livestock. Skeletal fluorosis causes certain bones to grow at irregular rates and to soften. In livestock, this causes their teeth to grow uncontrollably, soften, and fall out. This makes it impossible, for instance, for sheep to graze. They eventually starve to death. In the region, extensive cases of irregular tooth growth have been observed since the 1980s, with devastating effects on local pastoralist livelihoods in the following decades. Citing local Animal Husbandry Bureau statistics, an official in the Shadegesumu district, located between Baotou city and Bayan Obo within Baotou municipality, noted that the number of sheep in the area declined from 160,000 in 1964 to 16,000 in 1999 (Zhao 1999).

Skeletal fluorosis is also a devastating condition in humans, with crippling effects in its advanced phases (Tamer et al. 2007). In some cases, it causes the long bones to continue growing. Because the ligaments and muscle tissues remain the same size, they stretch to the point of tearing or snapping. In other cases, an increased fluoride load in the human body causes the bones to become more dense and brittle. This occurs when it binds with calcium ions in the bones to form an insoluble salt that, when cleared from the body, takes away part of the bone matrix. As a result, the early phases resemble arthritis and osteoporosis. As skeletal fluorosis advances, the ligaments of the spine and long bones calcify, hindering movement. Fractures occur easily, and cannot be treated using standard methods because of the brittleness of the bones. Recovery from bone fractures in these cases is extremely rare. Other effects include thyroid damage, ruptures of the stomach lining, and loss of motor control cause by spinal compression (Reddy 2009). There is no particular medicine to counteract fluorosis. The best treatment method is to find a new source of drinking water (Sharma et al. 2013).

The high incidence of these ailments over the past three decades has stimulated an emptying out of the grasslands around Bayan Obo and rural to urban migration, though this constitutes neither a clean escape nor access to a secure livelihood. The landscape surrounding the Bayan Obo mining district is marked with abandoned and crumbling houses formerly occupied by farmers and no-madic pastoralists. The official narrative delivered by the local director of the Land and Resources Bureau in Bayan Obo is that all farmers and herders were resettled and compensated. Officially, there is no grazing or agriculture within the mining district and all formerly agropastoral land has been converted to wind energy generation.

At first glance, this appears to be true. It is a desolate landscape under an im-mense sky, made all the more striking by the presence of hundreds of wind tur-bines. After traveling twenty kilometers away from the outskirts of Bayan Obo mining district, I encountered an elderly herder in a village built around a small spring. He boasted that the water was cleaner than anything around for kilometers, "far better than the pollution they drink in the city," and that enterprising urban residents periodically approached the community with offers to purchase water or build pipelines to the district. He explained that Bayan Obo used to belong to Mongolian nomads, and that the Bayan Obo mine was once a sacred mountain in local religious lore. Even as the mine expanded, nobody wanted to leave, but "first the animals got sick, then the babies, and then everybody else." Most left because the only alternative was death.[20]

Facilitating or forcing out-migration helped the local government contend with the economic and public health problems caused by human and livestock expo-sure to pollution. As additional measures, the local government in Bayan Obo implemented a ban on grazing animals and eliminated the census category for farmers and pastoralists, officially criminalizing and clearing the books of certain vulnerable populations to whom the local government might otherwise be ac-countable. Yet even from the window of the local Land and Resources Bureau director's office overlooking the grasslands outside of the small urban district of Bayan Obo, some tended herds of sheep and horses were visible among the mas-sive wind turbines. At both legal and illegal mines in the vicinity, hoof prints marked the gritty, sparkling mud accumulating along cracks in waste pipes and around the tailings ponds, indicating that livestock continued to graze even among industrial waste. Just as villagers around the tailing dam in Baotou had little means to leave, some herding families around Bayan Obo saw no alternative but to con-tinue their pastoralist lifestyle despite intensifying pollution. They are acquainted with the ravages of cancer and fluorosis, and also must periodically pay fines for violating the no-grazing policy. Those I interviewed stated that they preferred risk-ing "an early death beneath [their] own sky" to working as a day laborer on the margins of a far-off big city (Klinger 2013c).

All frontiers cut across space. In this case, imposing a border around IMAR and zoning Bayan Obo as a mining district cut across bodies and livelihoods, deterritorializing local ethnic others with competing claims to the land. The Mongolian pastoralists and their livelihoods remain as a stubborn fact contrary to the states' formalized borders, census categories, and zoning practices. Moving with their livestock among the wind turbines and mining infrastructure, these pastoralist families illustrate the very meaning of a territorial assemblage. As explained in the previous chapter, a territorial assemblage describes the coexistence of multiple competing territorial orders in one place. This particular territorial assemblage is composed of the local officials' periodic enforcement of their vision of the area outside of the urban district as a space where only wind turbines are allowed. This vision is literally overrun by pastoralist families who refuse to leave despite degrading living conditions. The semi-regular fines paid by pastoralists for violating the no-grazing policy provide a modest but reliable revenue stream for the local government. This suggests a possible compromise between two competing territorial orders, so long as the pastoralists and their livestock can bear the bodily burdens of pollution.

These territorial assemblages highlight the enduring sacrifices on which China's rare earth monopoly is built, as well as the incomplete nature of the frontier-taming project. This local complexity is key to understanding China's changing rare earth policy priorities, which are shaped in key ways by social mobilization against the environmental and epidemiological crisis unfolding on the ground.

Contemporary data on occupational health and safety issues for people working within the rare earth mining and processing industry was more difficult to obtain, especially in light of heightened sensitivities to foreign criticism, but there is ample data on the environmental and epidemiological effects surrounding production. Studies from previous decades published in China's academic journals provide a window into the human costs of China's rare earth production monopoly. For example, a longitudinal study of female workers in Baotou's smelting facilities found a rate of pregnancy complications 22.94 percent higher than the national average and a rate of congenital birth defects 20.89 percent higher than the national average (Liu 1996; Zhao 1994). The high incidence of respiratory cancers among Baogang workers has also been researched and reported on at annual conferences of China's Rare Earth Research Society since 1990 (Li 1990), and has been a core research area of Baogang's internal Public Health Bureau (Wang, Wanping, and Yulang 2002).

The toxicity of the land and water surrounding Bayan Obo and Baotou has been closely monitored by local environmental and public health bureaus for decades (Guo 2009). In the local documents repository at the Baotou Municipal Library and at other provincial and national archives, I found annual reports specific to particular industries, pollutants, and diseases dating back to 1972, as well as a wealth of academic literature on specific places and cases in Chinese print journals, many of which are referenced in this book. These academic studies have not been subject to censorship, but neither have they been widely disseminated beyond specialist audiences, nor have they been translated into English. There are three consequences to this.

The first is that there is little general understanding outside of highly specialized Chinese-speaking audiences about the human cost of rare earth extraction and beneficiation. This precludes substantive dialogue on how to address the human and environmental catastrophes of Bayan Obo and avoid recreating them elsewhere. Second, the paucity of material available in translation is mistaken for secrecy, and secrecy is mistaken for a lack of environmental monitoring, regulation, and remediation in the area. This leads to a third consequence, which is that the billions of yuan committed by the central government to support dedicated specialists working on these issues—and most important, the outcomes of their efforts—are simply not known because they are not publicized beyond specialized Chinese-speaking audiences. Yet a tremendous wealth of accumulated knowledge exists within China, built on four decades of observing, documenting, and attempting to manage this dangerous, yet necessary, enterprise.

This indicates that, contrary to the Western perception of China as an unregulated free-for-all for dirty industry, the situation has been closely monitored. The rare earth frontier in IMAR is characterized not by lawlessness, but by multiple overlapping and selectively privileged legal regimes. The devastation did not occur because political and industrial leaders were ignorant of the industry's effects, or only recently became aware of the hazards of radioactive waste or mining residues. The priority was, instead, to maximize exploitation of China's hinterland for the greater good—defined as China's economic development and national security—as well as for the global market (Chen 2010; Mancheri, Sundarasan, and Chandrashekar 2013). The consequence of this was the reconfiguration of Baotou into a national and global sacrifice zone according to the interests of key actors in the global rare earth sector.

Neoclassical economics would view this state of affairs as an unfortunate externality. When "externalities" degrade the landscapes and lives of ethnic others, they remain comfortably abstract. But there are limits beyond which the notion of externalities falls apart. Contaminating the water supply on which a tenth of the country depends, and sickening a local population to the point that illness and early death undermine regional economic development (Wang and Dai 2011; Yu, Le, and Li 2008) have prompted sweeping changes. The measures taken in production policy and practice are discussed in the next section.

From Peak Production to Strategic Resource Conservation and Regional Development, 2000–2010

> How to solve the problems of complex use in the beneficiation process as well as remediate the environment? First, change perspective. We must change production priorities and guidelines from "Metals First" to "diversification and environmental protection."
>
> —Ma, Gao, and Yu (2009, 90)

Transformations in policy, discourse, and practice at the start of the twenty-first century reoriented rare earth production toward strategic resource conservation and diversified regional development. These transformations resulted from the gradual culmination of research, activism, and broader political economic changes over time. The environmental and epidemiological hazards of rare earth elements have been known for decades. Therefore, the dramatic shift that unfolded over the first decade and a half of the twenty-first century can hardly be explained by a sudden revelation of facts or a flash of enlightenment on the part of key national leaders. Rather, the abundance of research on the environmental and epidemiological devastation of rare earth production in Baotou and Bayan Obo indicates that achieving peak production status was predicated on the reconfiguration of the region into a sacrifice zone. Most analyses of sacrifice zones do not account for what happens "after," when the place in question is revalorized as something other than an extractive frontier, but this is what is happening in Baotou.

Practically speaking, it is difficult to deal with the causes of pollution in Baotou and Bayan Obo: tailings ponds cannot be completely sealed, nor can the contamination be pulled out of the soil or groundwater (Wang et al. 2009). Recognizing these constraints, policymakers have instead focused on reducing the causes of pollution by controlling production and by removing some major sources of pollution from urban and agricultural areas in Baotou. Simultaneously, they have extended palliative measures such as hospital construction, expanding access to health care, and undertaking extensive economic diversification campaigns.

For China's central government policymakers, a straightforward way to rein in the environmental and epidemiological harms of rare earth production is to decelerate the pace of resource exploitation, production, and processing. In a market economy, this must be implemented through a host of complex policy instruments, including institutional restructuring, punitive measures, and propaganda campaigns. In China's context, this has been accomplished by reframing overarching political and economic development objectives according to these priorities. The frontier is not a permanent status. The fact that calling a place a frontier implicates a project to transform the space in question into something else is manifest in the ongoing transformations in Baotou and Bayan Obo.

The exploitation of Bayan Obo's geological endowments was framed as integral to regional industrialization and socialist revolution during the Cold War. The current policy to conserve these same endowments has been framed as integral to ongoing national development. The policy measures are sweeping, and organized according to numbered slogans favored by China's policy architects:

> The Eleventh Party Congress of Baotou Municipality enumerated the "Three Enhancements" that will be achieved in Baotou. These are: enhance the comprehensive economic strength of the region; enhance urban taste and quality; enhance the material and cultural living standards of urban and rural residents. Thus the Congress promoted the "Five Transformations": transform the region from an old industrial base to a new industrial base; transform Baotou into a regional center with sophisticated urban functions distinct from other relatively simple industrial cities; transform local structural mechanisms as needed to boost innovation-driven technological development by investing the necessary material resources; transform the development orientation from the sole pursuit of economic growth to give increasing attention to improving people's livelihoods; transform the growth mode from extensive to intensive. These transformations will be channeled through the "Six Projects Implementation": implement industrial restructuring and project upgrades; implement "ecological and livable city" construction projects; implement the urban and rural integration development projects; implement the "scientifically educated city" through human resources development projects; implement projects to protect and improve the people's livelihoods. (Li and Zhang 2012, 58)

If these policy measures seem vast and sweeping, it is because they are deliberately formulated to be applicable to the operations of entities as diverse as the Ministry of Culture and state-owned enterprises. It is typical of China's political process for the highest policy entities to formulate the ideological framework and policy objectives, and then for the range of policy actors at lower levels of government to interpret the objectives according to local conditions. For example, the Environmental Protection Bureau interpreted these measures in the following way:

> First is to emphasize pollution reduction by restructuring Baotou's economy in order to reduce emissions and make significant improvement to the economy. This is necessary in order to realize the "Primary Pollution

> Controls Plan in the Twelfth Five-Year Plan of Baotou Municipality"
> and the measures outlined under the "Baotou Municipality Working
> Program for Control of Major Pollution Emissions." . . . Second is to
> eliminate backwards production capacities . . . third is to strictly control
> access to the environment. . . . This is important to progress and to the
> improvement of the people's livelihoods. (Li and Zhang 2012, 59)

This might seem to be an unrealistically ambitious and comprehensive set of
goals to be undertaken in five years' time. It may even seem unlikely that the declarations of an environmental protection bureau would result in industrial restructuring, given that environmental protection and industrial organization are
managed by separate industries. This is where the productive fictions surrounding rare earth elements prove politically expedient. Reframing them as scarce
within China, necessary to strategic national interests, and also vulnerable to foreign expropriation allows policy and production priorities to pivot from maximum output and export to conservation and industrial innovation. The Twelfth
Five-Year Plan provided the following instructions: "Article 28 of the Emerging
National Strategic Industries Development Plan very clearly points out that we
must consider protecting and strengthening existing rare earth reserves while
growing our domestic rare earth industry to be a strategic task of utmost importance" (Li, Liu, and Zhao 2012, 196).

There, in plain language, is the central government strategy to change China's
position in the global division of toxic labor, and to transform domestic dominance in rare earth extraction and refining to international dominance in value-
added processing. The earliest statement of this intention can be found in the
2003 Policy on Mineral Resources. It has been repeated and expanded on in
China's subsequent white papers on rare earth elements, and in numerous other
policy documents.

In 2003, the central government issued an updated Policy on Mineral Resources. In contrast to laws issued in 1986 and 1993, the 2003 policy emphasized
the need to promote more technology-intensive mineral exploration and improve
environmental oversight. In January 2004, the export tax rebate initiated by the
Ministry of Commerce in 1985 was adjusted from 13 percent to 0 percent, and
the export tax rebate for yttrium and scandium compounds was adjusted from
17 percent to 5 percent. As of May 1, 2005, the export tax rebates were abolished.
Cancelling the rebates was part of a larger set of shifts in China's economic development strategy around raw materials. The purpose was to discourage exports
and increase imports of raw materials while building up domestic beneficiation
and components manufacturing capacity. Toward those ends, on November 1,
2006, the Ministry of Commerce announced that a 10 percent tariff on rare earth

exports would take effect on June 1, 2007 (Chen 2010). The Ministry of Commerce subsequently increased the tariffs to 15–25 percent on all rare earth products except praseodymium, gadolinium, holmium, erbium, thulium, ytterbium, and lutetium (Chen 2010).

These tariffs were preceded by a set of export and production quotas that had been announced in 1999 and went into effect in 2000. Most charts published internationally since 2010 begin with the year 2005, which gives the impression that China had been steadily intensifying the export quotas. Figure 13 illustrates that beginning in 2000 shows the more variable character of the quotas, indicating policy experimentation rather than the draconian imposition of limitations, as it has so often been portrayed in Anglophone discourse. The historical trends help show that quotas were not arbitrarily designated, but determined according to a set of calculations based on supply, market price, and strategic and diplomatic considerations.

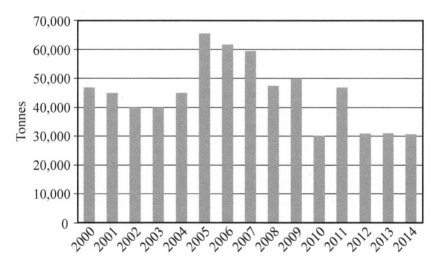

FIGURE 13. Rare earth export quotas in metric tonnes, 2000–2014. The sudden spike in quotas in 2005 is due to the extension of quotas to include joint ventures. The quota is not the same thing as the total amount exported. Because of robust black market activity, actual exports are estimated by comparing official export receipts from China with import receipts from downstream countries. Black market exports have been found to comprise 10–40 percent of the total annual volume of rare earth exports.

Sources: Compiled by the author based on data from Korinek and Kim (2010); Tse (2011); and Ministry of Commerce (2013, 2014).

Given the gradual introduction of these changes, notably occurring after China's accession to the WTO, this is hardly a revelation, although many in Anglophone and Lusophone literatures treated it as such in the aftermath of 2010. Continuing to "upgrade" China's production capacity means, in practice, moving the rare earth frontier somewhere else. In the years prior to the 2010 crisis, this has been viewed as critical to longer-term domestic sustainability, employment, and stability concerns. In other words: "Mining rare earth minerals employs hundreds of people, separating rare earth elements employs thousands of people, producing rare earth end products, however, employs millions of people" (Chen 2010 quoted in Lucius, 2014, 176). These policies were informed by three types of material conditions produced by China's de facto monopoly: first, the regional environmental and epidemiological crisis, to which local activists and courageous journalists forced government attention; second, the economic problems posed by industrial overcapacity in many rare earth-related enterprises, and; third, the seldom acknowledged problem of a global oversupply of rare earth elements.

The rise of Bayan Obo as the hometown of rare earths was useful as a means and ends for the central government to territorialize the Mongolian border. But this is not simply the case of a frontier boomtown that rises and falls with a single commodity. Baotou was built as a red hinterland, with diversified local industries and an extensive concern with reproductive politics intended to permanently settle and develop the region. The convergence of environmental and economic pressures prompted the government to rethink the frontier project at the turn of the millennium. It is not enough for Bayan Obo to serve as the world's rare earth frontier, with Baotou serving as the world's industrial hinterland for technological components. To advance the People's Republic of China's territorial agenda in light of twenty-first-century environmental and economic concerns, the hinterland must become a metropole. The strategy to achieve this transformation was to begin importing rather than exporting primary commodities, and to begin exporting rather than importing high technology and expertise. Subsequent chapters take up the manner in which this effort to transnationalize China's rare earth hinterland has helped drive the global rare earth frontier to previously impossible places.

It is important to understand these policy objectives as directly informed by the material conditions that resulted from China's rare earth monopoly rather than vaingloriously declared irrespective of political economic realities. Cold War caricatures of socialist politics continue to hobble international analyses of China's policies by presuming that official objectives are tone-deaf declarations of an out-of-touch Leninist regime (Canadian Security Intelligence Service 2013; International Business Publications 2012). To the contrary, these policy objectives have been built on gradual structural changes in line with the export tariffs of the previous decade.

Building on the macroeconomic changes precipitated in part by the quotas and tariffs, the next step in the policy strategy was to consolidate production control under Baogang (Guo 2009). In 2006, the central government announced a closure and consolidation campaign to reduce the number of private industries in the rare earth sector to twenty state or military owned enterprises by 2015. In 2010, the state issued a moratorium on new rare earth separation projects in order to reduce production overcapacity. In Baotou, privately owned and operated rare earth firms were directed to merge with or sell out to Baogang.

The central government also issued a mandate ordering all small-scale and informal mining operations to cease. However, key local officials opposed these mandates for the same reasons that central government officials had formulated them: national security and economic stability. As a result, the mine closure and enterprise consolidation policy followed the selective compliance pattern of many other national campaigns; it was most effectively implemented where the results would be visible to visiting officials, which is to say, from the paved roads. Along the highway in the northern regions of Baotou Municipality near the Bayan Obo mining district, dormant processing plants and abandoned workers' housing could be seen. But off the paved road, in areas accessible by dirt tracks leading off behind the hills, small-scale mines were still in operation in late 2013. The claim made by certain local officials refusing to implement the forced closure and consolidation was that the small mines were useful for local employment as well as for various nationalist objectives such as industrial and economic development.

Industrial consolidation in Baotou has been partial for several reasons. The first has to do with inertia in the built environment around Baotou. The city, provincial, and central government devoted significant resources to constructing a "Rare Earths High-Technology Development Zone" in order to attract private industry from around China and the world, including building a new residential district with luxury apartments to house a white-collar workforce. Their successes resulted in the establishment of sixty firms specialized in different aspects of the beneficiation and downstream production process. Some firms were private spinoffs of military industries and produced specialized components for advanced naval and weapons applications. With the more specialized firms overseen by individuals with influential connections, sales and mergers tended to be negotiated over private dinners with terms favorable to the seller. About thirty nominally independent firms remained in 2013, either because they were specialized in a particular aspect of the beneficiation process that Baogang officials judged too difficult or inefficient to take over, or because they were owned by current or former military personnel and thoroughly integrated with the local and global arms industry.

Despite this local compromise, Baogang remained responsible for allocating the raw materials coming from the mine at Bayan Obo. When there was not enough ore to supply all local industries, the informal mines provisioned certain private firms. This arrangement required no small measure of complicity from local actors. Bayan Obo is a closely monitored region. With the closure and con-solidation campaign entering into force in 2009, the local Public Security Bureau received considerable central government money to implement new surveillance measures. As a result, networks of surveillance cameras monitor all roads lead-ing into town and areas surrounding the mine. Another local official reported that every ore-bearing truck contains a GPS tracking chip and is monitored by satel-lite. Since there is only one paved road and one railroad linking Bayan Obo to Baotou, enforcement duties are less onerous than in other, better connected min-ing regions in the south of the country. These measures were reportedly put into place to crack down on black market activity, criticized for its environmental de-struction and market destabilization. Stabilizing prices and regularizing the rare earth economy by controlling informal production are measures that are now in keeping with China's short- and long-term environmental and national security interests.

But some local officials have a different perspective. In their view, the central government mandates are driven by environmental concerns, which carry less weight than the national security imperative to nourish the weapons manu-facturing industries with a steady supply of raw material. Furthermore, since central government axioms are formulated with the express purpose of being interpreted by lower level government officials in accordance with local condi-tions, officials supporting the ongoing operation of now illegal mines do not necessarily see their activities as contradicting national interests. One official in-terviewed said that such an arrangement was an appropriate application of the national mandate because keeping the arms industries well supplied was con-tributing to the fundamental and timeless mission of building a mighty nation.[21]

However, there is intense disagreement on the ground as to the best prac-tices. Local officials frustrated by the incomplete closure and consolidation pro-cess criticize their comrades as narrowly self-interested, profiteering from "selling out" China's precious natural resources to foreigners, and not caring about the severe environmental problems caused by ongoing illegal mining. One frus-trated official in the local bureau of the Ministry of Land and Resources in Bayan Obo characterized his complicit colleagues as possessing a twentieth century industrialist mentality about national security, in which the production of heavy machinery was the primary measure of national strength. By contrast, twenty-first century national security means enacting long-term measures to remove threats to strategic resource supplies—which include "chaotic" mining—and eliminating environmental destruction by rationalizing production and enforcing environmental protection regulations. In this particular case, this means consolidating production and processing under the umbrella of the state-owned enterprise.

Consolidating rare earth mining and processing into state-owned enterprises transforms the spatial politics of the industry and has important implications for ongoing environmental regulation. SOEs have their own environmental and public health bureaus, internal to the SOEs themselves. Because private firms were subject to the monitoring and audits of the local bureaus, the data gathered by the Environmental Protection Bureau and Public Health Bureau included pollution, occupational health, and safety data from private firms. Private firms that merged with the state-owned enterprise were removed from the jurisdiction of local government offices and became subject to the internal monitoring of the SOE, the findings of which are not shared with local government bureaus nor disclosed to the public. As Baogang assumed a greater share of mining and beneficiation processes, an increasing share of the rare earth industry has been exempted from public monitoring and regulation. The findings from the SOEs' internal monitoring offices are reported up the chain of command to State-owned Assets Supervision and Administration Commission instead of to the Ministry of Environmental Protection, Ministry of Land and Resources, and the Ministry of Public Health. This spatial delimitation of monitoring and regulation means that local environmental data does not include SOE data, nor is sharing such data legally required. Furthermore, SOEs are more difficult to petition for redress of grievances as compared to ministries, which complicates ongoing citizen-led campaigns for recognition and compensation (State Bureau for Letters and Calls 2014). This has deepened the knowledge gap between existing "policy silos" in China's government apparatus. As of this writing, if merging private firms into SOEs has reduced environmental degradation, the state has not yet disclosed it.

Conclusion

Tracing the changing geographies of the global rare earth frontier from approximately 1980 to 2010, this chapter examined the emergence of China's rare earth monopoly from the perspective of international political economy since the Reagan/Thatcher revolution and Deng Xiaoping's reforms. The first section showed how differing national and individual interests, brought together within emerging forms of neoliberalism, shaped global geographies of rare earth mining and beneficiation. The second section examined the physical effects of this shift in "the hometown of rare earths." Here, the rare earth frontier has literally cut through human bodies, degrading the muscles, bones, brains, and organs of people exposed to certain pollutants and needlessly shortening lives. The third section traced several domestic responses to the material conditions brought about by China's de facto rare earth monopoly.

The changes in Baotou and Bayan Obo over this period show how frontiers are temporary names assigned to dynamic places. After the IMAR frontier had been rendered into a red hinterland, it then became a global sacrifice zone, poisoning many of the people who had migrated from other parts of China in order to build it into the rare earth capital of the world. The environmental and epidemiological crisis in Baotou and Bayan Obo cannot be explained simply as a Chinese conspiratorial plot against the world, or vice versa, as it has been characterized in English and Chinese media, respectively. Rather, the devastation was the result of multiple policies and practices built on a tacit international and domestic consensus that viewed brutality and pollution as integral to everyday economic functions.

The consensus within China has shifted. Stemming the tide of destruction on which China's rare earth monopoly was built has become more important than growth at all costs. When environmental and epidemiological harms intensify to the point that the damage done by rare earth mining poses a serious existential threat to territorial hegemony, extractive relations change. The National Development and Reform Commission (2005) has expressed an explicit intention to trade China's global dominance in rare earth mining for value-added production and advanced research and development. Although this has met with resistance in areas where the local economy is largely dependent on rare earth extraction and beneficiation, and provoked Western ire in the post-2010 climate, the overall trend should be clear. The ongoing concentration of rare earth mining and processing in China does not seem to be in the interest of anyone: Western commentators emphasized the strategic and economic vulnerabilities presented by this arrangement, while Chinese commentators bemoaned the social and environmental costs of supplying the world from the Inner Mongolian hinterland.

This did not, however, signify a paradigm shift at a global scale. China's strategies to transform the domestic rare earth industry are meant to change its position within the global division of toxic labor by sending the problem elsewhere. Nor were the measures discussed in this chapter sufficient, in themselves, to create the 2010 crisis. The next chapter examines the 2010 events that awoke the world to its dependence on China's rare earths, transformed global resource geopolitics, and radically altered the geography of the global rare earth frontier. Despite the transformations that unfolded post-2010, many of the operative elements of the rare earth frontier have remained intact: using rare earths as a means to territorial ends; undertaking regulatory offensives to legitimate mining in historically autonomous or contested places, and; mobilizing the fictions of rare earths' supposed rarity to conquer new frontiers.

RUDE AWAKENINGS

In 2010, the rest of the world was shocked into sudden awareness of China's rare earth monopoly. What caused these rude awakenings? And how did they reshape the global rare earth frontier?

Trouble Downstream: The "Embargo" and the 2010 Crisis

In the first decade of the twenty-first century, the question of rare earths, global dependence on China, or the environmental costs of rare earth production detailed in the previous chapter scarcely made an appearance in international policy or media discourse. As had been the case for much of the twentieth century, rare earth elements remained the domain of specialized chemists, materials scientists, and niche investors. For most people, the rare earth frontier was not only remote; it was so insignificant as to be entirely unknown. That is, until rare earths were caught at the center of a high-profile dispute between China and Japan, which then evolved into an international trade dispute between China and major rare earth importers: the United States, Japan, and the European Union (EU).

The "Embargo"

The Senkaku, or Diaoyu, Islands lie in disputed waters off China's eastern seaboard, north of Taiwan and southwest of Okinawa, Japan. These eight uninhabited islands, with a total area of about seven square kilometers, are contested because of their geostrategic location, proximity to important shipping lanes, and abundant fish and undersea petroleum resources. China claims that they have belonged to China since ancient times, Japan claims that they belong to Japan since they were the first to formally survey the islands and erect a sovereignty marker in 1895, and the Taiwanese government varies its claims depending on which party is in power.[1] Tensions had already been periodically simmering since the discovery of oil deposits in 1970 and reached the point of erupting into episodic anti-Japanese demonstrations during the 2000s (Bao 2010; *China Daily* 2003).

On September 7, 2010, the Japanese coast guard detained a Chinese fisherman, Zhan Qixiong, who had strayed too close to the islands. Instead of complying with the procedural escort beyond the fifteen-kilometer range, Mr. Zhan reportedly rammed his boat into approaching coast guard vessels—a fact generally omitted from reporting on the subject. It is important to note that fishing activities around the island had generally proceeded despite the diplomatic dispute; both Chinese and Japanese fishers were entitled to fish in the area. International and Chinese media omitted the fisherman's behavior from their accounts, which conveyed the impression that Japan had detained a hapless fisherman as a unilateral act of escalation. Domestic Chinese media interpreted the detention of Mr. Zhan as an act of war. This particular conflict had nothing to do with rare earths, at least not yet.

In response, according to Anglophone accounts at the time, China halted rare earth exports to Japan. However, this was not an official policy or even a decision taken by national-level policymakers in Beijing. Rather, some military personnel in one of China's port cities colluded with port workers and local customs officials to retaliate by withholding shipments. Their motivation was "to teach Japan some humility" by reminding them of their economic dependence on China.[2]

Officials at the Ministry of Commerce publicly denied that any such disruptions had occurred. But contrary to international speculations about complex diplomatic maneuvering (Bradsher and Tabuchi 2010), Beijing issued this denial because they were caught unawares and were only alerted to the issue after the Japanese Customs Authority inquired about the rare earth shipments. Meanwhile, Beijing officials made urgent remonstrations at the local level that went unheeded for several weeks. Public statements from the central government disavowed any strategy to restrict rare earth exports beyond the decade-old quotas. Adding to the confusion, Anglophone analysts that do not differentiate between official declarations and specific local realities cited official statements from Beijing as evidence that the shipping disruption never actually occurred (King and Armstrong 2013).

Local port workers framed their actions as defending China's national interests according to local conditions,[3] which, as explained in the previous chapter, is an axiom of China's decentralized authoritarian governance (Xu 2011). In this particular case, local conditions were crucial to informing the action of military and port personnel. This port is on the coast of the East China Sea and, as such, it is one of the closer eastern seaboard ports to the islands. More important, it is also located north of Nanjing, where a living memory of the Imperial Japanese invasion and massacre is cultivated as a matter of public policy (Fogel 2000). Furthermore, Mr. Zhan was detained within a few days of the anniversary of the 1931 full-scale Japanese invasion of Manchuria. Thus Japanese claims to the islands were, in formerly occupied regions of China, interpreted through the historical lens of colonialism and war; seizing a Chinese national in waters that Beijing asserts are its own was framed in terms of a brutal history of humiliation suffered under early twentieth century Japanese expansionism. History, memory, and geography were crucial to how this issue developed: local military and port personnel saw themselves on the frontlines against a contemporary manifestation of Japanese aggression. Unaided, in their view, by the central government, they used rare earth shipments as a weapon. Central government officials had considerable difficulty persuading local officials and workers to resume shipments because the latter viewed it as an order to stand down, and essentially, to surrender to Japan.

As an intermediary step, port officials nominally resumed shipments in early October 2010, but subjected nearly every container bound for Japan to thorough inspections before loading them on ships, which had the effect of delaying shipments further (Yuasa 2010). Even though the fisherman was released on September 24, 2010 (Fackler and Johnson 2010), port workers did not restore normal export activity until late October or mid-November,[4] approximately two months after the initial incident (BBC 2010; Bradsher 2010b, 2010c).

Halting shipments was an isolated local decision. It was unrelated to export quotas, industrial consolidation in Baotou, or any other production control measure. But Anglophone media was quick to identify a conspiracy, so the act nevertheless had serious international consequences. Beijing did not know it was happening, denied it was happening once it learned, took steps to resume official shipments from this port in the interest of maintaining stability, and then denied the incident had ever occurred (Areddy, Fickling, and Shirouzu 2010; Richardson and Williams 2010). But the damage had already been done. The world had its first rare earth crisis in sixty years.

The Crisis and Its Aftermath

Accusations of an embargo prompted market panic, which drove prices to unprecedented highs. The sudden halt of official exports from China to Japan was interpreted as a significant disruption in the global supply of rare earth elements, given that at the time, Japan imported 40 percent of all rare earth elements produced globally, 97 percent of which came from China (Humphries 2013). Because shipments were only temporarily disrupted at one port, this episode was much less than if 40 percent of all available rare earth supplies had been withheld. This degree of nuance, which undoubtedly would have calmed market fears, was completely absent from any analysis on the issue. The myths of rare earth rarity were too potent to resist with reasonable inquiry. Fears of an embargo coupled with a rude awakening to the importance of rare earth elements and global dependence on China caused prices to increase by as much as sixty-one-fold for some elements in a very short time in the latter months of 2010, as shown in table 4.

In fact, the web of distress created by sudden fears of shortage exposed just how diverse the applications of rare earth elements are, and how much of modern life depends on them. Major downstream firms saw their profits cut into by as much as 35 percent between the third and fourth quarters of 2010 (Monahan 2012). Junior firms in the renewable energy sector were particularly hard hit. Rare earth elements are essential for the production of wind turbines and hybrid fuel cell batteries. Each two megawatt wind turbine uses roughly eight hundred pounds of neodymium and 130 pounds of dysprosium (Stover 2011), while each hybrid vehicle uses about thirty pounds of rare earth metals (Burnell 2010). With rare earths comprising only a small fraction of the thousands of inputs required to make these products, a 2,000 percent price increase was too much for many nascent green technology companies to bear. Some analysts claim that fifty-nine renewable energy companies in the United States were forced to file bankruptcy;

TABLE 4 Annualized quarterly volatility for selected rare earth oxides from late 2010 to mid-2012, US$/kg

	Q3 2010	Q4 2010	Q1 2011	Q2 2011	Q3 2011	Q4 2011	Q1 2012	Q2 2012
Lanthanum	16.37%	**374.47%**	**366.10%**	**348.22%**	**348.71%**	143.07%	140.42%	59.82%
Cerium	20.36%	**439.84%**	**425.33%**	**410.26%**	**412.68%**	139.05%	135.20%	65.63%
Praseodymium	46.15%	61.51%	59.72%	88.99%	94.00%	**108.49%**	**120.27%**	43.22%
Neodymium	37.13%	63.97%	64.58%	97.58%	84.79%	**122.07%**	**138.34%**	73.88%
Samarium	1.47%	**865.67%**	**840.27%**	**808.42%**	**803.10%**	122.45%	117.11%	45.00%
Europium	7.62%	5.85%	3.60%	77.71%	**285.42%**	**296.18%**	**303.59%**	**322.21%**
Terbium	48.30%	44.48%	45.35%	84.52%	**191.66%**	**204.88%**	**211.99%**	**222.02%**
Dysprosium	44.82%	48.99%	45.72%	76.68%	**199.73%**	**199.38%**	**218.60%**	**242.57%**

Sources: Bartekva (2014). Compiled by Molly Roy.

Notes: This table shows how not all rare earth elements were affected equally by the market response to the 2010 disruption. The most dramatic volatilities are indicated in bold.

others claim that the number is under ten (Lakatos 2014; Primack 2014; Stahl 2014). Department of Energy personnel interviewed in early 2014 did not have precise figures on the number of clean tech bankruptcies caused by rare earth price increases.

The story of the crisis broke in the West in an article by *New York Times* reporter Keith Bradsher, with a headline describing the disruption as an embargo. This was followed by a disavowal of the incident in the *People's Daily*, the English-language edition of Beijing's official mouthpiece (Bradsher 2010a; Liang 2010; Qi 2010). Given that international media and politicians needed to learn very quickly what rare earth elements were, it is perhaps understandable that the initial responses were knee-jerk and ahistorical. Commentators across the ideological spectrum speculated about the possibility of war with China, divorcing Deng Xiaoping's couplet[5] from its 1992 context to reason that rare earth elements would become the next oil, and therefore the next cause of global conflict (Brennan 2012; Chapple 2012; Coppel 2011; Dobransky 2013; Krugman 2010; Kudlow 2011).

But denial, even in the face of clear evidence, has proven to be an effective diplomatic strategy (Franken 2003; Kellner 2007; Woodward 2006). Since the position of China's central government was that the issue had not occurred, institutions whose legitimacy draws from taking state power at its word carried the message forward. There were special events in Washington, DC, that questioned whether the embargo had ever actually happened (Hao and Nakano 2011; King and Armstrong 2013). These were led by renowned experts in international relations who nevertheless did not have a practice of examining the facts on the ground.

Alastair Iain Johnston at Harvard University conducted one of the very few evidence-based analyses. His analysis was based on import data from Japan's Ministry of Finance, which found that there was little statistical relationship between import figures for rare earth commodities, and import declines at Japan's major ports. Johnston drew this conclusion after noticing that a third of the cases examined showed a decline in rare earth imports. In 46 percent of the cases Johnston looked at, rare earth imports to Japan increased from August to September of 2010 (Johnston 2013). Because two thirds of the cases examined do not reveal an irregularity in shipments from China, Johnston concludes that no embargo occurred. The customs receipts showed that Japan was importing rare earth elements during the time that the alleged embargo was taking place. The import receipts are presented as evidence that the embargo did not actually occur.

Yet all thirty-one Japanese companies that handled rare earth elements reported stoppage or disruption in rare earth oxide shipments from China (Agence-France Presse 2010). Furthermore, it is important to note that the legality of rare earths imported from China is almost never noted in recipient ports; black market rare earth elements have exceeded China's production quotas by 10 percent to 40 percent annually since 2007 (*China Daily* 2009; Els 2011; Stanway 2011; Topf 2013). A halt in official exports provided an excellent market opportunity to black market traders who, because of their lesser technological capacities, deal in oxides and not in alloys or other value added products (Bradsher 2010d; Reporter 2012).

Some further clarification is crucial here. Framing the question in terms of whether an embargo occurred obscures the nature of the event. Strictly speaking, an embargo is an official ban on trade or commercial activity with another country. Nothing that fits this strict definition occurred. The import receipts show only that Japan was importing rare earths from China; they do not indicate anything about the origin or legality of these imports, much less the government's official position on temporarily impeding their flow. Indeed, the black market trade in rare earth elements has been a persistent problem for China's policymakers, while also serving as a necessary pressure valve for international downstream buyers who would otherwise face supply shortages, much in the same way that clandestine mining is essential to supplying independent processing firms in Baotou.[6] China's Customs Administration monitors the volume of black market rare earths circulating in the global economy by comparing import data of recipient countries with export data from China. The discrepancy is interpreted as the volume of the illegal rare earths exports. Furthermore, the shipments were halted from one of many seaports that export commodities to Japan. There is no indication in my field research or available information on the matter that military or civilian personnel at other ports engaged in similar tactics.

Regardless of the actual flow of goods, the market response turned a temporary and by all accounts minor interruption into a full-blown economic and geopolitical crisis. Indeed, the market response during this period supports the contention that the market is driven by speculation, fear, and fantasy (Carrington 2015; Eichengreen et al. 1995; Taffler and Tuckett 2007). Such mass subjectivity, caricatured in market panics, shapes political thought and action in important ways. The mere suggestion of an embargo was sufficient to drive prices of certain rare earth elements through the roof, despite the lack of investigation into the veracity of the claims. Productively misreading the disruption in official exports as an embargo enabled the international community to build a case that China had violated the rules.

Although the crisis was short-lived, it transformed the global rare earth frontier by unleashing waves of global prospecting, speculation, and investment. By driving prices to unprecedented highs, the crisis generated a range of opportunities for mining companies in the rest of the world, as well as for researchers hopeful that market incentives would sustain funding for research into rare earth recycling and greener production practices. Furthermore, citizens, elected representatives, and prospectors in economically depressed regions of the United States and Canada saw the solution to their employment and tax revenue woes in bringing rare earth mining back home, so to speak. The immediate political economic and discursive responses that have been key to reshaping the contemporary global rare earth frontier are examined in the following sections.

Rational Responses? Green Nationalism, Global Exploration, and WTO Suits

The responses to the 2010 crisis were marked by incoherencies and contradictions, several of which are discussed in this section. Discursive responses in the United States and Brazil were characterized by the emergence of an environmentally inflected right-wing ideology that has been dubbed "resource nationalism," "neo-extractivism," and "green" or "environmental nationalism" (Baletti 2012; Hao and Liu 2011; Margonelli 2009). The global market responded to the sudden price increase with an unprecedented wave of speculation, investment, and exploration driven by the resolve to end dependence on China by opening up new sources on the global rare earth frontier. One analyst found that by mid-2011, less than a year after the crisis, there were 429 new projects outside of China and India being developed by 261 different companies in 37 countries (Hatch 2012). All of these ambitions were undermined by the World Trade Organization (WTO) suit brought against China by the United States, EU, and Japan. This reinstated the global political economic conditions that had undermined domestic industrial capacity in developed countries decades before.[7]

This incoherence should not be surprising, given that international anxieties with respect to China's global influence have grown in the past decade (Hoge and Hoge 2010; Klare 2008; Kurlantzick 2008; Xie and Page 2010). Existing fears primed people not only to accept the fictions of rare earth scarcity at face value, but also to cling to the fiction of China withholding its legendary endowments even after credible analyses began to emerge. The rare earth crisis of 2010 provided a touchstone for the Sinophobic anxieties that had been circulating in policy talk and scholarship over the past two decades. These ideas, many of them only partially accurate, helped stimulate the radical transformation of the rare earth frontier in subsequent years.

The new spatial politics of the global rare earth frontier are characterized by an extension of the frontiers of prospecting and production to previously off-limits regions, such as designated nuclear-free zones in Greenland, the heart of the war on terror in Afghanistan, and the putatively better-regulated Global North, specifically the Mountain Pass mine in California. This latter example contradicts the prevailing wisdom that large-scale mining interests move unidirectionally from developed to lesser developed regions in pursuit of the lower production costs captured in less regulated contexts. In ideological terms, this conventional wisdom has translated into an irreconcilable difference between environmental interests and extractive interests (Schroeder 2000; United Nations Development Programme 2012). The post-2010 geography of the global rare earth frontier indicates more complex dynamics at work.

Green Nationalism?

The responses to China's rare earth monopoly revealed that green nationalism takes many forms. The sudden shortage of these critical materials precipitated a new interest in the commodity chains of green technologies, which mainstreamed awareness of environmental degradation in Baotou and Bayan Obo (Hilsum 2009; R. Jones 2010; N. Jones 2013; Parry and Douglas 2011). Fossil fuel companies and nuclear lobbyists then used this information to argue against renewable energy generation, describing the devastation in Baotou and Bayan Obo as "clean energy's dirty little secret" (Epstein 2014; Fisher and Fitzsimmons 2013; Stover 2011). This, of course, ignored the fact that rare earths are critical inputs to renewable and nonrenewable forms of energy generation alike. National security analysts advocated developing "environmentally superior" rare earth production practices closer to home, including green sourcing and recycling in the name of protecting US interests (Kennedy 2013; Stoyer 2013). In both cases, toxic practices in China were racially coded[8] as both a "Chinese" problem and as a "dirty trick" played by China's central government in order to achieve dominance in rare earth production. If pollution could be defined as quintessentially Chinese, then going green could be a demonstration of US superiority. US parties typically opposed to clean technologies in favor of the petroleum-dependent status quo reversed their position once it appeared that ending fossil fuel dependence might be dictated by China's control of rare earth supplies. Viewed in this light, reviving US rare earth capacity could not happen quickly enough. What was needed was some sort of rare earth-specific "New Deal": "A few years ago, China showed its power, and cut the supply of rare earths to a trickle. The move sent the United States and other countries scrambling to end their reliance on China. . . . But the crucial element in escaping China's rare-earth rule isn't new mines, it's rebuilding the expertise and infrastructure to process the finicky metals" (Oskin 2013).

In a curious twist, conservative antigovernment congresspersons were authoring legislation to effectively nationalize the industry: "It is the sense of Congress that the United States should take any and all actions necessary to ensure the re-introduction of a competitive domestic and ally nation rare earth supply chain, to include the reintroduction of the capacity to conduct mining, refining/pro-cessing, alloying and manufacturing operations using domestic and ally nation suppliers to provide a secure source of rare earth materials as a vital component of national security and economic policy" (United States House of Representatives 2011c, 6).

Given the circumstances, right-wing policymakers found that a federal industrial policy was preferable to the possibility that the US policy choices might be constrained by China. Even though Congress acted relatively quickly to authorize the Department of Energy to establish the *Critical Materials Institute* at Ames National Laboratory in late 2010 (Bauer 2010), commentators bemoaned the long lead-time on policy changes and rare earth projects: "That still leaves a long gap when the green revolutions will rely on the economic and political judgment of China's exporters" (Heap 2010).

In the meantime, several US senators sought to curb China's rare earth dominance by using multilateral institutions to thwart China's mining investments within China and around the world. In a letter to Treasury Secretary Tim Geithner and Secretary of the Department of the Interior Ken Salazar, they

> urged Sec. Geithner to instruct the US representatives to the multilateral banks (including the World Bank entities) to oppose funding for any Chinese financed mining project in China or abroad. They also asked Sec. Salazar to use his authority to block any Chinese funded domestic mining project, until the Chinese end their anticompetitive practices in regards to REEs [rare earth elements]. US mining law recognizes that foreign investment in mineral exploration and purchase should be prohibited where a foreign country denies reciprocal privileges to US companies, and the Senators urged Secretary Salazar to invoke that power. (Casey et al. 2011)

No such action was taken. But this and similar requests were indicative of a strong sense that China's rare earth monopoly threatened US sovereignty and national security while ignoring the fundamental role of US policies in creating the situation in the first place. Anti-China discourses resorted to environmental comparison; the differentiating element between "us" and "them" became "our" Euro-American potential to be "environmentally superior." Given enough time and resources, the United States could "beat China" by being greener (Leifert 2010; Perkowski 2012). Major downstream buyers stoked this sensibility by suggesting that major firms would pay a premium for sustainably produced non-Chinese rare earth elements, and emergent rare earth companies attracted investors on promises to lead the new green wave: "Molycorp CEO Mark Smith plans a different kind of mining in a new age of environmental awareness. 'I don't want to produce another pound of product if we don't do it right environmentally. That's how serious we are,' Smith said. 'We don't want to be just environmentally compliant; we want to be environmentally superior'" (Kraemer 2010). This approach constituted a pillar of the US Department of Energy's (2011) Critical Materials Strategy: "In all cases, extraction, separation and processing should be done in an environmentally sound manner . . . research into more efficient and environmentally-friendly separation and processing technologies has the potential to boost supply from new and existing sources throughout the world, lowering costs while reducing the environmental impacts of mining and processing."

In a prescient investigative piece on the rare earth industry in the United States, journalist L. Margonelli described the charged climate around rare earth elements as "nascent green nationalism . . . a weird amalgam of environmentalism, economics, and national security":

> Consider the views of the industry analyst Jack Lifton—by no stretch your standard environmental activist ("I don't give a rat's ass about global warming"). To protect US industry from supply shocks, he has called on the government to mandate the recycling of strategic minerals. A "bottle bill" for cars, long dismissed as an environmentalist's dream, is just one possible outcome. Another could be a backlash of resource nationalism in supplier nations like China. As green nationalism's potent mix of idealism and fear changes the kinds of cars we drive, it also promises to change the course of globalization. (Margonelli 2009)

Broad declarations such as "chang[ing] the course of globalization" were typical of the postcrisis rare earths discourse. They reflected a recognition, however vague and unspecified, that the geography of the global rare earth frontier—and of resource geopolitics in general—was shifting in ways unforeseen in an international order built on China as the world's factory floor. There were several ways in which the hyperbolic discourses have been matched by hyperbolic behavior, especially with respect to the changing geography of global rare earth prospecting.

Global Exploration: Greenland, Afghanistan, and the Americas

The concentration of critical resource production in one or two places around the globe is hardly peculiar to rare earth elements. Yet the economic catastrophe that would result from a disruption in boron production, for example—the entire global supply of which comes from Turkey and the southwestern United States (Şebnem, Ayşe, and Işıl 2013; United States Geological Survey 2014)—has not inspired further exploration or investment anywhere else, much less in politically or geographically challenging environments. Mining for other materials critical to heavy industry and munitions, such as tungsten and antimony, are also concentrated in China (Carlin 2013; Shedd 2014). But this has not generated political or economic fervor comparable to that surrounding rare earth elements.

Clearly, asymmetry in the global production of critical materials is not sufficient to explain the wave of global exploration and prospecting that occurred after the 2010 crisis. Nor, as chapter 1 has shown, is the simple presence of rare earth elements sufficient reason for investment and exploration, given their relative ubiquity. Rather, the rare earth crisis provided an international rallying point for diverse anxieties concerning China's rise, which in some cases provided a renewed impetus to exercise frustrated territorial ambitions having little to do with rare earths. In contemporary rare earth politics, these anxieties have effectively leveraged bodies of geological knowledge old and new in the service of long-standing geopolitical ambitions.

Three factors explain the dynamics of global rare earth exploration following the rude awakening in 2010. First, the urgent need to source supposedly *rare* rare earths from anywhere other than China fit well with states' ongoing drives to solve long-standing frontier problems of their own, and in some cases to discipline historically autonomous territories. Second, the need to move the harmful effects of production away from centers of political accountability to the sacrificable hinterlands served state and imperial desires to effectively exercise sovereign national and international geopolitical power over and through the frontiers. Third, containing one aspect of China's growing global influence by developing supplies of these strategic materials in other parts of the world seemed likely to score political points with Euro-American powers. The strategic significance with which rare earth elements were imbued in this context was instrumental to generating the political will and necessary capital to pursue environmentally hazardous production in contexts that are well regulated or war-torn. Efforts to promote rare earth exploration outside of China, from Greenland to Afghanistan to the Americas, illustrate the immensity of the political, scientific, and regulatory undertaking required to transform the global rare earth frontier.

GREENLAND

Rare earth mining in Greenland was indirectly prohibited by a 1984 Danish declaration making all its territories a nuclear-free zone. This banned mining for radioactive materials. Because of their geological coincidence with uranium and thorium, this foreclosed any possibility of rare earth production in places under the Danish authority. Simmering Greenlandic desires for greater sovereignty seized the productive fictions surrounding rare earths emerging after 2010. Dormant bodies of geological knowledge from Cold War–era uranium expeditions were dusted off and circulated as evidence of Greenland's untapped potential. As the territory struggled to recover from the effects of the 2008 financial crisis and chafed under its ongoing dependence on Denmark, rare earth mining came to be seen as a ticket to prosperity, independence, and greater global geopolitical importance. In this light, the 1984 moratorium against mining radioactive materials was recast from a symbol of morally superior pacifist values to a hindrance to Greenlandic self-determination.

These sentiments culminated in a fiercely contested 2013 referendum. Belief in the promise of rare earth mining to buy independence from Copenhagen and accelerate the territory's global economic integration triumphed over conservation laws. Greenland had long been exemplary in its ecological and environmental protection laws, but the post-2010 climate surrounding rare earths gave previously marginal mining interests unprecedented political capital to accelerate policy changes that had been underway for several years.

In 2009, the Act on Greenland Self-Government gave the territory authority over its natural resources independent of Denmark for the first time in two centuries. In 2010, the Greenland administration began recruiting mining companies to explore the island's mineral potential (Boersma and Foley 2014). China's interest in Greenland's minerals prompted the EU to request preferential treatment in mineral concessions while also recommending that Greenland place restrictions on investment from China. Then prime minister Kuupik Kleist rejected this request, saying that "Greenland is open to investments from the whole world" (Briscoe 2013). In the end, the promise of sovereignty and a desire to integrate more fully into global circuits of power and accumulation, as (it is hoped) an independent state, triumphed over desires to preserve the social and ecological fabric of Greenland (*Arctic Journal* 2013a; Fletcher 2013; Gravgaard 2013). In October 2013, Greenland's parliament voted 15-14, with two abstentions, in favor of repealing the moratorium against mining radioactive materials (Faris 2014).

As the final vote reflects, authorizing large-scale mining of radioactive materials in Greenland was controversial (McGwin 2014; Olsvig 2013). But it is important to point out that it was the productive fiction surrounding rare earth elements—rather than interest in mining uranium, for example—that generated enough votes to repeal this environmentally and symbolically significant moratorium. Proponents viewed rare earth mining as the only hope for diversifying Greenland's economy away from dependence on fishing and subsidies from Denmark, while opponents feared the ecological devastation and social decay that comes with transforming small towns into large-scale mining operations dependent on migrant labor (Loewenstein 2014). The fact that foreign firms would be exploiting the resources and would be able to bring in their own laborers has caused further contention (*Arctic Journal* 2013b; Macalister 2013), as this could undermine the extent to which mining enables the territory to purchase its independence.

Beneath the question of whether or not to permit mining for rare earths and other radioactive material lay a deeper existential question. Greenlandic citizens were also grappling with the interpellation of their country as a new global resource frontier, opened up by a retreating ice sheet (Nuttall 2012). Such an economic designation conflicts with cosmopolitan sensibilities expressed in Greenland's newspapers, reflecting a geographical imaginary that places large-scale extractive activities in the Global South, outside of the European fold. These existential questions reached a deeper register in early 2016, following the Paris Climate Accord. The current Greenlandic administration announced that it would leave the accord, as it is looking to oil and gas reserves to fill the annual US$1 billion block grant that will be lost with independence from Copenhagen. Yet Greenland is one of the areas of the world most affected by climate change. Likewise, climate impacts in Greenland will have a profound effect on the rest of the world as the Greenland ice sheet melts and sea levels rise. In the face of these multiple forms of uncertainty, and averse to remaining within the European fold, Greenland's ministers are primarily concerned with generating revenue.

Greenland's quest for mining-financed autonomy has drawn considerable interest from Chinese investors, which has raised geopolitical concerns in continental Europe about growing Chinese influence in the Atlantic (Conley 2013; Desgeorges 2013). If Greenland were to permit mining in newly exposed areas, actors in the EU contended that European investors should be given preferential treatment. In Prime Minister Kleist's rejection of EU aspirations to transform Greenland into a hinterland for continental Europe's critical raw materials needs, he used language remarkably similar to that of policymakers in China by placing responsibility for rare earth sourcing with downstream countries. In an early 2013 address to Danish Parliament in Copenhagen, Kleist pointed out that each EU citizen discards an estimated seventeen kilograms of critical raw materials annually. Were access to rare earths truly the central concern, Kleist argued, continental Europe could make use of its own reserves valued in the hundreds of billions of dollars (Santos 2013). For Greenland, and for the EU, what is at stake in the debate over adopting policies amenable to foreign mining companies is not actually concerns about rare earths per se, but rather what the capacity to mine them might mean in terms of fulfilling—or undermining—dreams of sovereignty and regional hegemony. Long the remote and forbidding frontier, a world historical

convergence of climate change, geopolitics, and desires for territorial control compelled just enough of a voting majority to reenvision Greenland as a new global
mining frontier.

AFGHANISTAN

From ice sheets to warzones, speculative hyperbole abounds about these new rare
earth frontiers. Somehow, both Greenland and Afghanistan supposedly contain
the world's largest deposit. Of Greenland's potential, Greg Barnes, chief geologist
for an Australian mining company looking to exploit Greenland's rare earths reports: "It is the world's biggest rare earth deposit, it's probably got 50% of the
world's rare earth in it. This is one of the world's top 10 mines eventually we think"
(Barnes quoted in Fletcher 2013). With respect to Afghanistan, an internal Pentagon memo stated that the country "could eventually be transformed into one of
the most important mining centers in the world" (Mazurkewich and Greenaway
2010). As the 9/11 pretext for invading Afghanistan faded, the extractive agenda
became more pronounced. Ongoing security concerns came to be framed as investment opportunities: "The challenge is for an enterprising company to develop
this exciting deposit" (Coats 2006). In both Afghanistan and Greenland, domestic
and foreign proponents of rare earth mining have invoked the strategic imperative, the promise of sovereignty, and the tensions surrounding a particularly
salient aspect of China's power in order to generate political will for mining in
previously unexploited and ecologically delicate regions.

In Afghanistan, rare earth deposits received significant media attention in late
2010, following several years of relatively unpublicized prospecting funded by the
US Agency for International Development and undertaken by the United States
Geological Survey in collaboration with the Afghanistan Geological Survey (AGS).[9]
These expeditions published findings of reserves valued in the trillions of dollars,
not just of rare earths, but also of gold, copper, iron, precious gems, and other
nonferrous metals (Hansen 2012; Risen 2010).[10] In a special feature in *Scientific
American*, these resources were imbued with mythic qualities. They were framed
as a way to "beat" the Taliban: "vast deposits of rare earth and critical minerals
found in Afghanistan by US geologists under military cover could solve world
shortages and get the country off opium and out from under Taliban control"
(Tucker 2014). *Science* also published a story glamorizing militarized geological
exploration in a piece titled "Mother of All Lodes." Under armed protection,
"gutsy" US geologists "liberated" chunks of rare earth elements from remote regions of the country in order to assess Afghanistan's mineral riches (Stone 2014).

This is another case in which rare earth elements are imbued not only with
strategic significance, but also with the power to enable frustrated territorial
ambitions to succeed where a decade-long military occupation has failed.

These discoveries, however, are not new. Afghanistan's mineral wealth has been extensively surveyed by outside interests periodically over the past five decades, primarily Soviet.[11] The Soviet Union committed hundreds of billions of dollars to developing extractive infrastructure in Afghanistan, but the projects ended with the Soviet withdrawal in 1989. For contemporary use, much of the data gathered in the 1960s and 1970s simply needed to be translated from Russian. The US and UK coalition forces began supporting geological prospecting operations as early as 2003. Furthermore, small-scale "clandestine" mining operations continue to provide a modest but important income stream for many groups, not just the Taliban.

The US-led invasion created, perhaps counterintuitively, more favorable conditions for foreign prospecting in Afghanistan's subsoils, over which multiple domestic polities exercise claims to statehood (DiJohn 2010; Goodhand 2005). Potential investors view the US military presence as an essential security measure to support private sector operations. This is, perhaps, a more extreme variation of the importance of the state to violently impose the conditions for neoliberal economic practice. The explicit mandate of the US Department of Defense (DoD) to create conditions favorable for international business interests in this geopolitically vital Central Asian country supports this rather coarse expression of corporate entitlement to the "blood and treasure" of the coalition forces. Nevertheless, some key players in the international minerals and narcotics trade were emboldened by the excitement around rare earth minerals in the years immediately following the 2010 crisis. Some investors have even penned editorials calling for the US government to drop the pretense that the occupation should be about anything other than advancing Western economic interests underwritten by the US military (Chossudovsky 2005; Randall and Owen 2012; Scott 2010). If Afghanistan's riches can be made accessible to private sector firms, the logic goes, then the US-led occupation is "a war worth waging" indefinitely.[12]

Developing Afghanistan's mineral wealth, which includes subcontracting prospecting to the AGS and solidifying regulations favorable to international corporations, is the primary task of the Pentagon's Task Force for Business and Stability Operations (DiJohn 2010; United States Department of Defense 2011). As the Task Force's Director of Natural Resource Development put it: "By working with the Afghan Geological Survey on an airborne geophysical exploration program, we are taking an important step in preparing the Afghan government to conduct their own mineral exploration efforts. The goal of this training is to enable the Afghan government to give the best information possible to international investors" (Philippine News Agency 2011). Since 2012, investors have been openly arguing for an indefinite US military presence in order to secure their assets and temper the likelihood that China, Russia, and "others" might "cash in" on US and UK sacrifices over the past decade and a half (Benard 2012; Mehrotra 2013; Randall and Owen 2012). Numerous actors envision natural resource exploitation as the key to national development, but members of Afghan civil society suspect that international efforts to open up the country's resources have little concern for what large-scale foreign extraction might mean for the people of Afghanistan. These are historically informed suspicions. The history of the rare earth frontier is one of extralocal domination of local landscapes and lives, one in which local people matter only insofar as they aid extraction or get out of the way.

This is not to suggest that social and environmental organizations categorically oppose mining—quite the contrary. A coalition of domestic and international groups have been working hard to cultivate a sense of resource nationalism and to improve existing natural resource governance frameworks. According to Ikram Afzali, director of Integrity Watch Afghanistan and coauthor of a 2014 letter to British prime minister David Cameron signed by forty organizations: "We want to develop our natural resources, but from a position of pride and strength, not by lowering our standards and ignoring abuses" (Afzali 2014; Alliance for Peacebuilding et al. 2014). In the Anglophone press, it is not a question of whether mining should occur, but how and on whose terms. As in formerly occupied China, the priority is not necessarily to halt mining, but to consolidate national control over mines and infrastructure as a matter of national pride and development. Afghan civil society groups see mining as an engine for development and sovereignty, but only if Afghani rather than foreign powers set the terms.

Despite very different regulatory frameworks and development histories, both Greenland and Afghanistan are imagined by Chinese and Euro-American interests as global peripheries (Kjeldgaard 2003; Krieckhaus 2006) and two of the latest frontiers for exploration. Pro-mining actors in both countries resist the characterization in different forms while seeking to appropriate it for developmentalist ends. Each introduces a new dimension to the meaning of a frontier. The case of Greenland shows that the making of a rare earth frontier can require a regulatory offensive. In the words of Greenland's deputy foreign minister: "We are, in mining terms, a frontier country. But we are not a frontier country like frontier countries in Africa or South America. We are something very different—perhaps unique. We have evolved over 300 years a solid legal framework, a well-educated population, rules, democratic institutions and a strong society" (Deputy Foreign Minister Kai Holst Andersen as quoted in CER 2014).

From the perspective of conventional development theory, perhaps it is unsurprising that rare earth mining interests from global metropoles in Beijing, Copenhagen, and Melbourne should target these regions that are imagined to be on the global periphery. But the post-2010 wave of exploration was not limited to so-called peripheral or frontier countries.

THE UNITED STATES

The revival of the rare earth industry in the United States and ongoing prospecting efforts in both the United States and Canada (Machacek and Fold 2014) indicate shifting meanings of center and periphery that defined twentieth century world orders. Twentieth-century understandings of global development imagined the United States and Western Europe as "advanced states"; heavy and hazardous industry such as rare earth mining was something for "lesser developed" countries positioned lower on the "ladder" of democratic market development (Corbridge 1994; Johnson, Pecquet, and Taylor 2007; Redclift and Sage 1998). Efforts to bring mining back to areas that had previously exported dirty production to the developing world undermine linear conceptions of development and industrial modernization proposed by Rostow (1960) and advocated by policymakers in the Bretton Woods institutions. This illustrates the temporary character of the frontier. They do not move ever outward from global centers, as often imagined, but can shift, double back, and be conjured in the heart of areas long thought settled.

Indeed, the closure of the Mountain Pass mine in southern California followed the broader contours of the shifting international division of labor at the end of the twentieth century, defined by the global race to the bottom in pursuit of cheap labor and lax environmental regulations. But contemporary dynamics suggest that a qualitative shift has occurred. In the cases of Greenland, Afghanistan, and the United States, rare earth interests are not pursuing lax regulations per se, but are instead mounting regulatory offensives intended to render robust regulatory frameworks malleable to the needs of industry. If we understand neoliberalism as the reorganization of state functions to serve private accumulation rather than as a retreat of the state, this dynamic makes sense. In the US context, the 2010 crisis stimulated public soul-searching as to how such a strategically vital industry was allowed to go overseas. Commentators and policy-makers across the ideological spectrum characterized federal policymakers as "feckless . . . who did nothing while a rogue regime acquired a stranglehold on key materials" (Krugman 2010) and "dropped the 'rare earth ball'" (Goldman 2014). Although such an analysis mischaracterizes history and the protagonists, it nonetheless represents a demand for policy interventions that directly contradict fixed notions of the neoliberal status quo.

Following the 2010 crisis, a range of elected officials representing economically depressed postindustrial regions of the Western United States lobbied for the development of some sort of federal industrial policy to revive domestic rare earth mining and processing (United States House of Representatives 2011a). The proposals ranged from eliminating mining regulations in the United States (Tanton 2012) to creating federal funds to support the revival of a strategically vital industry (Dillon 2010). The former was a disastrous proposition proffered by the scientifically illiterate and the latter was a proposal that would violate US WTO agreements. Over thirty bills concerning rare earth elements were introduced to the US Congress between 2010 and 2014 (Grasso 2013). Although the legislative framing of the issue varied among proposals, they all called for state intervention in the global rare earth market in order to repatriate production. The National Defense Authorization Act for Fiscal Year 2013 contained clauses requiring the secretary of defense to assess the rare earth materials supply chain to determine which materials were critical to national security. In the case of a positive finding, the secretary would have been required to formulate a plan to ensure long-term availability within a three-year timeframe. This resolution would have also mandated that the secretary of defense develop a plan to establish a domestic source of neodymium magnets used in defense applications that had only been produced in China following the closure of Magnequench in 2003 (see chapter 3).

The Pentagon responded to this directive with a report on "the positive changes in rare earth supply chains" precipitated by "global market forces" which should incentivize the development of "economic and environmentally superior" domestic production capacity (United States Department of Defense 2013, 25). This is an example of the ahistorical and agent-less perspective on global political economy that was critiqued earlier in the chapter. "Global market forces" only take shape through the actions of specific actors and institutions at specific times. Those within the US mining industry as well as Congressional representatives from both parties characterized the DoD's standpoint as willfully ignorant of the investment required to rebuild US industrial capacity on par with China, let alone to develop new and superior green technologies. Those familiar with the industry contended that without "government life support," it would not be possible to rebuild the US supply chain.[13] The failure of federal US legislative efforts reveals that competitive pricing remains the sourcing priority for the DoD, even insofar as critical materials are concerned. In the absence of carefully crafted government incentives, competitive pricing regimes currently depend on a geography of rare earth production that devastates landscapes and lives to keep prices low.

The positive changes in the rare earth supply chain referenced by the DoD refer primarily to the temporary reopening of mining facilities in the United States. Following the 2010 crisis, the Molycorp mine[14] at Mountain Pass mine was revived with a US$1.55 billion investment to overhaul production, upgrade the facility, and eliminate the causes of past environmental disasters. But the promise of Molycorp as a newer, better, greener, and above all more American rare earth producer deflated as prices continued their downward trend in 2013 and beyond. In 2011 and 2012, Molycorp acquired a number of firms with processing capacities in China and Eastern Europe. To compete with China's exports, the firm continued to subcontract the dirtier and more complex aspects of rare earth processing to overseas facilities (Gordon, Wilson, and Dickson 2012; Sims 2011).

Subcontracting the more hazardous aspects of rare earth processing did not help Molycorp achieve anything that could be called green production in the United States. A surprise inspection by the Environmental Protection Agency in October 2012 found that leaked materials containing lead and iron were present in storm water on the plant site. Investigators also found several containers of hazardous waste that were improperly closed and labeled (Danielski 2014; Steinberg 2014). Curious documentary filmmakers interviewed some truck drivers in the daily convoy of loaded semis leaving the plant, noting that the volume of truck traffic seemed incongruent with the reported quantities of ores produced. One driver reported that he was hauling radioactive wastewater from an onsite accident to be dumped in the Pacific Ocean (Espilie 2014).

As the global price of rare earth elements fell in the years after the crisis, the stock price for Molycorp fell accordingly, from its 2011 high of US\$76 per share to less than a dollar in late 2014 (Xu 2014). The company declared bankruptcy in 2015 following an SEC investigation of its disclosures, a lawsuit related to engineering deficiencies at the Mountain Pass facility, and a class-action lawsuit alleging that former CEO Mark Smith overstated the importance of rare earth elements to shareholders and the public (Elmquist 2012; Pearson 2012; Wayne 2012). These issues have been attributed to the challenge of carrying out a hazardous business in a well-regulated context and a volatile market dominated by cheap goods from China. But this, too, is only part of the story. The rise and fall of Mountain Pass shows that it is not simply the presence of rare earth deposits, or even an existing industrial architecture that is sufficient to explain why some areas are mined and others are not. Multilateral policy actions pursued by the executive branch of the US government worked to restore the global political economy of rare earth elements to the pre-2010 status quo, which was characterized by the inability of US firms to compete with China's cheaper rare earth exports and forestalled the return of the rare earth frontier to North America.

WTO SUITS

In the 2001 Protocol of Accession to the WTO, China agreed to progressively eliminate all export duties with the exception of a list of products listed in Annex 6.[15] However, Article XX of the General Agreement on Tariffs and Trade (GATT) provides that member states can impose exceptional regulations that would otherwise be in violation of WTO rules. Specifically, member states are permitted to restrict the production and trade of commodities when doing so is "necessary to protect human, animal or plant life or health" and "relating to the conservation of exhaustible natural resources if such measures are made effective in conjunction with restrictions on domestic production or consumption" (GATT 1994). Broadly, Article XX would seem to apply not only to rare earth mining, but also to a host of other extractive industries within China. The concentration of extractive industries for dozens of critical materials in China generated a cascade of environmental and health problems that have been extensively documented by domestic and international researchers. In brief, these are: the environmental degradation caused by expanding mining and basic processing; the epidemiological crises unfolding in communities near and downstream of mining operations; and compounding problems of inadequate technology, lax environmental regulation, and lack of legal recourse for people laboring in dangerous conditions. The serious occupational health and safety problems of the mining sector had made mining the deadliest job in China, and China the country with the most mining deaths in the world (Areddy 2012; Mehmood 2009).

When the rare earth crisis unfolded in late 2010, the United States, Mexico, and EU were already engaged in a raw materials dispute with China concerning other critical materials[16] on which the latter had imposed export duties and quotas in the name of environmental conservation and public health. China successfully defended its practices in the claims brought by the United States and Mexico when they failed to establish "sufficiently clear linkages between the broad range of obligations contained in the provisions of covered agreements allegedly violated by China's export measures" (WTO 2013). However, the panel also found that China's measures were not developing in a WTO-compliant fashion, mainly because the panel interpreted Article XX as applicable only to temporary provisions, and China's measures appeared neither temporary nor equal in impact to domestic and international firms. The panel ordered China to bring its measures into compliance. China appealed, and won. The Appellate body disagreed with the panel's finding that a trade restriction must equally impact domestic industries. Nevertheless, China's General Administration of Customs removed several of the export duties and export quotas on some of the raw materials in question (WTO 2013). Rare earth elements had not been mentioned in the suit.

In March 2012, the United States, EU, and Japan brought another suit against China for the same issues pertaining to thirty measures covering 212 commodities related to tungsten, molybdenum, and rare earths. China claimed that the quotas were linked to the conservation of exhaustible natural resources and were therefore necessary to reduce mining-related pollution. The United States, EU, and Japan disagreed, claiming that the restrictions were designed to protect

Chinese industries producing downstream goods and to force international firms to relocate to China, thereby introducing more value-added processing technologies to the country. In October 2013, the WTO ruled that China failed to provide sufficient evidence that the production and export quotas were necessary for either environmental conservation or national security. Furthermore, the panel found that conservation provisions under Article XX do not allow members to adopt measures that result in one country controlling the international market for a particular natural resource. Since China produced 97 percent of the global supply of rare earth elements at the time, any action to control rare earth production unavoidably exercised controlling influence on the global market. Thus, the complainants alleged that rigging the global market was the motive behind the rare earth export measures, and the majority of the panel agreed. China's Ministry of Commerce continued to issue export quotas through 2014 on the basis that the panel did not uphold its obligation to conduct an objective assessment of the matter, and appealed the decision in April 2014. In August 2014, the Appellate body acknowledged various ambiguities between China's Accession Protocol and GATT, but nevertheless upheld the panel's ruling that China's export quotas were not justified under the conservation provision. It made no mention of the human health provision (WTO 2014).

Experts, such as Liao Jinqiu (quoted in Tu 2012), note that the environmental costs of rare earth production and processing had not been integrated into the commodity price. This is one reason why China was able to undersell foreign competitors. The solution to the grave environmental and epidemiological situation described in the previous chapter has been to further consolidate the rare earth industry in order to better control extraction and processing.

Building on the closure and consolidation campaign of 2006, in 2014 the State Council approved an additional measure to be taken by China's Ministry of Industry and Information Technology in order to control sources of mining-related pollution. The measure mandated further consolidation of all mining and processing companies into six large groups, each led by either a state-owned enterprise or a large private domestic firm. At the helm of the "Big Six" is a revamped Baogang, renamed China North Rare Earths Group. State-owned companies run two other regional groups: Aluminum Corporation of China (Chinalco) and China Minmetals Corporation. Three groups will be organized under now private firms: Xiamen Tungsten Co., Ltd., China National Nonferrous Metals Industry Guangzhou Corporation, and Ganzhou Rare Earth Group. The objective, "to build up a highly concentrated rare earth industry" (Caijing 2013), is intended to reduce overall exploitation of rare earth resources and facilitate the construction of several vertically integrated military, IT, and renewable energy industries. Industry restructuring proceeded in fits and starts, as influential heads of private companies and local officials could resist mergers and closures if they judge them to be against local interests. As discussed in chapter 3, consolidation in Baotou had been partial for several reasons, ranging from the inertia of the built environment, to entrenched power relations, to officially sanctioned clandestine or black market activity judged by some local officials to be in the national interest.

Officials in the Ministry of Industry and Information Technology found the WTO suits regrettable because "as a rule, the WTO allows members to take necessary measures to protect resources and environment, and considers it fair if export restraints are accompanied by simultaneous restrictions over domestic production and consumption" (Xinhua 2012). The WTO did not agree, and China rescinded all export quotas pertaining to rare earth oxides in January 2015.

The WTO victory against China was hailed in the United States as a political victory against China and an economic act of "shooting oneself in the foot" as far as domestic industry was concerned (Dinwoodie 2013). The businesses that benefitted from the WTO suit were, most notably, downstream consumers of processed oxides and value-added components. Those businesses harmed by the WTO suit were those attempting to produce oxides and various components outside of China. The Office of the United States Trade Representative (2014) stated that the victory supported American businesses, workers, and rule of law because the export restraints artificially increased world prices while lowering prices for Chinese producers. In this analysis, the quotas unfairly enabled China to produce downstream products more cheaply while incentivizing US firms to move their operations, jobs, and technologies to China, as though this were a new phenomenon.

Capital flight from the West had happened three decades prior, facilitated by Deng Xiaoping's reforms and the Reagan/Thatcher revolution, and was hardly the central issue in the debate. For rare earth producing firms outside of China, prices and shares dropped with the quotas (Paul 2015). The two primary US policy responses—to let the market shore up domestic industry and take action through the WTO—worked at cross-purposes. No single piece of legislation had passed to support domestic mining in the United States. One high-profile failure came in July 2014, when the bipartisan Securing Energy Critical Elements and American Jobs Act lost by nine votes on the basis that it created a government handout and unnecessarily expanded the role of government in industry. Opponents to the bill argued that the "federal government should open access to the thirteen states where rare earths are known to lie" instead of offering subsidies to US industry (Wegmann 2014). The effect of this has been to create a climate more favorable to better-equipped foreign prospecting interests in the United States (Epley 2014; Gee 2014). This outcome shows that the US federal government is—in practice—primarily committed to maintaining a global regime of free trade favorable to transnational interests above and beyond domestic resource security and technological competitiveness. Despite anxieties surrounding China's rise, and despite the brief flurry of green nationalism, the prevailing approach was to attempt to restore the pre-2010 status quo in order to maintain a steady, low-cost rare earth supply for large purchasers and military contractors such as General Electric and Lockheed Martin. This came at the expense of domestic jobs in the United States, and the landscapes and lives in Baotou.

Amid all of this, a phalanx of junior mining companies emerged in the Americas, asserting their competitiveness on the basis of developing more sustainable production practices. In the discourses of green nationalism following the crisis, figuring out how to produce greener rare earths seemed as important as finding non-Chinese sources. Liberalized exports effectively eliminated the market conditions under which more projects outside of China would be economically viable (Matich 2015). Some predicted that liberalized exports would drive other firms out of business, thereby restoring China's monopoly over rare earth elements absent a long-term investment to continue producing these elements outside of China. Others pointed out that China's export quotas were rarely filled anyway, so this was likely just a symbolic move to signal compliance with multilateral institutions (Wilson 2015). Yet as of this writing, the primary projects examined in this book are proceeding regardless of market conditions. With the nullification of short-term economic argument, territorial agendas and speculative fantasy assumed a greater role in sustaining the regulatory offensives necessary to continue exploration and mining initiatives on the new rare earth frontiers. There was one exception.

Greener Rare Earths: Uncelebrated Breakthroughs in Brazil

Postcrisis concerns generated some compelling technological consequences that were at once breathtaking in their promise to transform the way rare earth elements are produced, and heartbreaking in the manner in which they highlighted the limits of green nationalist rhetoric. The appearance of a viable, more sustainable alternative source of rare earth elements has thus far only served to highlight the practical disinterest in greener rare earth production on the part of downstream industries. But for a short time between late 2010 and 2013, there was a strong sense that downstream purchasers would pay a premium for non-Chinese rare earths. Furthermore, the political climate in the Americas was such that no one wanted to be seen as stooping to China's level of environmental degradation. The idea was powerful enough to stimulate some firms to take unprecedented steps to clean up their operations, as with the case of the Molycorp mine at Mountain Pass, California, which in addition to resorting to illegal dumping, did spend US$2.4 million per year on environmental monitoring and compliance during its brief revival. The idea also stimulated some firms to invest in the development of new technologies and greener industrial capacity, as in the case of the Brazilian Mining and Metallurgy Company (Companhia Brasileira de Metalurgia e Mineração; CBMM) located in Araxá, Minas Gerais, Brazil.[17]

Betting on the willingness of downstream firms to pay a premium for more sustainable and, above all, "non-Chinese" rare earth elements, CBMM fast-tracked R&D on extracting rare earths from the tailings of its niobium mining operations. The company has a lucrative near monopoly on niobium, currently supplying 85 percent of global demand for its products.[18] With the rise of the Mountain Pass mine at the time of CBMM's founding in 1955, and then China's dominance of the global rare earth market, CBMM focused solely on niobium. Rare earths have been accumulating in their mine tailings since the 1960s.

According to CBMM representatives, it was "China's interference in the global market" in 2010 that catalyzed and increased the velocity of their technological development to deal with accumulating mine tailings. Because separation technologies must be developed specifically for each deposit, the apparent demand for non-Chinese rare earth sources inspired the firm to "create a new bottom line" in sustainable rare earth production in 2011. It is worth quoting at length a member of CBMM's rare earths research and development program in order to illustrate the global paradigm shift that might have been:

> We believe that the world needs a supplementary source of rare earth elements other than China. This question of China controlling more than 90 percent of rare earths is not going to happen any longer in the market.
>
> I think there will be space for values in the market greater than those attained through the cheap China price. In truth, the China price is a grand illusion. As soon as you . . . factor in the environmental costs, the social costs, the so-called competitive low-cost will rise to a more normal base price. Today, the base price is very distorted. But Europe, Korea, Japan the United States, they *will* pay a higher value in order to have access to different rare earths players other than China. There is no doubt about this.
>
> Do you think that Japan is going to rely on China eternally? Will Europe or the Americas trust in China forever? No. This is the pillar of our strategy. We understand that there exists a real possibility, there exists a real opportunity for new players, and we want to be the primary ones.

We have developed a technology in-house and a strategy that respects the fact that with China on the scene, our status for the time being is to be almost everyone's Plan B. But major potential partners have assured us that they are interested in our rare earths. Everything that we are developing is up to the same technological, environmental, and social standards that we maintain with our niobium operations.

So we are seizing an opportunity, actually, to improve our niobium production. We don't need to commit environmental or social transgressions in order to seize an isolated market opportunity here and now to make some sales. No. This is not what we are doing. We are going to have a cost, and consequently a price, that would only with great difficulty be equal to China's, but we are going to have a guaranteed supply with the proper social and environmental commitments in order to demonstrate the superiority of our product.[19]

In a relatively short time, CBMM developed the technology to separate rare earth elements from their existing mine tailings and produced high-purity oxides. In 2012, the company invested US$430 million to expand production facilities to produce three thousand tonnes of rare earth oxides annually. In 2013, the company invested another US$24.7 million to double the capacity of rare earth production. The plant uses recycled water from mining operations to produce rare earth oxides in a way that reduces their overall waste footprint, and operates under the most stringent international industry standards for environmental and occupational health and safety in a region that is not considered environmentally sensitive.[20] With these measures, the rare earths produced by CBMM are the "greenest" in the world, and that is not just because the Bayan Obo and Mountain Pass mines set such a low bar for environmental health and safety.

But the CBMM strategists overestimated the environmental values of major purchasers on the global rare earth market, at least for the time being. While their production breakthrough has the potential to transform the way rare earths are produced by effectively eliminating the need to open new mining sites, the WTO victory against China restored much of the global status quo and removed sustainability and security concerns from the priorities of major consumers and regulators, as the flow of low-cost rare earths from China resumed. This had the effect of decoupling "green" concerns from those of nationalism in general, and foreclosed for the time being the possibility that recycling and reclamation might displace new mine exploration. But because rare earth elements are not CBMM's primary business; because they take seriously China's intention to become a net importer;[21] and because they are convinced that global anxiety over China's rise will eventually supersede the allure of low prices, CBMM has adopted a long-term approach to the issue and is determined to produce rare earth oxides regardless of global market conditions. In their analysis, the WTO victory undermined countermeasures to avoid another rare earth crisis. By mining its internal waste frontier, the firm intends to be prepared for the next crisis.

Conclusion

Greenland, Afghanistan, the United States, and Brazil: each of the cases explored in this chapter illustrates a different aspect of the rare earth frontier. They show how histories of domination couple with shifting regimes of technological change, changing environmental conditions, and competing territorial ambitions to produce distinct local instances of the global rare earth frontier. In particular, each of these cases shows how the productive fictions surrounding rare earths generate new spatial politics. Rare earth mining is not necessarily attracted to lawless places; in Greenland, Afghanistan, and the United States, mining interests have mounted regulatory offensives in order to modify existing laws. In Greenland, rare earth elements are seen as a means to assert independence from Denmark by positioning the country as an attractive frontier for global mining investment. In Afghanistan, they are seen as a way to rationalize a historically autonomous region by imposing foreign investment-friendly extractive relations in a way that echoes US and British efforts during the colonial era to source rare earth elements from India and Brazil. The US case shows vividly that the shifting geography of the global rare earth frontier is neither linear nor teleological in the sense of dirty industry radiating outward from developed countries, never to return. It further shows that the toxicity of the rare earth frontier is not a uniquely Chinese problem. Proponents of reopening the Mountain Pass mine were somewhat naïve in their green nationalism, as though reopening a mine in the United States would necessarily mean greener practices. The case of Brazil shows that it is only by re-conceptualizing the global rare earth frontier to include sites of waste, and investing in the industrial capacity to process it, can the rare earth frontier actually be made greener. The failure of the initiative to launch shows that the 2010 crisis radically altered the geography of the global rare earth frontier while keeping its more fictive and destructive elements intact.

Although the discourses of green nationalism were framed in antagonistic terms against China, all international reactions to the 2010 crisis—with the exception of those advancing the WTO case—were actually complementary to the policy objectives of China's export and production quotas. At the annual meeting of the Central Government Economic Working Committee held on December 28, 2010, Yao Jian, a top official in the Ministry of Commerce stated: "To maintain the international rare earth supply is the common responsibility of countries around the world. I hope other countries will further the development of rare earth resources. Countries should find a way to cooperate to open up new resources, as should they strengthen cooperation in energy-saving technologies for global industries, especially high-tech products for superior provision of rare earths" (quoted in Han, Lei, and Yuan 2010).

However unintentional or contentious, these aspirations were realized to varying degrees by efforts to open new mining sites and develop greener production practices. A period of greater global awareness of the acute environmental and epidemiological harms of rare earth production stimulated criticism of diverse technology supply chains. This inspired many to call for more sustainably produced rare earth elements and compelled some key players to develop environmentally superior production practices. The WTO suit restored much of the pre-2010 status quo, undermining both the revival of the Molycorp mine at Mountain Pass and the entrée of more sustainably produced oxides into circulation. This has not, however, slowed the ventures in Greenland or Afghanistan. Extractivist interests succeeded in rolling back codified pacifist values in the case of Greenland, and leveraging US military operations in the name of Western business in the case of Afghanistan. The regulatory changes in these two cases occurred at the expense of local people and environments, but the case of CBMM in Brazil presented a possible alternative scenario.

CBMM's breakthrough provided a glimpse of alternative futures of rare earth production. A future in which no new mines are opened would mean that heavy metals and radioactive materials would stay underground instead of contaminating surrounding soils, waters, and human beings. A production model in which existing mine tailings are reduced would mean that existing sources of heavy metals and radon gas would likewise diminish. Most significantly, this production model would mean that the question of who must be displaced in the interest of exploiting the minerals beneath their feet would cease to be an inextricable part of the mining prospectus.

Had CBMM's practices for creating a "new bottom line" not been ignored, the extractive imperative dividing Greenland's populace between environmental and sovereign integrity, the neoimperial inflection to Anglo-American intervention in Afghanistan, and the ongoing contamination of the southern California desert might have diminished significantly. Had major Western purchasers been serious in their commitment to source environmentally superior, non-Chinese rare earth elements, perhaps we would be witnessing a paradigm shift in global rare earth politics. Perhaps in the future global discourse will move towards making productive use of the wealth of rare earth resources latent in mine tailings across the globe, and sustainable production of rare earth elements will become a point of international cooperation. And perhaps, for the first time in over a century, the flow of radioactive wastewater, toxic gases, and heavy metals into soils, water, and bodies around the world will begin to slow.

From simple market logic, the price effects of the WTO rulings against China should have slowed prospecting efforts in controversial places. From an ecological modernization perspective, the subsidized entrance of environmentally superior production technologies should have presaged a global paradigm shift. Even from a logic of enlightened self-interest, the fact that China's interests in becoming a net importer directly aligned with state and national interests across the globe to become rare earth exporters should have generated a more substantive change in the global political economy of rare earth mining and processing.

Clearly, these theories cannot explain the fact that rare earth prospecting has continued in Afghanistan and Greenland, nor can they explain why, in Brazil, an entirely different set of actors has been working to open up high-risk mining concession in the extreme northwest of the Brazilian Amazon, even amidst CBMM's potentially paradigm-shifting breakthrough. As the final two chapters demonstrate, the driving interests behind the contemporary geography of the global rare earth frontier are not just economic; they are also geopolitical and entangled with longer-term territorial agendas. To understand what is happening on the farthest reaches of the global rare earth frontier, we must depart from the logics of a world governed by laws of supply, demand, and scarcity.

5

FROM THE HEARTLAND TO THE HEAD OF THE DOG

Madam President . . . I wish to address another subject that is crucial for the development of our country, a subject that concerns the reaffirmation of our sovereignty especially with respect to scientific and technological autonomy in more advanced areas of innovation. In addition to the gains in basic research in a very new area of technology, our country can benefit economically if we develop the knowledge necessary to transform some rare minerals that we have right here in our territory into the inputs for complex and sophisticated products . . . I am referring to those which are called "rare earths."

—Senator Luiz Henrique da Silveira (2013)

There are well-regulated rare earth producers established in Brazil's industrial heartland. So why are state, corporate, and indigenous actors campaigning to mine a remote and protected region of the Amazon?

In the wake of the 2010 crisis, public and private sector actors sought to "reglobalize" Brazil's rare earth industry to restore it to pre–World War II importance in global political economy (Lapido-Loureiro 2013). In particular, the state-owned geological research enterprise Mineral Resources Research Company (Companhia de Pesquisa dos Recursos Minerais: CPRM), jointly with the National Department of Mineral Production (Departamento Nacional de Produção Mineral: DNPM), received the largest federal budget in their history in order to reinvigorate their mandate to generate basic geological information about Brazil's thirteen million square kilometers of territory. This information was intended for the public and potential investors and meant to facilitate strategic resource extraction (Jones et al. 2011). In late 2011, the Ministry of Science, Technology and Innovation and the Ministry of Mines and Energy held Brazil's first seminar on rare earths in Rio de Janeiro, in which government geologists declared that Brazil possessed the largest known rare earth reserves of any country in the world (Lima 2012). As noted in chapters 2 and 4, each of the major finds examined in this work are purported to be the largest in the world. Although the content of these claims must be treated with skepticism, the work done by such claims in producing frontier space requires closer examination. In Brazil, these claims shape contemporary iterations of historical struggles between indigenous peoples, multiple state interests, and corporate and artisanal miners.

The objective of Brazil's first and subsequent rare earth seminars was to disseminate information on domestic rare earth reserves to diverse audiences. It was hoped that promoting the country's rare earth endowments would recruit investors as well as generate political will to organize a vertically integrated rare earth production chain in the country (Ferraz 2011). Here I examine diverse efforts over time to transform an explored site into a new rare earth capital. As with the complex territorial campaigns and immense shifts in global political economy that constituted the establishment of Bayan Obo as the rare earth capital of the world, the case of Brazil shows that much more than simple accidents of geology or invisible hands of the market drive the geography of the rare earth frontier. Since the 2010 crisis, multiple regulatory offensives have unfolded in order to make rare earth mining possible in Brazil's remote and protected regions. In 2012, the Senate established a temporary commission to elaborate a special regulatory code to facilitate domestic rare earth mining (Federal 2013). Framed in official discourse as strategically necessary to Brazil's technological innovation and increased international influence, rare earths are referred to as "bearers of the future" (*portadores do futuro*)[1] (Diniz 2013), as crucial to national defense (Portales 2011), and as essential to the national sovereign development of Brazil (Henrique 2013).

Even before the liberalization of China's rare earth exports in January 2015, the Brazilian rare earth frontier displayed a spatial paradox that defied simple market logics. From 2011 to 2014, several rare earth frontiers emerged within a broader national developmentalist framework, two of which are presented in this book. The first concerned the aspirational challenge of procuring rare earth elements from existing, partially separated mine tailings at a state-of-the-art niobium production facility. This facility is situated on well-established infrastructure networks in the Brazilian heartland within a regional regulatory context favorable to extractive industry. The challenge, in this case, was to develop the technological capacity to remake the wastes generated by niobium beneficiation into a new rare earth frontier. With the major technological, regulatory, and infrastructural issues resolved, the remaining challenge for CBMM was to find downstream buyers willing to pay a premium for a sustainably produced, non-Chinese source of rare earth oxides. Industry analysts explained the failure of the sustainable rare earths initiative strictly in terms of price competition: if CBMM could beat China's price, then they would dominate the global market.[2]

But there is more to the story. Simultaneous with CBMM's initiative, other mines opened and state geologists reexamined samples collected in the 1970s from a sacred hill in the northwestern borderlands. Called Morro dos Seis Lagos (Hill of Six Lakes), this site is located squarely in indigenous territory on the border with Colombia and Venezuela, in an ecological park on the eastern edge of the municipality of São Gabriel da Cachoeira (Simões 2011). This region, nicknamed Cabeça do Cachorro (Head of the Dog) because of the shape formed by national borders, is far from the only promising deposit in Brazil. It is perhaps the farthest—in political, ethical, and logistical terms—from the possibility of industrial development. Nevertheless, a small group of geological researchers, backed by state, military, and powerful investor interests, dusted off forty-year-old archives identifying a deposit that reportedly contains record concentrations of rare earths that could supply global demand for the next four hundred years (Lima 2011). The infrastructural constraints alone negate this proposition, as illustrated in figure 14. This has not tempered the intense interests in mining rare earths in this region, however.

FIGURE 14.   Location of Araxá and Cabeça do Cachorro. The map shows the relative density of Brazil's paved road networks on a national scale as well as the position of two of Brazil's rare earth frontiers.

Source: Image by Molly Roy.

The basic paradox at work in the Brazilian geography of the rare earth frontier is this: even though reportedly abundant, technologically and environmentally superior high-purity rare earth oxides are already under production in Araxá—a site with well-developed physical and regulatory infrastructure—there is still a tremendous push to exploit rare earths and related elements in an inaccessible, legally protected, and politically inhospitable site located in the northwestern Brazilian Amazon.

Paradoxes cry out for some form of resolution (Proctor 1998). As with the cases of Afghanistan and Greenland, the debate concerning rare earth mining in Cabeça do Cachorro takes actual market concerns merely as a point of departure. In these new rare earth frontiers, longer-term struggles over territory, sovereignty, recognition, and geopolitics are (re)enacted through competing claims over these critical element endowments. In practice, these competing claims are exercised over the question of whose geological knowledge is recognized as legitimate, and which extractive practices are sanctioned by the state. Although the international community tends to see mining and rainforest conservation as diametrically opposed, among domestic stakeholders, including a diverse set of indigenous activists, the question of mining in Cabeça do Cachorro is not framed in terms of whether it should occur,[3] but rather by whom and under what political economic conditions.

Because these struggles are not about whether state-sanctioned extraction takes place, but rather, by whom, certain elements characterizing the quest to tame the rare earth frontier in Baotou are recognizable here. Competing transnational actors surveyed the region with the ambition to consolidate territorial or geopolitical power by capturing the mineral and spiritual riches of this place, while struggles over who shall bear the mandate to mine these particular deposits are fused to dreams of sovereignty, desires for greater recognition, and the need to control the conditions of one's own development. That such an intense struggle is unfolding over the meaning and control over rare earth extraction in this region despite the economic and logistical infeasibility of such an enterprise indicates that there is much more at stake than the putative scarcity of rare earth elements, or even the quest to develop a national rare earth production chain. The struggle over the terms of mining in Cabeça do Cachorro is a clear instance in which historical continuities shape contemporary debates, changing the meaning of the latter through reenactments and reinterpretations of the former. In other words, rare earth elements are not the sole cause of contemporary struggles to determine the fate of the region. As in other cases, not even the presence of potentially minable deposits is sufficient to explain the conflict over the fate of the region's mineral endowments. Rather, it is because these particular deposits are strategically valued by powerful interests, which have interpreted rare earth mining in terms of longer-term territorial ambitions, that Cabeça do Cachorro emerges at all. It is neither history nor geology nor territorial politics alone, but a combination of the three.

Conflicting notions of sovereignty and security lay at the heart of the struggle over who shall mine rare earth elements in Cabeça do Cachorro. The stakes are amplified by ongoing geopolitical contests, three of which are the focus of this chapter: indigenous and artisanal struggles for greater political and economic recognition; multiple failed efforts on the part of extralocal powers to control transborder flows of people and commodities, and; border militarization in the context of the US global militarism. Because of these factors, this frontier, remote in distance and feasibility, has garnered greater political interest and international exposure than the more promising frontier manifest in the technological breakthroughs in Araxá. These asymmetrical interests in mining are better explained by the way in which rare earths shape longer-term struggles in Cabeça do Cachorro, rather than by any actual scarcity of rare earths themselves. This is another case that illustrates how the rare earth mining is a means to broader territorial ends.

Although the historical continuities are important, there is something fundamentally new about the debates concerning resource extraction in Cabeça do Cachorro that is scarcely visible in Anglophone discourses. Namely, the manner in which indigenous mining proponents are engaging in struggles over the fate of various legal conventions currently prohibiting mining on their constitutionally recognized lands. Contrary to prevailing representations in academic and advocacy work on the relationship between Amazonian peoples and mining, there neither is a unified "Indigenous" perspective nor is the debate drawn along the lines of indigenous opposition to mining as colonial invasion.

To some extent, the prevailing notions are understandable. The weight of centuries of historical memory, etched deeply with the racialized hierarchies of colonial brutality, overwhelm more nuanced perspectives on indigenous peoples and mining in this place referred to as the "end of the world."[4] But the struggles around mining in Cabeça do Cachorro demand a more complex view of indigenous politics in the Brazilian Amazon.

In particular, the diversity of positions on extractivism and the alliances among actors generally portrayed as opponents are crucial to understanding the stakes of rare earth mining in the region. The sometimes deadly conflicts over who is entitled to mine, and in what manner, are centered on the meaning and circulation of geological knowledge unfolding over multiple temporal scales. All aspects require due consideration to make sense of this point on the global rare earth frontier. This chapter uses the production of geological knowledge and the struggle for control over the definition of the rare earth frontier as the lenses through which to bring these complex dynamics into focus.

Competing Sovereignties: Indigeneity, NGOs, and the Military

Cabeça do Cachorro has been positioned as the frontier of many ambitious powers since the 1600s. Over the centuries, extractive interests have colluded with imperial and state powers in order to discipline, rationalize, and exploit the subsoils of this place, all with varying degrees of failure. Cabeça do Cachorro is situated on the upper reaches of the Rio Negro River in a historically contested border region between Brazil, Colombia, and Venezuela. Slavers, missionaries, rubber companies, and treasure hunters large and small have attempted to wrest the region's wealth from local hands. Violence periodically flares between indigenous people, outlaw geologists, police forces, and traffickers of various sorts. Seventy-eight percent of the population belongs to one of the twenty-six indigenous groups[5] possessing pre-Colombian histories in the area extending back at least three thousand years (Neves 1998). Their collective land rights were legally recognized by the federal government in 1996 with a demarcation of five continuous protection areas totaling approximately 10.6 million hectares, as shown in figure 15. No one, not even the military, is allowed to enter indigenous lands without permission. Today, the region is governed by a working alliance between indigenous federations, the Brazilian military, and a rotating cast of civil servants.

The region can only be accessed by air or river transport. Although indigenous and environmentalist victories in the 1980s and 1990s deterred many large-scale extractive interests, in early 2011 the Brazilian government announced that it would encourage exploration for rare earth deposits in the Amazon. President Dilma Rousseff personally invited the firm Companhia Vale do Rio Doce to identify new mining sites (Gozzi 2011), which inspired other enterprising geologists to follow suit. Since then a troubled alliance among military planners, state geologists, small-scale miners, indigenous activists, and federal government officials has coalesced around a struggle to rewrite the hard-won state and federal laws that have categorically prohibited mining in Cabeça do Cachorro for the past three decades.

At present, an executive order, a constitutional amendment, and several acts of congress would be necessary to repeal the constitutional measures forbidding mineral exploration and extraction in this sensitive and protected region. This is a separate matter from the conservation measures discouraging regional infrastructure construction. Long-standing contests over national sovereignty, indigenous citizenship rights, and basic livelihood security in Cabeça do Cachorro lie at the core of the impetus to exploit rare earth elements in such a challenging context. As in other sites examined in this book, material concerns over rare earth supply security are secondary to the imagined geopolitical spoils to be won by controlling their extraction in these remote frontiers.

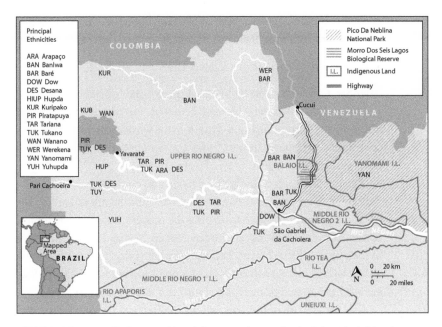

FIGURE 15.   Detailed map of land demarcation and principle ethnicities in Cabeça do Cachorro.

Sources: Ricardo and Ricardo (2006). Image by Molly Roy.

This has generated some strange and perilous alliances. At a conference organized by the commanding general of the Amazon Military Command (Comando Militar da Amazônia; CMA), senior military officers, state planning officials, geologists, artisanal miners (*garimpeiros*),[6] and Indigenous mining proponents assembled in Manaus on April 26, 2014 to discuss the subject of rare earths and strategic geopolitics. In an open comment session following a series of lectures on these subjects,[7] the entanglement between rare earths and competing visions for the territory was laid bare in an exchange between veteran geologists, an Indigenous garimpeiro, and the commanding general of CMA. To start, a senior geologist who had participated in the preliminary geological explorations of Morro dos Seis Lagos during the military dictatorship said the following:

> I am here weeping before you because it has been thirty-nine years since I graduated with a degree in geology . . . forty years ago, we went up the Rio Negro doing our research, finding the resources that would develop our nation and FUNAI[8] chased behind and said "No, this is Indigenous land." But I have confidence that you, here, who defend our sovereignty [gestures toward audience primarily composed of three and four-star

generals], are going to wipe the tears from our eyes that come from seeing so much poverty amidst the riches that we have conditions to develop.

This inspired a lead geologist with the local DNPM to say the following:

After São Gabriel became a biological reserve and Indigenous territory, we had to cease all mineral research activities. But what we have there, I tell you—the greatest niobium deposit in the world, with rare earth concentrations previously unheard of—is a *national patrimony*. It would bring tremendous returns to all of us Brazilians.

The demarcation of Indigenous lands and conservation areas is financed by European and North American and Japanese banks. The Brazilian government is coordinating the process, but all of the money comes from international environmental entities and banks that want to lock up Brazil's resources, because they know that they are not going to fare well if they have to enter into competition with a developed Brazil! We have let this go on for too long! Too long! [Applause]

Enough with all of this concern for the Indians! Enough with creating conservation areas! The Indians are not interested in conservation areas; they are interested in mining and development! [Addresses Indigenous activists in the audience] International NGOs [nongovernmental organizations] are lying to you! They don't want you to know that beneath you are tremendous riches! They want you to continue being poor, to continue being simple! The international NGOs are never going to let you grow! They want to keep you stupid, illiterate!

They are making Indigenous areas and putting you there like you are some kind of animal and they aren't giving you a single legitimate economic activity. It is a degradation of your very lives! You want to work and develop, but it's prohibited! You can't mine, you can't develop tourism, you can't do anything. Because everything that would be good for you is prohibited. [Applause].

In response to this, an Indigenous garimpeiro took the floor and introduced himself as a member of the Tukano people, a military veteran, and the organizer of a small mining cooperative:

Folks, when I hear you talking about how the world is not letting *us* develop, it is very hurtful. *We* are as *Brazilian* as you. Our condition is not the fault of the world.

We are the most mineral rich country in the world, and we don't have our own affairs organized. We are dependent on China! This is pathetic, folks. Today: no, it should have been done yesterday; the law that permits us to mine on Indigenous lands must be regularized. But it is not the world, it is we Brazilians who make the laws. These laws need to be changed, and we are the ones to change them.

People go out claiming to represent the Indians. They say Indians don't want small-scale mining [*garimpagem*]. They say Indians are puppets of politicians. Politicians say Indians are puppets of NGOs. And when this country doesn't have progress people blame the Indian. They say: the Indian is interfering with our development. The Indian is not interfering with anything. The Indian wants progress! The Indian wants to help secure the frontier!

. . . I have formed a mining cooperative because I have hope in Brazil. We hope to obtain a concession to have a small mining operation in my region so that we Indigenous people, ourselves, can bring minerals to augment the nation's economy, which we need so badly. So I hope you will understand me when I say that the NGOs are hindering us in our efforts to move forward. Expel the NGOs from our lands! We want progress.

This speech garnered a standing ovation. Next came the general's closing remarks, in which he projected a map showing the largest deposits of rare earths and other high value mineral commodities in the Amazonian region. He then super-imposed a map indicating Indigenous lands and conservation areas, which covered many of the largest deposits (see figures 16A and 16B). "It would seem that Indians really like minerals," he said, which provoked laughter. He continued:

You know, I made this same joke at a presentation in São Paulo and somebody denounced me to the Public Ministry for disrespecting Indians. I hope you will forgive me; I made such an ironic statement because the Indians, you, my Indigenous brothers, are tools. You are tools being used in this process that [the DNPM geologist] characterized very well. Mining is the basis of sovereignty, yet we are preoccupied with human rights questions about things that happened forty years ago, and in the meantime, we have completely incapacitated ourselves from enacting concrete solutions to our problems.

No one had mentioned human rights or the military dictatorship during the conference, so the general's unprovoked reference is a telling indication of the ideological continuities underlying military concerns toward the Amazon.[9] Furthermore, the exchange indicates how perilous the Indigenous mining position becomes in the face of competing territorial agendas. The general's response to the indigenous activist shows that despite a declaration of loyalty, a claim to military fraternity, and a condemnation of NGOs, the Indigenous are seen as obstacles to corporate or militarized extractivism. Indigenous control over mining in Indigenous lands does not qualify as a "concrete solution" and does not satisfy the military's vision of sovereign control over the Amazon. Only the domination of Amazonian landscapes and lives through large-scale industrialization will suffice. So long as Indigenous counterparts propose anything other than acquiescence, they are seen as tools of international NGOs lacking the capacity to control their own affairs.

There is a way in which international environmental and Indigenous rights campaigns, on the one hand, and indigeneity scholarship, on the other, align with the general's dismissal of Indigenous claims to mine within their own territory, on their own terms, for their own economic gain. The relationship between people and nature in indigeneity discourse often frames Indigenous agency in terms of a "more balanced" relation to nature that has not entirely let go of the "noble savage" tropes (Redford and Stearman 1993). As a result, there is virtually no conceptual space for the figure of the Indigenous miner. With remarkably few exceptions (Graulau 2001; Lahiri-Dutt 2011), mining does not even enter into

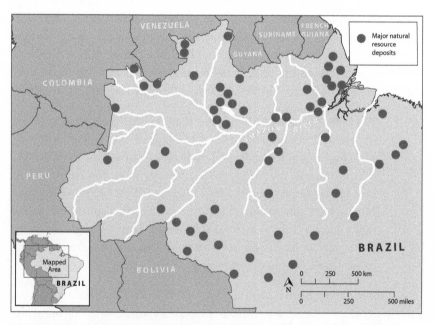

FIGURE 16A.   Adaptation from a widely circulated slide presentation used by the General of the Amazon Military Command. The first slide indicates several areas of major known mineral deposits in the Brazilian Amazon.

Source: Image by Molly Roy.

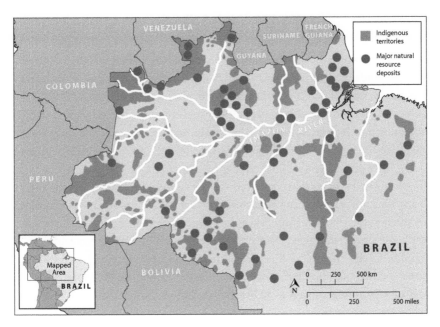

FIGURE 16B. Adaptation of the second slide showing mineral reserves overlaid with Indigenous lands and conservation reserves.

Source: Image by Molly Roy.

consideration as a legitimate economic activity for Indigenous peoples: not in the international conservation movements, not in the majority of literature on indigeneity, and not in the perspective of the Brazilian military.[10]

Exiled from transnational advocacy networks, but not recognized as being wholly Brazilian, the exchange between the Indigenous miner and the general indicates the limited agency attributed to contemporary indigeneity. Furthermore, whether and how Indigenous people in São Gabriel da Cachoeira actually identify with notions of indigeneity circulating globally and in domestic Brazilian discourse is a question seldom asked of Indigenous activists in the region. In international conservation discourse, the intense duality of mining as destruction wrought by outsiders versus the image of Indigenous people as environmental stewards[11] does not account for the difference between corporate and small-scale mining activities. Indigenous mining proponents are dismissed as "coopted Indians" (*Índios cooptados*) and find themselves forced into an impossible space of silence on the matter of basic citizenship and usufruct rights. But silence would amount to acquiescence to the frontier-taming visions of distant powers, allowing no space for indigenous agency. So silence is not an option.

As written, the law differentiates between horizontal and vertical land use rights. Indigenous people are legally entitled to the use of everything above forty centimeters depth (to allow for cultivation) contained within their federally recognized lands. Everything beneath this depth is reserved as property of the federal government and part of the national patrimony (*Patrimonio da União*). The current status quo relies on an explicitly spatial compromise: Indigenous people are entitled to horizontal land use to sustain themselves, while the state asserts exclusive control over vertical territory. Legally, nothing can be extracted from below forty centimeters without paying royalties to the federal government. But the permitting process is prohibitively expensive and garimpeiros face the risk of being violently displaced by large-scale mining companies should they register their holdings.[12] Further complicating the issue is the geological incidence of rare earth elements in the region. Many rare earth deposits on Indigenous lands in Cabeça do Cachorro are found at, or very close to, the surface of riverbanks and streambeds. Since "mining" these types of deposits requires little to no digging, it is actually ambiguous as to whether this constitutes an exploitation of horizontal or vertical territory.

The state denial of vertical land rights to Indigenous people and the particular geological incidence of rare earth elements on Indigenous lands in Cabeça do Cachorro has generated some confusion, as illustrated by the ordeal suffered by veteran local leader in the Indigenous rights movement Mr. Santos. In early 2011, Mr. Santos noticed bluish green stones and clays at the mouth of a riverside cave within his lands. He sent a sample to a geologist friend in Manaus for analysis, who confirmed an unusually high concentration of rare earths. This geologist subsequently informed a buyer who quoted a price that was noticeably higher than what Mr. Santos recalled his other acquaintances ever receiving for their mineral shipments. This was in 2011, when global rare earth prices were still quite high, although Mr. Santos was unaware of the reasons at the time. He arranged to fill a four-meter long boat with raw material and transport it to Manaus to sell.[13] He, his friends, and his family joined in the effort:

> There was nothing clandestine about this. My friends and family gathered and loaded rock in the plain light of day, and I took the boat down river, stopping in the city [of São Gabriel] for lunch. I secured the boat on the beach in front of the federal police post and when we returned from lunch the federal police asked me: "Mister, is this your boat?" I said yes. They asked me: "Where is your fiscal note for these materials?" And I said: "Fiscal Note? What fiscal note? This is from my land." The Police said: "You robbed these resources from Indigenous lands?" And I said, "Robbed, what's this? I am from that land, I found

these rocks, and my relatives helped me gather them." The police informed me that removing minerals from Indigenous lands was a crime. They seized my boat, arrested me, and sent me to jail in Manaus, where I was charged with illegal mining and environmental crimes. I was sentenced to six months in jail and assessed a fine for picking up stones on my land. Now I, a fighter [for Indigenous rights], invested in the reform of the legal system of our country, and the father of two young daughters, must return to Manaus several times each year to report to the Justice. I conducted my business in the plain light of day, and now I am a criminal.[14]

Mr. Santos and his family had planned to use the money from mineral sales to purchase a computer for his daughters and to take care of medical expenses needed by some community elders. In a region where social relations were largely nonmonetized, certain essential goods and services nevertheless require cash. Exchanging a boatload of rocks for a good price seemed to be an entirely legitimate means to acquire currency. Mr. Santos was not the first to experience this. A newspaper report from 2001 reported a similar affair, in which a member of the garimpeiro cooperative Cooperíndio was intercepted by the federal police in Manaus with a boatload of amethyst and tantalite (Brasil 2001). The cooperative has since politicized its activities, deploying white Brazilian garimpeiros to engage with allies in the DNPM and pro-mining interests in the CMA in an effort to shape national mining policy to protect smallholder interests.[15]

At stake in this debate over who mines in what way is the question of who has the right to define and conquer Brazil's geological frontiers. This is, at its core, a question of citizenship, which at present only extends to large-scale corporate actors and from which Indigenous people and garimpeiros have been excluded because of the costly barriers to legal mining. Put another way, Indigenous people do not have the means to conquer their own geological frontiers in ways that are recognized by the state. The sense that an unknown entity could take the earth from beneath one's feet is a chronic source of insecurity among Indigenous people, who cite mining entitlements as a matter of fuller citizenship rights within Brazil. Currently, only wealthy prospectors and established firms can afford to comply with the law as it is written. This suggests that the law was written for a specific notion of citizenry that excludes both garimpeiros and the Indigenous.

The legal codification of extractive entitlements extends state-backed corporate power to this enduring frontier and porous border region. This preference for large-scale corporate mining over small-scale indigenous extractivism illuminates the states' attempts to selectively (dis)empower certain agents to execute its territorial agenda. This is a striking example of how history shapes the present by reinscribing the opposition between Luso-Brazilian colonizers and Indigenous peoples. In this case, the struggle unfolds over who is entitled to exploit the nation's geological patrimony in order to control this point on the global rare earth frontier. This racial codification of legal and illegal mining erases any possible constructive alliance between the figures of the indigenous miner and the Luso-Brazilian garimpeiro.

Indigenous mining activists as well as white garimpeiros argue against these obtuse identity politics informing current state policy with logical and legal propositions. Those interviewed have identified two primary issues behind the state's rationale for criminalizing their activities. First, the prohibitively expensive permitting process means that garimpeiros are not conducting mining in a way that allows the state to collect royalties. Second, the clandestine nature of their activities means that they cannot organize investment or other programs to implement environmentally superior technologies. They expressed a firm desire to formalize their operations in order to reduce the social and environmental danger that plagues their currently lawless trade. In other words, they understand the problems and propose solutions: legalization, regularization, and state prioritization of small-scale mining activity as an antipoverty measure. These efforts to engage in policy reform complicate the position of the garimpeiro in Amazonian politics, which is perhaps why they are only selectively acknowledged in broader Anglophone and Lusophone discourses on the region.

In general, garimpeiros are cast as villains in Amazonian conservation and Indigenous rights politics. They are condemned for destroying the environment, invading Indigenous lands, and for wreaking murderous havoc wherever they go (D'incão 1994; Guimarães 2010; Hoefle 2013, MacMillan 1995; Slater 1994). There are important truths to these characterizations, but they are too simple. It does not allow for the possibility of Indigenous and garimpeiro cooperation. It racially codifies mining activity by presuming that only whites conduct mining on Indigenous lands and that Indigenous people are opposed to mining because of some innate quality of their indigeneity. Such a characterization forecloses the possibility that Indigenous people might also be garimpeiros, engaged in mining of their own volition, as part of their own livelihood practice, on their own lands.

Because Indigenous people are only permitted to use minerals found within the first forty centimeters of subsoils on their land for traditional and spiritual purposes, Indigenous mining proponents argue that the law is designed to keep them in a state of nature. By confining their usufruct rights to the realm of ceremony, the protections do not allow them to advance or live as modern citizens of Brazil. The Federation of Indigenous Organizations of the Rio Negro, (Federação dos Organizações Indigenas do Rio Negro; FOIRN) in particular has been at the forefront of condemning the current legal regime as primitivist, classist, and racist insofar as it only makes it possible for those who can afford the up-front permitting costs to comply with the law, while criminalizing customary activity regardless of how necessary small-scale mining might be for local livelihood security in the context of a globally integrated Brazil.[16] But because proposed revisions to the federal mining code would repeal the moratorium on indigenous lands while also outlawing mining for all but major corporations (Coêlho 2013), Indigenous activists and politicized white garimpeiros are engaged together in a multipronged fight against corporate interests in the Brazilian Senate. Their goals are to legalize and grant statutory protection to small-scale mining operations on Indigenous territory, to prioritize indigenous governance and prohibit large-scale corporate mining by foreign firms. Only then, they argue, will they rise from their status as criminals at worst and second-class citizens at best.

The struggle represents a claim to citizenship and belonging that is outside the categories typically ascribed to both Indigenous and extractive interests. The claim for greater Indigenous autonomy over Indigenous lands is actually a claim to belonging to the Brazilian nation. André Baniwa, elected deputy mayor of São Gabriel da Cachoeira expresses it thusly: "Being Indian means knowing one's own culture, traditions, and maintaining one's identity without failing to know the Brazilian state [and] the culture and tradition of the nation" (Baniwa 2009 quoted in Guzmán 2013, 50). The struggle over the mining code, as it unfolds between Indigenous leaders and the state, is characterized by a demand on the part of Indigenous garimpeiros to participate in national development through small-scale mining on one's own lands and on one's own terms.[17] Indigenous mining proponents also argue their case in geopolitical terms, maintaining that greater control over their local resources would empower them to exert greater control over the Brazilian border while also capturing a greater share of the global rare earth frontier for the benefit of the country.

One Tukano member, Mr. Barreto, has had a small-scale mining project that he began in 2013 to promote as a model for a sustainable development policy: "The activity would seek profits, not from the point of view of the capitalist world but from the measure of sustainability using traditional techniques and Indigenous conceptions and taking into account the relationship with nature. We do not want large companies and large corporations doing the work"[18] (Barretto quoted in Farias 2013a).

However, both environmentalists and their political opponents in the Senate have misapprehended this sentiment. On the one hand, certain politicians pointed to Indigenous desires to mine their own lands in order to justify the liberalization of mining on Indigenous lands writ large, in exactly the terms Indigenous activists oppose. On the other hand, the international environmental conservation community has been vehemently opposed to changing the mining code on

Indigenous lands under any terms whatsoever. In their global campaigns to pressure the Brazilian Senate to vote against procorporate, anti-indigenous mining measures, international NGOs have privileged a simplified narrative of Indigenous interests. The hope is that this will serve to keep the federal government and allied military and corporate interests out of indigenous lands. But the stakes of this form of representation are high. Military interests see allied indigenous and international NGO efforts as an affront to national sovereignty and security, and have responded by putting forth Ministerial Order 7957. This would allow state ministries to request the use of military force against Indigenous people who oppose large-scale development projects. This is an especially illustrative example of the way in which law reflects and reproduces power relations in society, rather than objectively arbitrating among them.

In a 2014 tour sponsored by Survival International, the renown Yanomami shaman Davi Kopenawa spoke to diverse audiences around the world to generate international opposition to mining on Indigenous lands (Bayer 2014).[19] This international campaign against Amazonian mining furthered the racial codification of extractivism insofar as the problem was framed in terms of white miners invading pristine Indigenous lands populated by people who had no interest in mineral exploitation. While it is true that capitalist mining is understood as a degradation of the land and a violation of the Indigenous cosmos (Kopenawa and Albert 2013), small-scale mining is not understood in these same terms (Graulau 2003; Hinton, Viega, and Beinhoff 2003).[20] Unfortunately, the international campaign traffics in the primitivist vision of Indigenous peoples against whom activists and inhabitants in Cabeça do Cachorro have been fighting. There is, of course, fierce debate within indigenous communities regarding the practice and proliferation of mining. The point is that there exists a plurality of perspectives.

The current success of the international campaigns comes at the expense of more nuanced proposals intended to give Indigenous people greater control over their lands, such as those advanced by FOIRN and local activists. This has generated intense resentment against international NGOs among Indigenous interviewees and mining proponents, but the antipathy goes both ways. Environmental activists dismiss Indigenous pro-mining arguments as not representative of the real concerns of the community, suggesting that Indigenous garimpeiros have been tainted by greed or brainwashed by military propaganda. This erases the fact that many of the mining proponents interviewed have been involved in the Indigenous struggles since the bloody days of the 1970s and 1980s.[21] The purpose of these struggles, from the outset, was to establish Indigenous control over Indigenous lands.

In the view that emerged from the concentrated gathering of diverse mining interests at CMA headquarters in Manaus, conservation and Indigenous protections were cast as an injury to Brazilian sovereignty insofar as they constrict the power of the state to conquer the Amazonian frontier in the name of rare earths. The relationship between centralized Lusophone power and the people living on the Amazonian frontier has changed little over the past few centuries: Indigenous entitlements to the land are a hindrance that would be best dealt with by the imposition of large-scale mining activities and the obliteration of existing local territorial orders. Not even an Indigenous investment in the extractive effort, expressed directly to military officials and state geologists in the name of national development and frontier security, could unseat this entrenched frontier narrative.

International NGOs position Indigenous mining proponents as coopted by corporate and military interests, but federal policymakers and military officials dismiss Indigenous mining proponents of the sort referenced above as indistinguishable from the traitors who have aligned themselves with international NGOs against the Brazilian state. Under the latter framing, both NGOS and Indigenous mining proponents demanding special protections against outside interests are accused of stalling national progress by "locking up" the strategic minerals of the Amazon in conservation reserves and Indigenous lands. Thus, Indigenous mining proponents find themselves doubly exiled in political discourse on the matter: from their transnational communities composed of Indigenous activists and international NGOs and from the position of recognized citizenry of Brazil. The dual nature of this exile is distinctive to contemporary rare earth politics, but the troubled relationship between Indigenous inhabitants and outsiders intent on remaking indigenous lands into a frontier to be conquered and exploited has an extensive history. Therefore in this case, another feature of the contestation over the rare earth frontier is the meaning, legitimacy, and scope of the legal protections provided by citizenship. Although these tensions are not unique to contemporary rare earth politics, they are nevertheless constitutive of them. As in Afghanistan, Greenland, and China, mining rare earth elements is the latest iteration of longer-term struggles to create and conquer a frontier, and to determine who belongs and who is trespassing.

Amazonian Frontier: Four Centuries of Geology, Imperialism, and Nation Building

The dynamics of the April 2014 encounter at CMA are rooted in several centuries of extralocal efforts to territorialize Cabeça do Cachorro regardless of existing local territorial orders. In effect, rare earth politics provide a contemporary iteration the struggle between extralocal interests and their multiple narratives that characterize the frontier as unproductive, empty, and problematic. Historically, as now, these frontier narratives clash with local desires to exercise resource-based economic agency. From imperial conquest to modernist nation-building projects, geological prospecting has been a key tool through which multiple powers have sought to discipline the northwest Amazonian frontier into a resource hinterland and a source of geopolitical power. The history of surveying and exploring the upper reaches of the Rio Negro is characterized by a rotating cast of aspiring hegemons with extractivist designs on the region.

Europeans first explored the upper reaches of the Rio Negro in the expedition of Pedro Teixeira in 1639 as part of a mandate by the Portuguese crown to consolidate control over the Amazon. Teixeira's expedition consisted of forty-five canoes, seventy soldiers, twelve hundred archers, and conscripted Indigenous rowers (Miranda 2007). Along the way, Teixeira's expedition fought Dutch and English interests that had set up forts on the Amazon river and seized control over their Indigenous slave trade operations (Salvador 1627). The Spanish Jesuit Cristóbal de Acuña accompanied Teixeira in the hopes of finding Indians to catechize to expand the territorial scope of their missionary activities. Acuña also served as the expedition's chronicler, noting the abundance of precious stones visible along the rivers and on the surface of the soil. In his writings, he declaimed riches that appeared entirely disregarded by native inhabitants who apparently had no idea of their value (Acuña 1641).

Both religious and economic interests drove Portuguese territorial expansion on the upper reaches of the Rio Negro. In 1690, the Portuguese crown sent Carmelite missionaries to convert the Indians and solidify control over Amazonian frontier regions disputed with Spain. Their territorial strategy was typical of religious settlements of the time and crucial to state formation in frontier regions (Chernela 2014). Where they went, missionaries set up a convent, a farm, and a village, effectively creating the first imperial footholds in new frontiers. Established in 1695, São Gabriel was the first Portuguese settlement on the upper reaches of the Rio Negro. In order to sustain themselves, the Carmelites expropriated land and labor from the Indigenous inhabitants and began buying and selling Indigenous peoples with riverine slave traders (Hemming 1978).

This was a time of intense competition between multiple colonial powers intent on capturing the territorial and mineral bounty of the Amazon. Echoes of these early contests ring through contemporary debates over the Amazonian frontier, where it is routinely presented as a treasure nearly lost. For half a century, Spanish forts remained further to the north, where the Orinoco connects with the Rio Negro via the Casiquiare waterway. However, the 1750 Treaty of Madrid divided South America between Spanish and Portuguese powers; Portugal ceded part of the Río de La Plata and much of present day Argentina in exchange for control over a greater portion of the Amazon. To enforce the treaty, in 1761 the governor of the Captaincy of São José do Rio Negro at Barcelos[22] organized several defensive patrols of the upper reaches of the Rio Negro. The patrols found that the drainage of the Orinoco into the Rio Negro provided a fluvial passage between the upper Amazon and the Caribbean Atlantic. Representatives of the Barcelos Captaincy promoted the possibility of developing a trade route to move timber, gold, and silver from this inland frontier to Europe. Two years later, frustrated by apparent imperial disinterest in the Captaincy's grand plans, they shifted the framing of their cause from trade to territorial threats. They made the case up the chain of command to the Office of the Secretary of State for Marine and Overseas Affairs of the Portuguese Imperium that improper defense of this region could jeopardize Portuguese control over the deep jungle (Bernardo e Mello 1763). The threat of losing control over the Amazon struck a fearful chord that resonates through to the present.

Alarmed at the prospect of losing the Amazonian frontier, the Office for Marine and Overseas Affairs responded by providing financing and manpower for a construction expedition. The Portuguese crown enlisted the services of German Military Engineer Phillip Sturm with instructions to attend to the solicitations of the Captain of Barcelos (Nabuco, Sampaio, and Rodriguez 1903). Sturm recommended the construction of two forts, one in present day São Gabriel da Cachoeira, and another upstream to mark the northern extreme of the Portuguese dominion near Cucuí, called the Forte de São José de Marabitanas. From these forts, Portuguese soldiers could expel Spanish settlements and defend against possible incursions via the Orinoco (Santos 2008). Despite its remoteness, Portuguese officials were convinced by the Barcelos contingent that Marabitanas would be the first part of the Amazonian frontier to be attacked by competing colonial powers (D'Almada 1785); therefore, militarizing the deep jungle was necessary.[23]

Despite Portuguese, and later Brazilian anxieties over incursions into its Amazonian claims, non-Iberian powers did not explore the Orinoco-Rio Negro-Amazon River network until World War II, when US rubber supplies were seriously jeopardized by the break in relations with Germany. At the time, US chemical companies had no idea how to produce synthetic rubber, and Japanese imperialist designs on Southeast Asia threatened supplies from Malaysia (Dean 1987). Meanwhile, trading routes from the Amazon River delta and the Atlantic were being "severely harassed by submarine attacks" (Engineers 1943, 1) that threatened US supplies essential wartime minerals and other primary commodities supplied by Brazil (Pecora et al. 1950; United States Geological Survey 1942–47).[24] In response, US military and rubber interests looked to the Amazonian frontier. The US Army Corps of Engineers (1943) cooperated with the governments of Venezuela, Colombia, and Brazil to survey the 1,842 mile long Orinoco-Casiquiare-Negro Waterway with the purpose of identifying alternative shipping routes and "to further the understanding and development of that region by the sovereign governments."[25]

The survey team, composed entirely of US military personnel, conducted hydrographic and climatological surveying, aerial photography, and mapping of the route from the Caribbean to São Gabriel da Cachoeira. The final reports of this expedition were shared with regional governments, providing the first body of geological and geographical data to grant twentieth-century substance to the frontier narratives. On mineral wealth, the Army Corps of Engineers report noted that "the natural resources of the region are considerable, but, because of its comparative remoteness, not easy of development" (US Army Corps of Engineers 1943, 20).

This report provided baseline data for the first systematic Brazilian geological survey of the region, undertaken during the military dictatorship (1964–85). In the 1960s, a group of geologists participating in the preliminary border surveys observed a "chimney-like" hill protruding out of the northwestern Amazonian landscape outside of the city of São Gabriel da Cachoeira. This geological feature tends to be rich in ferrous and non-ferrous metals. It is a remnant of subterranean magma plumes that harden and remain as an eroded cylinder after the surrounding material has eroded away. As discussed in chapter 2, local nomads described the analogous formation at Bayan Obo as a "yurt."

In 1973, on the estimation that local mineral resources could sustain a regional industrialization project, the Twenty-First Engineering and Construction Company of the Brazilian Army relocated to São Gabriel da Cachoeira (Brasileiro 2006). This relocation occurred contemporaneously with the notorious military assault on Communist guerrillas in Araguaia, which had the effect of refracting the dictatorship's Amazon policies through the Cold War prism of dealing with communist threats (Portela and Neto 2002; Vecchi 2014). Anticommunist industrialization therefore defined the dictatorship's approach to Cabeça do Cachorro. As part of its relocation agenda, the Brazilian army trained a troop of Indigenous mercenaries in the upper reaches of the Rio Negro in case Communist guerrillas should occupy the region and interfere with plans to build an industrial mining hinterland to fuel Brazil's development (Ricci 2014). Although no major industry was ever built, it was in service of this right-wing nation-building agenda that DNPM conducted the RADAM[26] geological survey of Morro dos Seis Lagos and São Gabriel da Cachoeira in 1975, during which the "chimney-like hill" was identified as the Morro dos Seis Lagos non-ferrous metals deposit (Corrêa, Costa, and Oliveira 1968), and is now described as the largest niobium and rare earth deposit in the world.

Geological survey was central to consolidating military dictatorship control over the Amazonian frontier, therefore it served as a cornerstone to national policy projects. The survey was organized across ten ministries and spearheaded by the Brazilian Air Force. It took place between 1970 and 1985, during which time the entire Brazilian territory was surveyed with aerial radar and mapped with over thirty-eight thousand logged flying hours (Souza and Cavedon 1984). The first phase (1970–75) covered 54 percent of the national territory, comprising primarily the Amazonian frontier and parts of neighboring countries (Momsen 1979). It later expanded to cover all of Brazil in the second and third phases: 1975–80 and 1980–85 (Archela and Archela 2008). This was a project of unprecedented scope. The radar technology allowed Brazilian surveyors to collect their first imagery unaffected by cloud cover (Momsen 1979). The Morro dos Seis Lagos carbonatite complex was recognized in 1975 after geologists published a report on the radioactive anomaly detected on the site (Hassano, Biondi, and Javaroni 1975). That year, CPRM conducted preliminary geophysical and geochemical explorations that consisted of drilling four samples between 110 and 255 meters in depth. The samples were only partially analyzed at the time. One veteran geologist interviewed reported that after they completed their initial reconnaissance of the radioactive anomaly at Morro dos Seis Lagos, US spy planes based in Panama conducted clandestine geological reconnaissance of the place. According to the story, Brazilian mapping of the Amazon rainforest apparently aroused the distrust of the United States with respect to Brazil's reportedly peaceful nuclear ambitions (Brazilian Nuclear History 1947–89; Cruz 2014; Souza and Cavedon 1984). It is not the truth or falsity of the account that is informative. Rather, it is the sense among Brazil's geologists, reflected in military policy, media, and popular culture that the country's Amazonian resources are coveted by the United States, which would like to establish greater control over the region. Brazilian exploration of the Amazon has been carried out under the perceived threat of US intervention.

To mitigate against the threat of foreign expropriation, a 1979 law placed matters of mineral exploration in frontier regions under the jurisdiction of the National Defense Council to significantly restrict access to the region's geological endowments. Without military authorization, mining activities "in areas indispensable to the security of national territory, especially in the frontier regions and with respect to activities related to the preservation and exploitation of natural resources of any type" could not proceed (PL 6634/1979). This law was enshrined in the Constitution of 1988 along with a moratorium on activities intending to occupy, fragment, or exploit Indigenous lands or their mineral wealth, unless specifically approved by Congress on a case-by-case basis (Cognresso Brasileiro 1988, Articles 231 and 232). By overlaying frontier security, mining prohibitions, and the integrity of indigenous territories, the constitution effectively froze the status quo of state control over the frontier. If the state did not have the immediate capacity to conquer this frontier, then neither could any other extractive interests be permitted, whether Luso-Brazilian, Indigenous, or foreign.

The constitutional prohibitions on mining have had the effect of intensifying extralocal desires to exploit the region, especially among veteran architects of bygone state development failures who had built grand plans to capitalize on immense mineral wealth now "locked up" in Indigenous lands. Among these actors, there is a sense that the only thing holding back Brazil's development is a legal fluke resulting from the federal government's inability to stand up to foreign interests laundered through Indigenous movements. Because of this informal yet pervasive racial codification of citizenship, the sophistication with which Indigenous mining proponents defend rights over their lands deepens their symbolic exile from the Brazilian nation. Their visions of mining their own lands as a key to unlocking greater citizenship rights and recognition in Brazil have no place in a governance structure that only permits Indigenous agency to be exercised in traditionalist terms, and that views indigenous livelihoods as part of an ongoing frontier problem to be solved.

Beneath the racial codification of extractivism, the current disputes over rare earth mining in Cabeça do Cachorro are essentially struggles between industrial[27] and artisanal[28] mining. The former is generally understood as heralding the catastrophic penetration of capitalism into human communities, bringing war, death, and the death of nature. It is often counterposed to the latter, or to that which industrial mining approaches seek to delegitimize: small-scale, family mining operations that complement other livelihood activities to provide a supplemental income to other subsistence activities drawn from the land and regional trade. Both forms of mining, of course, are motivated by profits. In the latter, mining is one family or community enterprise in which children, women, and men participate. It is extremely difficult to extract surplus value or tax revenue from such an arrangement, and so it is vilified. In Cabeça do Cachorro, industrial mining interests have relied on knowledge expropriated from local communities in order to access geological riches. Historically, the divulgation of geological secrets occurred informally, but also revealed the fragility of ties between Indigenous communities and outsiders of all kinds. Given the pitched battles for territorial control over the region, the handling of localized geological knowledge has been a matter of mortal consequence for Indigenous people and artisanal miners (Wright 2005). The struggles of the 1980s, in particular, shape contemporary rare earth politics in important ways. They are related in detail here.[29]

In late 1979, a group of Baniwa was heading over the border along the Serra dos Porcos to participate in Colombian gold mining when they discovered gold within their own lands. They informed people in their communities. News traveled up and down river, and Indigenous peoples from Colombia, Venezuela, and Guiana came to mine. In the first years, there were no whites, only Indigenous garimpeiros. By 1983, traders came to São Gabriel da Cachoeira to exchange goods for gold, and white garimpeiros joined the operations. Indigenous peoples set up a robust regulatory and permitting system to control the number of white garimpeiros allowed and collected a small tax on their findings, but it was never recognized by the government, military, or corporate actors.

This experience has informed their legal campaign with respect to rare earth mining. Several Indigenous interlocutors viewed rare earths as a "new gold."[30] Media commentators often used the same term, entrenching the idea. As a result, Indigenous peoples' approach to the question of rare earth mining on their land—whether for or against—is directly informed by a living memory of the violence and attempted seizure of their lands by an alliance among missionaries, mercenaries, mining companies and the military.

By 1984, the upper reaches of the Rio Negro was the site of one of the most intense searches for gold in the history of the Amazon. Private firms began sending agents and requesting licenses from the DNPM. Two of these firms—Paranapanema and Goldmazon—conspired with the Governor of Amazonas to help state surveillance efforts in exchange for control over all mining activities in the region. The "New Tribes Mission," which had built basic infrastructure to support the evangelical activities headed by the American Sophie Müller, offered mining bosses the use of their remote landing strips and basic supplies.

Because the mining companies, the state, and foreign missionaries conducted themselves as though no regulatory systems governed extractivism in the region, despite the permitting and taxation system implemented by the indigenous people in the previous year, their foray into Cabeça do Cachorro was based on a world that did not exist. Where they encountered a reality that conflicted with their ideas of the frontier as empty, ungoverned, and unproductive, bloodshed resulted. In early 1985, representatives of the mining companies recruited Baniwa to carry heavy machinery, gasoline, and other supplies through thirty kilometers of jungle from the river to the primary mining site in the Serra dos Porcos. Upon arrival, representatives ordered the Indigenous to clear forest and to set up camp.

According to Wright's (2005) extensive ethnographic data gathered among the Baniwa, it was not clear to the Indigenous porters that the representatives were from large mining companies intending to expel garimpeiros or that the newcomers would attempt to expropriate Indigenous mines that had been operating for several years. When the Baniwa demanded payment for their porter services, mining personnel threatened them with obscene violence and ordered them to leave. In response, a group of sixty leaders from nearby villages prepared for war. In full battle regalia and armed with arrows and rifles, the Indigenous leaders surrounded the agents and offered them the options of leaving immediately, or staying and fighting. The corporate agents left immediately, leaving the operations to the Indigenous.

The mining companies retaliated by conspiring with corrupt personnel working for FUNAI (National Indian Foundation) to hire mercenaries to go after the Indigenous leaders and intimidate them into signing agreements opening their lands to outside mining interests. Alarmed by this, representatives from fifty-four communities wrote a letter to the President of FUNAI in Brasília, Nelson Marabuto.[31] They demanded the recognition of their rights to their lands; the removal of all outside mining interests; and the explicit proviso that, in the future, these communities would mine according to their own terms and with full discretion over technical assistance, production, and profits.

Indigenous peoples leveraged different forms of resistance—from armed to political—to regain control over their subterranean resources. Delegations traveled to Manaus and Brasília to petition higher levels of government to recognize their land claims and expel corporate mining agents. They held local meetings with mining representatives, community members, and government personnel. Many of these official efforts were fruitless in the short term as government and industry continued to "dribble the ball"[32] in terms of who was responsible for what in Indigenous territory. In late 1985, FUNAI president Álvaro Villas Boas sent a working group to research the mining and conflict situation and propose a series of measures to alleviate the tensions. The working group lacked any authority to implement its recommendations. One of the few concrete legislative changes that emerged was a territorial assertion on the part of the federal government stating definitively that the Serra dos Porcos was within Brazilian territory and that Colombians were prohibited from mining in Brazil. Colombians were not the problem, as far as the locals were concerned, so this did little to address the daily concerns of Indigenous peoples dealing with the ongoing intrusion of mining companies onto their land. Erasing Indigenous concerns, the federal government interpreted the conflict through the lens of geopolitical interests rather than a dispute among different groups of Brazilian citizens. Episodes of gruesome violence flared between Indigenous people and corporate miners, which survivors from both sides described as war.

From an Indigenous perspective, this represented the latest in centuries of white attempts to steal indigenous lands. From a state perspective, the conflict was an embarrassing illustration of government incapacity to rationalize and integrate the farthest reaches of its territory. From the military perspective, this conflict showed why integration schemes such as those that had failed in the 1960s and 1970s were needed once again.

In fact, simultaneous to the conflict, the Brazilian military had been developing a program—the Northern Trench Project (Projeto Calha Norte; PCN)—to militarize the Amazonian frontier following the failures of the massive integration projects of the 1960s and 1970s. Under the twin banners of development and sovereignty,[33] it was conceived in order to provide a justification and operating budget for maintaining a military presence in Cabeça do Cachorro despite growing evidence of military abuses of Indigenous peoples (Instituto Socioambiental 2001; Ricardo and Ricardo 1990).[34] This was a frontier-taming proposal *par excellence*. Submitted to President José Sarney in June 1985, it proposed to better integrate the region into Brazil, develop the local economy, and assimilate Indigenous people into the Brazilian population, which—it was hoped—would eventually eliminate the need for any specially demarcated indigenous lands.

The proposal sat unaddressed for months. As with the preindependence proposal from the Captaincy of Barcelos to the Portuguese crown, it took the invocation of an immediate threat to the region to receive a response from on high. On November 6, 1985, the international press exploded with news of a siege by the Colombian Armed Movement M-19 on the Colombian Supreme Court in Bogotá. Although the attack had taken place roughly a thousand kilometers away from São Gabriel da Cachoeira, and although the M-19 movement claimed less than two thousand members who were overwhelmingly based in urban areas in Colombia, military personnel capitalized on popular fears of jungle guerrillas in Brazil. A propaganda campaign cynically mixed accounts of Indigenous attacks on corporate mining personnel and broader cultural anxieties about attacks on Brazilian territory by circulating rumors that guerillas were building up an arsenal on the border in order to invade and capture Brazilian gold (Hayes 1986). The propaganda worked. President Sarney approved the project in December 1985.[35] In subsequent cooperation with representatives from the United States' Drug Enforcement Administration, the Brazilian military determined that the upper reaches of the Rio Negro would serve as a "test case" for their Amazonian occupation, surveillance, and assimilation program.[36]

Although the guerilla attacks on Brazilian territory never materialized in 1985, the specter of FARC (Revolutionary Armed Forces of Colombia) incursions has been periodically invoked to justify the military presence in Cabeça do Cachorro. Guerilla leaders have been clear that they do not wish to engage in armed conflict with the Brazilian military,[37] but neither does the Brazilian Ministry of Defense wish to appear that it is lagging on the offensive against FARC, lest the US military decide that its needs to increase its presence in the region (Filho 2006).[38] This dramatically underscores the conventional geopolitical needs driving ongoing state attempts to industrialize the region. Tearing up the impenetrable green and rationalizing the space with large scale rare earth mining operations would powerfully demonstrate not only Brazilian control over the region, but also advance a specific, anti-indigenous notion of Brazilianness.

Perhaps this helps explain why nothing in the Constitution or related laws expressly prohibited DNPM from issuing mining concessions in Indigenous lands so long as applicants followed the proper permitting procedures. Furthermore, the moratorium did not nullify existing mining concessions in the region, but rather placed them in a state of suspension until such time that the mining code could be revised to permit mining on Indigenous lands. Such a proposal came in 1996 in the form of Senator Romero Jucá's PL 1610. Under discussion for over two decades, this law would overturn the constitutional mining moratorium in Indigenous lands and allow prospecting firms the same rights of exploration as permitted on nonprotected lands, without providing a process for Indigenous inhabitants to contest or modify mining concessions on their territories. Banking on the success of PL 1610, the number of mining concessions acquired in São Gabriel da Cachoeira increased from thirty-six before the ratification of the Constitution to 401 just after the turn of the millennium (Ricardo 2013).

Large-scale industrial mining interests tend to present themselves to state counterparts as killing two birds with one stone. Not only would Amazonian mining activities contribute to national development, the requisite logistical support infrastructure would also help manage the multifarious threats to the Amazonian frontier. Whether these threats are believed to be military, guerrilla, or environmental, their periodic invocation of a vulnerable frontier in political and popular discourse effectively generated mass consent for the largest defense procurement deal in the history of Brazil.

At just over US$1.4 billion, the Amazonian Surveillance System (Sistema de Vigilância dos Amazonas; SIVAM) is the world's largest environmental monitoring network. It was conceived during the presidency of Fernando Henrique Cardoso (1995–2003) and came online in 2002. One of its principle objectives is to monitor and collect data about natural resources in the Amazon in order to further the modernization task of completing an inventory of national riches (Nascimento and Sá 2008). Using a combination of ground and aerial radar, this "system of systems" was built to monitor drug trafficking; illegal mining, ranching and deforestation; agrarian conflicts; and invasions on Indigenous lands.[39]

Implementing SIVAM required a twenty-first century remilitarization of the Amazonian frontier in the name of surveillance, which reached beyond the developmentalist ends of the PCN. In 2004, the Brazilian army relocated the First Strategic Brigade of Niterói in Rio de Janeiro to São Gabriel da Cachoeira. This was presented as part of a new orientation of the Brazilian military in response to poorly defined "new" international pressures (Filho and Vaz 1997; Messias da Costa 2013). But in fact the remilitarization of Cabeça do Cachorro does not seem to reflect any profoundly new thinking about the Amazon in general or the upper reaches of the Rio Negro in particular. Perhaps it is most accurate to describe the military as reviving its foundational strategies to address what has long been perceived to be the greatest national vulnerability. The push to mine rare earths in the region is in many ways the latest iteration of the struggle to conquer the northwestern Amazonian frontier.

Clearly, the drive to exploit the region long precedes the identification of rare earth elements. Furthermore, the very possibility of mining them in Cabeça do Cachorro was not seriously considered until decades after they were discovered. The histories of São Gabriel da Cachoeira from the 1600s to the late twentieth century show successive attempts on the part of imperial and state actors to rationalize the region through the production of cartographic and geological knowledge. Where that knowledge was contested, violent geopolitical struggles resulted. These bodies of knowledge about the Amazonian frontier have been generated and deployed by Imperial Portuguese as well as military actors from the US and Brazil in the hopes of capitalizing on the region's resources and assuage broader territorial and geopolitical insecurities. Indigenous claims to the land and resources have only been recognized since the late twentieth century. These too are under threat.

In 1985, ten years after the RADAM-BRASIL survey of São Gabriel da Cachoeira, state geologists calculated that Morro dos Seis Lagos contained a reserve of 81 billion tons of niobium at 2.898 percent concentration, which was fourteen times the known global reserves at the time (Justo and Souza 1986). The deposit was also found to possess remarkably high concentrations of rare earth elements as well as vanadium, beryllium and zirconium (Radambrasil 1976). In this period preceding the proliferation of rare earth-based information and military technologies, these findings did not stimulate much practical interest (Rossini 2012). After the successful demarcation of Indigenous lands in 1987 and biological reserves in 1990,[40] all further state geological research was halted. Excitement over the region's rare earth and other technology metals is a phenomenon that emerged after the 2010 crisis awoke Brazilian researchers to the importance of rare earth elements.

Contested Hinterland: Extractivism, Geopolitics, and Militarism, 2010–2014

The question of geological mapping of Brazil is far from being resolved. What is clear, however, is that Brazilian geology is still unknown compared to other mining nations. This represents an excellent opportunity for businesses especially focused on exploration, in search of potential activities in frontier regions such as the Amazon . . .

—Jones et al. (2011, 10)

The RADAM-BRASIL samples were taken out of storage for further analysis in the immediate aftermath of the rare earth supply crisis. In December 2010, a joint research group between the Federal University of Rio Grande do Sul and the National Council of Research on the Mineralogy and Geochemistry of Mineral Deposits reopened investigations into the Morro dos Seis Lagos deposits. This initiative was overseen by CPRM doctoral fellow Mateus Marcili Santos Silva, who had participated in large-scale hydrographic mapping initiatives and was therefore able to gain access to samples through his colleagues at CPRM. But the project was dealt an unexpected setback when Silva died in a car accident in July 2011 ("Obituário" 2011). Analysis did not resume until nearly a year later.

Subsequent analyses carried out by members of this research group using the samples that were made available to the late Mr. Silva found concentrations of heavy rare earth elements between 5 and 10 percent (Giovannini 2013), which is exceptional compared to other sites that successfully garnered investment based on concentrations of between 0.9 and 2.2 percent.[41] In addition to the resources at Morro dos Seis Lagos, there is an abundance of rare earth elements, coltan, vanadium, and other elements now considered "strategic" or "critical" present in alluvial deposits and clays on Indigenous lands elsewhere in Cabeça do Cachorro. This new information revived Senator Romero Jucá's infamous PL 1610. In June 2013, President Dilma Rousseff sent a mandate to Congress to formulate a new mining code that would be "favorable to business and to productive investments that would strengthen a new cycle of development in our country, but all with gains for society, for workers, and for the environment" (quoted in Bustamente et al. 2013).

This became PL 5807 of 2013, and was anxiously awaited by the mining sector, which had hoped that the charged geopolitical climate surrounding rare earth elements would facilitate lifting the moratorium against mining on Indigenous lands (Bustamente et al. 2013). Corporate mining interests lobbied Congress with claims that the nation was at a critical point to determine its future development and standing in the global economy. Industry was quick to buy influence among politicians. For example, mining companies contributed at least R$1.8 million to the reelection campaign of PL 5807's sponsor, Senator Leonardo Quintão, which comprised 37 percent of his entire 2013 campaign budget (Souza 2014). Efforts to force through the bill on the pretext of national urgency failed, however, as the price of rare earth elements stabilized and allegations of bribery and corruption surfaced (Instituto Socioambiental 2013). As of this writing, the bill has yet to be voted on. A substantially similar bill, PL 5263, was introduced in June of 2016 to liberalize mining in Indigenous lands, but does not govern radioactive materials. Therefore it is unclear what this might mean for rare earth elements, given their common association.

Since 2010, some DNPM and CPRM geologists have taken a proactive stance on exploiting the rare earth riches of Cabeça do Cachorro which, despite extensive survey data, is problematized by state geologists as the least known geological region of Brazil (Santos 2003). By maintaining that Indigenous and international NGO efforts to keep the region undeveloped have undermined the scientific and economic potential of Brazil, geologists have found a welcoming audience among the Ministry of Defense. The Ministry of Defense takes care to include the most outspoken among state and Indigenous mining proponents in high-profile military events across the country, such as the gathering described in the first section of this chapter. Reopening Indigenous lands to large-scale mining operations would enable the military to extend its agenda of industrialization-driven consolidation of the northwest Amazonian frontier.

After the 2010 rare earth crisis, the budget for the PCN increased precipitously, from R$250 million per year to R$770 million in 2013.[42] PCN Director Brigadier Dantas attributes this to "the growing confidence Senators have in our initiative" as well as "growing awareness of the need to protect São Gabriel da Cachoeira's resources from Colombian and Venezuelan *garimpeiros* by integrating the region into Brazil."[43] Here again, the extractivist threat was internationalized at the expense of Indigenous claims to frontier resources.

The post-2010 circulation of geological knowledge reinvigorated efforts to intensify Brazilian state control over a frontier region that had been, in practice, more integrated with the mineral, cocaine, rubber, and contraband markets in Colombia. In 2014, rumors again circulated about alleged plots against two separate Brazilian army bases. As with most other rumored plots in the past decades, these attacks never materialized.[44] Invoking FARC to amplify the perceived threat of foreign garimpeiros and to justify expanded military presence in the Amazon serves two purposes. The first is to engender broad public support for the ongoing militarization of the Amazon, and the second is to ward off potential encroachments by the US military.

Brazilian military officials express discomfort with the enduring US military presence in Colombia (Marques 2007) since this represents a major global occupying power fortifying its military facilities near Brazil's porous Amazonian border. This has compelled the military to formulate offensives in the region in order to enclose possible spaces in which the US military might decide to intervene. In one such case, the CMA reframed the fluid transborder economies as a violation of Brazil's sovereignty and undertook joint training exercises with the air forces of Colombia, Peru, and Venezuela. To simulate illicit trafficking, they deployed small airplanes to fly at low altitude along the border of Brazil and participating neighbor states. They practiced radar detection and mid-air interceptions (Marques 2007). The practical motives of these exercises are unclear, given that the majority of the commodities, legal or otherwise, are moved in small quantities, boat by boat, in an "ant-like" (*formigando*) fashion up and down river. Both military and federal police interlocutors in São Gabriel da Cachoeira professed a fundamental incapacity to monitor these small-scale riverine movements on a regular basis.[45]

There is no place for Indigenous economic agency in the national imaginary: at worst, Indigenous activists are viewed as traitors, and at best, they are seen by the state as incapable of defending themselves against foreign incursions. The historical vulnerability of Indigenous populations to imperial and state violence has been reframed as a fundamental national vulnerability against an imagined external threat, and against which the Indigenous cannot be trusted to defend the country.

What is new in this contemporary era of neo-extractivism is the intensity with which the need to defend the Amazon has caught hold of the Brazilian popular imagination. Once thought to be a punishment, posts in the Amazon are now among the most sought after among military officers (Marques 2007). If anything has changed about the older feelings toward the Amazon, it is a sense of morally and technologically empowered purpose emerging out of several years of broader national political economic ascendancy. But the contemporary attitude of the military and the federal police toward Indigenous inhabitants has changed little from the colonial era. Indigenous people are still discursively framed as the counterpoint of civilized Brazilians, lacking the initiative to work, improve themselves, or even practice good hygiene (Castro 2003; Marques 2007).[46] In these discourses, Indigenous people do not count as a living presence on the Amazonian frontier, as illustrated by the military strategy to "vivify" the frontier with military colonies:

> Thanks to the PCN there has been an accentuated vivification on the frontier zone, based in the presence and deployments of the Special Frontier Platoons. If it weren't for the PCN, what would we have in this Amazonian vastness? . . . For the strategists, the Special Frontier Platoons are today little points of "national civilization" holding our frontier together, with the hope that in the future they will transform themselves into human agglomerations, small towns, small cities, municipalities. (Nascimento and Sá 2008, 41)

However, these antiquated attitudes must now reckon with the fact that Indigenous polities are well organized on the regional, national, and international scale. Chastened by the well-publicized disgraces of the colonial and mid-century extermination practices in the name of greater geopolitical control over the Amazon (Ricardo et al. 2014), the Brazilian military has been compelled to change its approach to the frontier. The necropolitics of the past, in which state-directed campaigns of mass death aimed to rid the frontier of those with competing territorial claims in the name of development, have given way to the "100 percent transparent"[47] biopolitics (Foucault 2007) of the present. This is evident in the language of assimilation and vivification used to describe state and military proposals to develop the frontier.

Beneath this changing approach is an underlying continuity of objectives. For planners in the federal state and military, large-scale industrial mining remains the ideal means for rationalizing Cabeça do Cachorro. Rendering these continuities palatable to local citizens obliged the military to undertake an extensive propaganda campaign in order to vilify small-scale mining operations. This campaign precedes the excitement over rare earths, but also sets the terms in which the debate unfolded over who is entitled to these strategically valued resources. Switching its politics from *necro* to *bio* further obliged the military to reconfigure public perceptions of difference between civilized Brazilians and Indigenous "others." With respect to mining, this involved recodifying legitimate and illegitimate extractive activity in terms of enlightened patriots and environmental criminals traitorous to Brazil. Corporate mining interests were codified as "us": legitimate and law-abiding, while garimpeiros were codified as "them": outlaw others bent on destroying Brazil's natural wealth.

One key program through which this project unfolded was in fact SIVAM, the massive satellite surveillance program. The SIVAM Social Communications Advisory Team developed the mascot of a young, light-skinned Indigenous boy wearing athletic shorts and a Yanomami-esque haircut named Sivamzinho (Little SIVAM). The promotional materials provided the following explanation: "This nice little Indian is Sivamzinho, mascot of the SIVAM project. He's the number one friend of the children of the Amazon" (SIVAM 2008 quoted in Guzmán 2013, 111–12). Over one million pieces of pedagogical materials, such as pencils, rulers, posters, calendars, and notebooks, were distributed to schools in the Amazon, particularly those sites targeted for the construction of SIVAM infrastructure. Another five hundred thousand notebooks distributed in Amazonian schools showed Sivamzinho raising the Brazilian flag and singing the national anthem.

This is another case, like that examined in Inner Mongolia, wherein the conventional and critical geopolitics intersect in the formation of desirable frontier subjects, to "generate a legion of Sivamzinhos" by connecting with Indigenous children's "love of the land" and "spirit of adventure" across the Amazonian frontier (SIVAM 2008 quoted in Guzmán 2013, 111–12). The military developed this explicitly raced and gendered social mobilization project to generate proper popular feelings toward the world's largest environmental surveillance and law enforcement regime. The geopolitical stakes are clear: if the Indigenous presence must be accommodated, then it is imperative to convert Indigenous people into assimilated citizens on whom the Brazilian state could depend to accept the project of rationalizing the Amazonian frontier into a national hinterland. This would occur through efforts to eliminate small-scale resource exploitation in the interest of eventual industrialization. Toward this end, a series of comics depicted Sivamzinho reporting wrongdoers such as garimpeiros and Anglo-European poachers—note the configuration of otherness—to the federal police, and then speaking directly to the reader about his love of the Amazon and his resolve to fight against those who destroy it. There are no comics in which Sivamzinho vilifies mining companies, military bases, or missionaries.

This may not seem unusual for a military-backed propaganda campaign. Viewed in historical context, it is one of many bold attempts to erase a history of violence visited on Indigenous peoples by states, missionaries, militaries, and mining companies. The subject of mining is especially fraught, where corporate mining is positioned with national development while small-scale mining, as in Inner Mongolia, Afghanistan, and elsewhere, is vilified and presented as anathema to the nation. The power of these discourses is visible in the difficulty with which indigenous interviewees expressed themselves on the topic of mining. They found it difficult to even talk about mining without also navigating the anguish generated by memories of violence in the 1980s. Yet the public discourse overwhelmingly attributes the horrors of the period to small-scale production rather than state-backed corporate mining. Therefore those arguing for indigenous controlled rare earth mining must also address the connotations of treason and illegality with which artisanal mining has been imbued. In the China case, local informal miners and sympathetic officials could use the same language as the national government to assert the legitimacy of their operations in providing raw materials to downstream industries. In the Brazil case, there was neither the political concept nor the immediate industrial need to help indigenous mining proponents advance their claims. These ambiguities weighed heavily on the psyches of Indigenous mining proponents interviewed in São Gabriel da Cachoeira, who were doubly exiled from transnational environmental conservation communities, on the one hand, and from the Brazilian national community, on the other, by virtue of their desire to mine their own lands.

The conditions of impossibility in which Indigenous mining proponents find themselves calls into question the depth of the critiques of the "noble savage" in academic and other discourses. If we are serious about deconstructing fixed notions that equate Indigenous people with partial and problematic notions of wilderness stewardship, then we need to be prepared to recognize critical instances of agency as exercised by Indigenous people—even if such actions upset established notions of the relationship between Indigenous people and mining.

However, as shown by the multiple attempts to coopt indigenous claims to mining-based agency, mining on indigenous lands is a proposition fraught with hazards. These hazards are entirely distinct from the obvious market and logistical constraints to mining rare earths in this region. Diverse actors, each of whom sees these elements as the key to unlocking diverse possible futures, strategically value rare earth elements in different ways. Indigenous mining proponents saw them as the means through which to finally capture legitimate recognition from the state. The state saw them as a means to fulfill three territorial ambitions at once: territorialize the remotest Amazon, promote national development, and capture international recognition. These aims to bring about Brazil's future through rare earth development were racially coded. Federal government and military interviewees did not equate an increase in Indigenous power with the consolidation of state power on the frontier. Quite to the contrary: any advance in indigenous power—especially autonomous economic power—was discussed as a threat to the security of the territorial state. Rare earths, discursively cast as "strategic elements" and "bearers of the future," highlighted this antagonism. For whom would rare earths bring the future? And what sort of future would that be?

Conclusion

> I repeat, Madam President, rare earths are a matter of national sovereignty, whether to provide advances in knowledge, whether for the multiplicity of its uses, including in the defense and oil industry. Therefore, we need a strategic policy to foster its production and prioritize entrepreneurial boldness to process deposits into products that are capable of nourishing the most advanced industries existing in the world today. This is a crucial issue for the future of our country.
>
> —Senator Luiz Henrique da Silveira (2013)

This chapter argued that the paradoxes currently characterizing rare earth exploration in Brazil cannot be explained by a simple economic calculus. Longstanding contests over sovereignty and citizenship rights are at the heart of the impetus to exploit rare earth elements in the challenging context of the northwestern Amazon—despite the ongoing production of more sustainably produced rare earth oxides in the heartland. This shows that despite the discourses of neo-extractivism in Brazil in the context of rising BRICS hegemony, strategic resource concerns continue to be powerfully shaped by (post)colonial desires to control the Amazonian frontier. Military-industrial mining interests in Brazil saw themselves as fighting the battle for rare earths on three different scales: competing with China internationally, defending a certain vision of sovereignty over the Amazonian frontier regionally, and suppressing indigenous claims to economic sovereignty locally.

Although the actual scale of rare earth extraction at the time of my research was miniscule—ant-like, in local parlance—the stakes were nevertheless high. In this place where there is a decades-old moratorium on commercial mining activity, a multitude of national, local, and transnational actors have undertaken regulatory offensives in order to legitimate their particular vision of rare earth exploitation. At stake in these multiple regulatory offensives to change existing Indigenous and environmental protection laws is control over defining the future of the Amazonian frontier. These definitions determine whether rare earth extraction translates into greater geopolitical control as envisioned by the military, or greater citizenship rights and fuller participation in the Brazilian economy as envisioned by garimpeiros and Indigenous activists. The feasibility of rare earth exploitation in such a place is secondary to the ways in which multiple actors imagine it might support their varied territorial agendas.

The dreams of rare earth elements or, more precisely, the dreams of power and sovereignty imparted by control over their extraction, are entangled with ongoing struggles over the meaning of the Amazonian frontier and entitlements to geological patrimony. Generals, investors, geologists, environmentalists, and Indigenous activists all conceive of rare earth elements in Cabeça do Cachorro as essential to their respective visions of the future. Amid these competing visions, there is also the recognition that regardless of how rare earth elements are extracted, whoever captures authority over their extraction stands to gain tremendously. Given this, it is possible to see how research on the high geological incidence of rare earth elements in the region intensified competing territorial contests in the upper reaches of the Rio Negro. This chapter has presented some of the lived and material consequences of such fantastic dreams, where the rare earth frontier reaches to the far northwestern Brazilian Amazon. The political power of fears and fantasies is especially salient in the next chapter, where some are attempting to extend the rare earth frontier beyond our Earthly confines.

EXTRAGLOBAL EXTRACTION

Rare earth elements are more abundant and accessible on Earth than on the Moon. Even if they were rare on Earth, hundreds of tonnes of reusable rare earths are discarded annually in mine tailings and old electronics. So why are civilian, military, and private sector actors working to subvert international law in order to mine rare earth elements from the Moon?

As of this writing, no one is actively mining the Moon. However, at least six national space programs, fifty private firms, and one graduate engineering program, are intent on figuring out how to do so. The fictions of rare earth scarcity emerging from the 2010 crisis gave lunar mining proponents something specific around which to frame their cause to conquer this prospective rare earth mining site. Trafficking in fictions, fears, and erroneous assertions made in the face of established facts concerning both international legal conventions and the global rare earth supply, lunar mining proponents made significant progress in subsequent years in moving the question of off-Earth mining from the fringes into the mainstream of political discourse and public consciousness across the globe.

The final case examined in this book, the Moon, appears at first glance to differ from other cases in its abiding concern with the future. But all cases examined thus far have been built around visions of future wealth and power—visions that carried sufficient force to rewrite laws and rearrange landscapes or, at the very least, shift debates from whether rare earth mining should take place to how and on whose terms. There are many ways in which the geographies of rare earth production and consumption reflect the best and worst of our contemporary social relations. In this respect, the Moon is no different. What is new is its relevance to global resource geopolitics and the preponderance of public and private sector actors jockeying to be the first ones to exploit the Moon in the name of the "greater good." This chapter examines who some of the primary actors are in the race to mine the Moon, and how they attempt to negotiate the technological, economic, legal, and ethical hurdles to lunar mining.

It has been well over two decades since our everyday lives have integrated with outer space in increasingly intimate ways. Our satellite-linked smart phones are one conspicuous example. Yet we have been slow to understand outer space as a site of social and political significance. Examining the Moon in terms of contemporary territorial struggles refracted through rare earth politics is one way to begin thinking about the space beyond our atmosphere in more immediate and concrete ways.

In basic market terms, the business of maintaining a steady global rare earth supply has very little to do with the current struggle over lunar territory. Discursively, however, they have played a key role in redefining the "final frontier" from a space of international scientific collaboration to the latest arena for geopolitical proxy wars. As such, rare earth elements serve as a lens through which to examine the complex array of actors and interests comprising the contemporary race to claim the lunar frontier.

Understandably, the sheer immensity of outer space and the relative newness of the possibility of expanding mining frontiers to the Moon challenges established ways of thinking—even those that exhort us to think globally—which have considered anything beyond our atmosphere to be beyond global concerns. Yet the same conceptual tools that aid our understandings of the creation and destruction of frontiers in China, the Amazon, and elsewhere, can also clarify our relationship to the Moon. Specifically, three key concepts that have helped us understand the relationships among far-flung sites along the global rare earth frontier can be adapted to guide our thinking about mining the Moon. First, as with the Mongolian Steppe or the high Amazon, outer space is not an empty or undifferentiated immensity free for the taking. Frontiers are seldom as empty as aspiring conquerors would claim. Where frontiers are not populated with people, they are imbued with collectively held meanings. Outer Space is enshrined in one of the most robust international treaty regimes to date as the patrimony of all humankind. Although outer space is infinite, the basic fact remains that some places are nearer to Earth than others, and therefore valued differently.

This leads to the second point, which is that our current discussion of mining the Moon is in fact a matter of the political economy of natural resources. Struggles over the question of enclosure, accumulation, production, and distribution are key ingredients in the latest race to the Moon. As on Earth, resource abundance alone does not resolve struggles for control over and access to mineral resources, their benefits, or the politics of sacrifice on which most mining enterprises are built. Quite to the contrary: resource abundance—and our valuation of a given body of resources—creates the very struggles off-Earth mining proponents promise to overcome.

This brings us to the third point. Just as the frontier is made in part through human imaginings of it, the contemporary production of outer space is coextensive with Earth-bound resource geopolitics. The Moon is an extended battleground for ongoing struggles over resource production and consumption, the relationship between the public and private sector, and the future of territorial politics here on Earth. The Moon, as a rare earth frontier, was conjured as such following the 2010 crisis. As with other rare earth frontiers examined in this book, it was a frontier of another sort before it became a rare earth frontier. In all cases, frontiers are not permanent spaces but instead are conjured by specific interest groups in order to be conquered (Tsing 2005). The fact that rare earths are essential components to many of the technologies used in space exploration is interesting, but not, in itself, sufficient reason to divert resources from developing more sustainable production and consumption practices here on Earth.

In the immediate aftermath of the 2010 crisis, rare earths appeared to be one of the few elements that were sufficiently valuable and useful in small enough quantities that they could be mined on the Moon for use on Earth. Since then, prices have declined and the industry has struggled with oversupply. Yet high-profile investment and regulatory offensives aiming to enable private sector lunar mining have proceeded despite economic, legal, and technological constraints. This indicates that there is much more at stake than alleviating terrestrial resource scarcity. In fact, the scarcity thesis does not stand up to serious scrutiny. To expose what is actually driving the campaign to enclose lunar space, we must examine which interests have invoked lunar rare earth mining in order to gain political and economic traction. As with the Cold War geographies of rare earth extraction elaborated in chapters 1 and 2, competing imperialisms lie at the heart of the latest race to the Moon. Like the struggles over mining Afghanistan and the Amazon, mining rare earth elements on the Moon is not only about capturing resources. It is about enclosing space to preclude uses that might not be subject to the state, corporate, or military control of (aspiring) geopolitical hegemons. As with other sites examined in this book, mining the Moon requires a regulatory offensive that is both controversial and contested.

This sets the Moon apart from the deep seabed, with which it might otherwise share conceptual and technological similarities insofar as it is fundamentally inhospitable to humans. Seabed mining is governed by a robust regulatory regime.

Accessible resources on the ocean floor, outside of special economic zones pertaining to sovereign states, are regulated by the International Seabed Authority, to which 159 countries are party, including the United States, China, and EU member states. The Authority sets limits on the area of land that can be explored, requires contractors to report on their activities, and most significantly, requires them to relinquish their areas of exploration back to the Authority after eight years (International Seabed Authority 2013).

What is new about the contemporary moment is the preponderance of the private sector, driven by crises of over-accumulation alongside great strides in technological development. This chapter examines the key actors in the race to redefine the Moon according to visions of rare earth mining, and looks at the driving rationales and geopolitical stakes of (un)making the lunar frontier for these purposes.

Conjuring the Lunar Frontier

> Frontier culture is a conjuring act because it creates the wild and spreading regionality of its imagination . . . A distinctive feature of this frontier regionality is its magical vision; it asks participants to see a landscape that doesn't exist, at least not yet.
>
> —Tsing (2005, 68)

"Our nearest neighbor" (Cave 1944), the Moon, is the first stop on the race to conquer off-Earth El Dorados. "Only two days away," the untouched "lunar treasure chest" is reportedly packed with resources "desperately needed on Earth" (Clark 2004; Day 2009). "Our offshore island" (Ostini 2011), as it has been called, lies firmly within the reach of expanding circuits of accumulation (Robinson 2004) and the corresponding ambit of resource geopolitics (Le Billon 2004).

Centuries before reaching the Moon entered the realm of serious possibility, elite and speculative actors attempted to lay claim to it. One German claims that his family has owned the Moon since 1756, when the Prussian King Frederick the Great granted his ancestor the Moon as a token of gratitude (CNN 2000). A Chicago man registered ownership of the Moon with the Recorder of Deeds and Titles of Cook County, Illinois in 1949 ("Chicago Man" 1949). A few entrepreneurs have become millionaires by selling lunar real estate in single-acre parcels, as indicated in figure 17 (Davidson 2007). Despite existing legal conventions explicitly prohibiting the appropriation of any part of outer space to the exclusion of others, in practice the current race is driven by visions of the economic and geopolitical power to be gained by being the first to capture off-Earth resources.

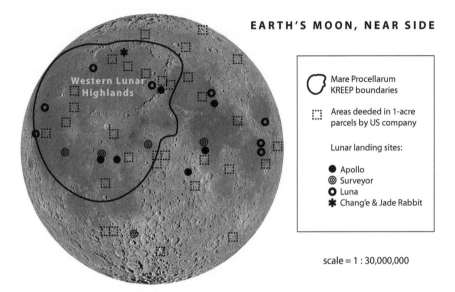

EARTH'S MOON, NEAR SIDE

Mare Procellarum
KREEP boundaries

Areas deeded in 1-acre
parcels by US company

Lunar landing sites:

● Apollo
◎ Surveyor
○ Luna
✳ Chang'e & Jade Rabbit

scale = 1 : 30,000,000

FIGURE 17.   The Mare Procellarum KREEP region is high in rare earth elements, thorium, and iron. Lunar landing sites constitute historic landmarks and may be subject to legal protections as sites of archeological, historical, and world cultural heritage (White 2013). The legal validity of the deeded areas sold by US companies has not been established, given the current treaty agreements enshrining outer space as the common heritage of all humankind.

Source: Image by Molly Roy.

The question of mining the Moon did not begin with the rare earth crisis in 2010. The earliest dreams of lunar mining emerged among a specialized group of space engineers, military brass, legal experts, and space mavens when the first lunar rock samples were brought back to Earth in 1969 (Wakita, Rey, and Schmitt 1971). This fired the imaginations of renowned science fiction authors and former astronauts alike.[1] But it was not until the 2010 crisis that the possibility garnered broader commercial, political, and popular attention. Promising to capture the elements "desperately needed on Earth" (Moon Express 2013), startup space mining companies collected billions in investment and government technology-transfer contracts. Indeed, the fiction of rare earth scarcity has been instrumental in the latest space race, which on closer analysis appears to have little to do with actual rare earth supply concerns.

Yet these fictions bear a loose relationship to geological fact. Like the other cases examined in this book, the race to this particular frontier is not built entirely on speculation, insofar as there are actually rare earth elements to be found on the Moon. Recalling the discussion of the geological formation of rare earth elements

from chapter 1, the rare earths found in the lunar KREEP[2] deposits, shown in figure 17, (Shervais and McGee 1999) formed after a Mars-sized object cataclysmi-
cally collided with Earth. Immense quantities of debris from the crust and mantle
were sent flying into space. Some gathered its own mass and gravity, yet did not
entirely escape Earth's gravitational pull. Spinning through space, locked in orbit around Earth, and glowing molten from the heat of the collision, this material
eventually consolidated into the Moon. The lower gravity of the Moon left an
ocean of magma trapped between the mantle and the crust. This lunar magma
cooled very gradually, resulting in the formation of higher concentrated deposits
of rare earth elements than typically found on Earth (Heiken, Vaniman, and
French 1991). Like other cases, geological exploration advanced the territorial
agendas of expansionist powers. Cataloging mineral resources on the frontier
would provide the means and the later justification for sacrificing local landscapes
to further distant geopolitical ends. However, the case of the Moon differs historically in one key way. Before any exploration took place, the international community agreed in 1967 that all of outer space, and all knowledge generated by
exploration thereof, belongs to everyone on Earth.

As with other points along the global rare earth frontier, speculative hyperbole abounds with respect to the Moon. In a BBC interview on the occasion of
the successful landing of China's lunar rover *Jade Rabbit* in 2013, China Academy of Sciences professor Ouyang Ziyuan stated, "the Moon is full of resources—
mainly rare earth elements, titanium, and uranium, which the Earth is really
short of, and these resources can be used without limitation" (Rincon 2013, 2).
Earth is not actually short of rare earths or platinum group metals, nor can lunar
resources be used "without limitation" as though outer space were the ultimate
no-consequences terrain of extractivist freedom. Nevertheless, this sentiment has
been seized by the public relations staff of private firms, who incant it as though
it were fact. When asked specific questions about the necessary economic conditions under which mining rare earths on the Moon could be feasible, business
development personnel retreat to truisms over Earth's dwindling resources, in-
variably cast against the infinite abundance of the cosmos. They also tend to invoke the multiple threats posed by China's control over rare earth production as
well as the country's progress in space exploration as compelling reasons to support private companies' bids for the Moon.

Behind these discourses of potentially infinite resources lies a fundamental geographical principle: some are closer than others. Therefore, outer space geopolitics are similar to Earth's. Some sites, routes, and resources are more valuable than
others, particularly those closest to Earth. The Moon is closest to Earth; therefore
it is the most valuable piece of potential extraglobal territory.[3] One legal scholar
notes: "Space may be vast, but many of the most valuable resources—especially

those convenient to Earth—are limited. Our Moon is one example. It may be one of the most promising sites for mining, energy capture, and spaceship refueling, but a limited amount of usable land exists, with an even more limited quantity of usable water . . . in truth, every resource is limited. The question, then, is who, if anyone, should have the rights to the riches of space?" (Reinstein 1999, 64–65).

International treaties in force for fifty years are clear that no one has exclusive right to anything on the Moon, but by persisting in asking the question, privatization proponents have succeeded in unsettling the matter. They are part of a broader discursive project to recast the Moon as under-utilized commons. On February 2, 2015, the UK-based Institute of Physics published an article reporting that with "an abundance of rare-earth elements hidden below its surface, the Moon is a rich ground for mining" (Corfield 2015). A few days earlier, *Astrobiology Magazine* ran a story titled "Earth's Moon May Not Be Critical to Life," reporting on a five-year-old scientific paper that challenged the theory that the Moon was essential to stabilizing Earth's orbit (Cooper 2015; Lissauer, Barnes, and Chambers 2011). Television and Internet news programs ran stories in January 2015 with the provocative titles like "Is mining the Moon economically feasible?" citing an analysis completed by Cornell professor of astrophysics Ian Crawford and published in *Progress in Physical Geography*. Crawford's cautious conclusion that the concentrations of rare earth and platinum group elements "might become of economic interest in the longer term" was drowned out by scarcity myths in the mainstream media (Crawford 2015; David 2015). Also in January 2015, the *International Business Times*, the *Diplomat*, and several other news sources reported China's plan to mine the Moon for rare earths and to "colonize" it with a lunar base by 2030 (Lang 2015; Mortier and Finnis 2015; Osborne 2015).

Mining the moon is illegal under international treaties to which all space-faring states are party, but that has not deterred policymakers, engineers, venture capitalists, and social scientists from exploring the possibility in legal, technological, and conceptual terms (Heim 1991; Helmreich 2009; Planetary Resources 2012; Stone 2007). Most prospective mining schemes employ two types of metaphors, oceanic and colonial, to conceptualize human ideas of, and activities within, the "complex, four dimensional materiality" of outer space (Steinberg 2013). A decade ago, geographer Fraser Macdonald (2007, 594) proposed considering outer space in terms of the "rediscovery of the sea . . . not as an undifferentiated emptiness between the land, but as a culturally configured site of knowledge and power." Lunar mining is not so different from the deep ocean insofar as both are relatively unexplored, pose steep technological and capital barriers to entry, beckon to expansionist tendencies, and are governed by robust regulatory regimes (Steinberg 2013; Valentine 2012). In both cases, the abundance of rare earth elements can be accessed only if the technological and political obstacles to extracting them are overcome (Amah et al. 2012; Jeandel et al. 2013; Kato et al. 2011,).

The analogy between outer space and the deep ocean has commercial, legal, and political implications. Many of the legal notes concerning outer space rely on conventions governing the seas and deep seabed (Guner 2004; Heim 1991; Sadeh 2011). Space entrepreneurs and legal scholars draw parallels with the Law of the Sea, but with differing motivations. Entrepreneurs and their spokespeople tend to conceive of international waters as a free-for-all, where "no one really owns the water but any company or country can mine the resources . . . there is a strong legal precedent and consensus of 'finders keepers' for resources that are liberated through private investment, and the same will be true for the Moon" (Naveen Jain quoted in Hennigan 2011). Judging by the frequency with which this sentiment has been repeated by the press and certain politicians over the past few years, it has proven to be a compelling fiction that has helped produce new forms of knowledge around the collective patrimony of Earth's resources.

However, neither resource extraction nor maritime salvage law are governed by the principle of "finders keepers." The attempt to construct a parallel between what is actually an archaic, medieval maritime law and contemporary conventions that concern lunar resources is at best uninformed and at worst deliberately misleading. In fact, there are a number of internationally agreed-on rules and institutions for ocean exploration, including the International Seabed Authority, which regulates extraction in international waters. While certain space entrepreneurs naively claim that the legal constraints do not matter because what they are doing is for "the benefit of all humanity,"[4] other scientific and legal experts see the urgent need for an international regime governing the proper conduct and use of private exploration and mining technologies in outer space (Reilly 2013; Salter and Leeson 2014).

These various viewpoints boil down to a difference between the principles of "first rights" and *res communis*[5] (Laver 1986). Proponents of the former are most vocal in Anglophone discourse, arguing that res communis is impracticable because it cannot be enforced (Landry 2013, Hickman 2012) or is "antithetical to the economic development of outer space resources" (Fountain 2003, 1753) because it "fails to create an adequate incentive for space exploration and colonization" (Gruner 2005, 306). Indeed, some fear that res communis threatens any sort of human progress in outer space: "If humanity hands control of the exploitation of space over to an international political body in an effort to use space development as a wealth-redistribution mechanism, the entire project is likely to fall on its face and there won't be any wealth to redistribute" (Reinstein 1999, 98). Conversely, proponents of res communis argue that, like Antarctica and the deep sea, agreements built around this principle would ensure that such areas are used only for peaceful purposes while guaranteeing their long-term preservation (Heim 1991; Riederer 2014). Furthermore, without a system of checks on the first space-faring nations, global wealth and power inequalities created during the colonial era will simply be reproduced in the space age (Frakes 2003; Raclin 1986; Reinstein 1999). Therefore, it is morally imperative to consider the interests of non–space-faring states when formulating space property law. In the current treaties in force, collective action and fairness hold priority over competition and "first rights."

The primary legal document governing the behavior of states in outer space is the 1967 United Nations (UN) Treaty on Principles Governing the Activities of States in the Exploration and Use of Outer Space, Including the Moon and Other Celestial Bodies (OST). Signed by 167 countries, the OST represents a Cold War–era "gentlemen's agreement" between the United States and Soviet Union to establish boundaries to the arms race. At the time, the extension of armed conflict into outer space would have bankrupted both countries. Led by these two powers, the international community established a treaty against placing weapons of mass destruction in Earth's orbit or on the Moon. The OST frames outer space as the "common heritage of all mankind [*sic*]" and prohibits assertions of national sovereignty "by means of use or occupation, or by any other means" (UN 1967, Article II). Although the OST has no explicit clauses governing private capital, it includes the following stipulation: "States Parties to the Treaty shall bear international responsibility for national activities in outer space, including the Moon and other celestial bodies, whether such activities are carried on by governmental agencies or by non-governmental entities, and for assuring that national activities are carried out in conformity with the provisions set forth in the present Treaty" (UN 1967, Article VI). Built on the mutual distrust between the United States and the USSR, the subsequent paragraph charges parties to the OST to "inform the Secretary-General of the United Nations as well as the public and the international scientific community, to the greatest extent feasible and practicable, of the nature, conduct, locations and results of such activities. On receiving the said information, the Secretary-General of the United Nations should be prepared to disseminate it immediately and effectively" (UN 1967, Article XI).

It is because of decades of international compliance with this provision to make all findings public that private space enterprises have robust selenological data at their disposal. Without decades of publicly disseminated research, it would be impossible for prospective miners to express confidence that they will encounter sufficiently rich deposits of the elements they seek. Selenological knowledge is part of the common heritage; therefore the obligation to collaborate in exploration of all types is enshrined in Article I: "Outer space, including the Moon and other celestial bodies, shall be free for exploration and use by all States without discrimination of any kind, on a basis of equality and in accordance with international law, and there shall be free access to all areas of celestial bodies" (UN 1967).

Interlocutors in the United States have pointed to this passage in isolation to suggest that it is not obligatory for private sector firms to share mineral resources captured for economic gain. Furthermore, actors in the private space industry have leveraged the fact that the OST explicitly mentions state parties in this provision to argue that the treaty does not apply to them.[6] But their attempts at the language of exception land them in the language of colonialism. For instance, Eric Schmidt, executive chairman of Google and key investor in private sector space industries, stated: "The pursuit of resources drove the discovery of America and opened the West. The same drivers still hold true for opening the space frontier" (Planetary Resources 2012). It has been established that the American West was governed by sophisticated land use customs that European settlers refused to recognize or honor.[7] By insisting on the fiction of the Americas as wild, empty, or uncivilized—and violently suppressing claims to the contrary—European settlers forced a reality to match their frontier fantasies. However, some argue that exploiting the Moon is fundamentally different from previous resource-driven waves of human expansionism. The late Indian political journalist and filmmaker S. Balasubramanian (2011), commenting on the initiative of his compatriot Naveen Jain, explained: "It is not the same as the colonization of The Gold Coast (Ghana) by the Portuguese and British, subjugating the native residents, or of the D R Congo by the Belgians for diamonds . . . Moon is just as free and uninhabited as the Antarctic (just a bit further away), and several countries have pitched their tents and hurled their flags in the latter."

This perspective incorrectly assumes that the brutalities and injustices of (post) colonial extractive regimes are confined to the area in which extraction occurs. Resisting such a palimpsestic narrative, other states, particularly those whose space programs were not as advanced as the United States or the former Soviet Union, saw quite clearly how unregulated space exploration would reproduce colonial relations. Spearheaded by Ecuador, which had been frustrated over the lack of recognition of its sovereign claims to the orbital space above national air space (Beery 2011), several developing countries worked through the UN (1979) to draft the Agreement Governing the Activities of States on the Moon and Other Celestial Bodies (hereafter Moon Treaty).[8]

The Moon Treaty strengthened and clarified measures related to lunar resource extraction from the perspective of social progress, development, and equality among all states. It also expressed a commitment to allow all countries to benefit from any space exploration "irrespective of their degree of economic or scientific development" with "due regard paid to the interests of present and future generations as well as the need to promote higher standards of living conditions in accordance with the Charter of the United Nations" (UN 1979, Article IV.2). In particular, the Moon Treaty reinforces the res communis principle by declaring that the Moon shall be used exclusively for peaceful purposes, for "promoting international cooperation and mutual understanding . . . as wide[ly] as possible" (UN 1979, Article IV.3).

With sixteen party states, excluding major space-faring powers, many Anglophone commentators have called it a "dead" treaty. However, it required only five signatories in order to enter into force in international law, which occurred in 1984. Since the 2010 crisis, three new developments indicate that the treaty is assuming greater prominence. First, Austria, the fifth country to ratify the Moon Agreement, passed domestic legislation in 2011 to attract private sector satellite companies to set up shop within its borders as part of a strategy to help the country achieve space exploration (Listner 2011). Second, Turkey's accession in 2012 further suggested a growing strength in numbers around egalitarian use of lunar resources (Beldavs 2013; United Nations Office for Outer Space Affairs 2012). Third, in both the United States and Canada, private citizens have sued multiple levels of government to have their private claims to lunar property recognized (Daly 2012; Jaggard 2009; Sablotne 2012). Courts have thus far dismissed these cases as "frivolous" and characterized the litigants as "paranoid." However, these cases raise an interesting issue: these private claims could have been summarily dismissed with a reference to the Moon Treaty, but if the courts of nonsignatory states were to comment in any way on existing conventions, that would set a legal precedent, which would provide signatories with significant leverage over non-party states (Listner 2012).

With respect to the production and the dissemination of selenological knowledge, the Moon Treaty further stipulates that because all states have an equal right to conduct research on celestial bodies, any samples obtained during research activities must be made available to all countries and scientific communities for research. But what is especially germane to lunar mining advocates is the elaboration of the OST statement that "the Moon and its natural resources are the common heritage of mankind [sic]" (UN 1979, Article XI.1). The Moon Treaty defines this to mean that:

> Neither the surface nor the subsurface of the Moon, nor any part thereof or natural resources in place, shall become property of any State, international intergovernmental or non-governmental organization, national organization or non-governmental entity of any natural person. The placement of personnel, space vehicles, equipment, facilities, stations and installation on or below the surface of the Moon, including structures

connected with its surface or subsurface, shall not create a right of ownership over the surface or the subsurface of the Moon or any areas thereof.
(UN 1979, Article XI.3)

Of further relevance is the obligation to share. In order for each party to the
agreement to be assured that the activities of other parties in the exploration and
use of the Moon are compatible with treaty provisions:

All space vehicles, equipment, facilities, stations and installations on
the Moon shall be open to other States Parties. Such States Parties shall
give reasonable advance notice of a projected visit, in order that appropriate consultations may be held and that maximum precautions
may be taken to assure safety and to avoid interference with normal
operations in the facility to be visited. In pursuance of this article, any
State Party may act on its own behalf or with the full or partial assistance of any other State Party or through appropriate international procedures within the framework of the United Nations and in accordance
with the Charter. (UN 1979, Article XV.1)

Paragraph five of the same Article charges signatories to "hereby undertake to establish an international régime, including appropriate procedures, to govern the
exploitation of the natural resources of the Moon as such exploitation is about to
become feasible."

In anticipation that eventual exploitation would occur, Article II Paragraph 7
explicitly outlines the structure and purpose of the regime, which is to enforce:

(a) The orderly and safe development of the natural resources of the Moon;
(b) The rational management of those resources;
(c) The expansion of opportunities in the use of those resources;
(d) An equitable sharing by all States Parties in the benefits derived from
    those resources, whereby the interests and needs of the developing
    countries, as well as the efforts of those countries which have
    contributed either directly or indirectly to the exploration of the Moon,
    shall be given special consideration.

These treaties lack a robust enforcement mechanism. The OST and the Moon
Treaty may represent a broad international consensus, but who is to stop those
intent on violating these agreements? Currently free from clearly enforceable social and environmental accountability requirements (Listner 2012), the Moon
would seem to represent the ultimate terrain of capitalist freedom (Brennan and
Vecchi 2011). Framing it as the common heritage of all humankind intensifies
rather than attenuates this appeal, because it fits so neatly with the frontier

concept. For any space, including areas of outer space, to be conceived as an underutilized commons eligible for enclosure and privatization, it must first (appear to) belong to no one. While the explicit prohibition on claims of national sovereignty placed important limits on a Cold War-era imperialist turn in the space race, enshrining outer space as belonging to everyone was tantamount to decreeing that it belongs to no one (Beery 2011). As postcrisis events have shown, legally designating outer space as a commons left it vulnerable to processes of enclosure. Frontiers are imagined as belonging to all and to none. To all, because in theory anyone with the means can conquer them, and to none, lest that conquering be contested.

On February 3, 2015, the United States Federal Aviation Administration (FAA) issued a memorandum coordinated with the Departments of State, Defense, and Commerce, and the National Aeronautics and Space Administration (NASA) stating that they would support US companies' claims to lunar territory through their existing protocols for licensing space launches. Reportedly meant to help businesses protect their assets, the permit would allow private companies to set up a "habitat on the Moon, and expect to have exclusive rights to that territory—as well as related areas that might be tapped for mining, exploration, and other activities" (Klotz 2015). The reliance of the private sector on the state to secure their actions on the Moon keeps global space politics concentrated at the level of the nation-state, indicating some important continuities with the Cold War-era space race. Despite hedging around government action, and despite the very clear conditions set forth in current international law, private space industries claim that "it's a very wild west kind of mentality and approach right now" and that they do "not see anything, including the Outer Space Treaty, as being a barrier to our initial operations on the Moon," including "the right to bring stuff off the Moon and call it ours" (Thorton and Richards quoted in Klotz 2015).

Both the colonial frontier imaginary and the continuation of Cold War–era race and gender politics are prominent features of the contemporary race to mine the Moon. The emergent private space sector is much more homogenous than NASA, for example, which has made great strides in the past three decades toward correcting the racist and misogynist hiring practices of the early decades. Race and gender matter when it comes to decision over resource use: *Who* is vested with the authority to determine *how* the Moon is explored, and *for whose benefit,* is inseparable from broader societal dynamics of race, class, and gender discrimination. Uneven progress on this front means that the colonial conquistador mindset enjoys a prominent place in space entrepreneurship. To explain his motivations to mine the Moon, one entrepreneur smiled and said: "You could say it's my culture. My culture has a history of taking risks, exploring and civilizing uninhabited lands!"[9]

Mining Rare Earths on the Moon: Who, How, and Why?

> Most of the elements that are rare on Earth are believed to have originated
> from space, and are largely on the surface of the Moon. Reaching for the Moon
> in a new paradigm of commercial economic endeavor is key to unlocking
> knowledge and resources that will help propel us into our future as a space far-
> ing species.
>
> —Moon Express (2015b)

This quote, taken from the promotional materials of Moon Express, a prominent
firm in the race to mine the Moon and a contender for the Google Lunar XPrize,
conveys confusion about geology and an inaccurate description of the incidence
of lunar resources in relation to Earth. As the lack of scientific precision or fac-
tual discretion over public pronouncements suggests, the latest race to the Moon
differs fundamentally from those in the past. In the first lunar race in the 1960s,
it was unthinkable that rocket launches would be subject to market competition, or that Moon rovers would be developed by small teams working in startup
offices funded by billionaires. After all, the first lunar race was characterized by
herculean efforts of global superpowers, imbued with great dignity and scientific
gravitas, and carried out under top secret conditions. But times have changed.

In the aftermath of 2010, the commercial space mining sector successfully
leveraged the prospect of freeing the rest of the world from China's rare earth
"stranglehold" to advance its claims on the state. Although there are many minerals on the Moon that could hypothetically serve various Earth-bound endeavors
(Lucey, Taylor, and Malaret 1995; Taylor and Jakes 1974; Taylor and Kulcinski
1999), at this current technological moment rare earths are the only elements
that have been promoted as possible to bring back from the Moon because they
are sufficiently useful in small enough quantities. Proponents took the
2,000 percent price increase of late 2010—still several orders of magnitude below
any hypothetical break-even price for lunar mining—as an indication of their
potential future value. That is why rare earths have been the target of contenders
for the high-profile Google Lunar XPrize, which will reward the first two private
companies to place a robot on the Moon with US$30 million (XPrize Foundation
2015), and also why they were a focus of the data gathering mission successfully
undertaken by China's *Jade Rabbit* lunar rover in 2013 (Rincon 2013).

For many in the contemporary space race, commercialization is the only feasible way to return to the Moon (Bakhtian, Zorn, and Maniscalco 2009; Eckert
and Barboza 2005; Peek 2014).[10] In the early 2000s, NASA began to programmati-
cally pursue subcontracting arrangements amid threats to its operating budget

and growing right-wing hostility toward certain government expenditures. It employed three broad strategies to facilitate private companies' participation: holding competitions around a particular space-faring challenge and offering large cash prizes; supporting a broad institutional framework that privileged these firms by providing infrastructure, liability indemnification, and favorable tax policies, and; offering specific development and demonstration contracts in order to generate market certainty (Culver et al. 2007). While this approach is meant to foster private-sector investment in the space program, it also means that subsequent profits from this immense public outlay remain in the private domain, with concrete outcomes available only to public programs for a price. This is positively framed as a public-private partnership, which at best allows the government to "achieve certain forms of basic infrastructural maintenance" while opening up "new streams of revenue and profit" for the private sector (Beery 2012, 25). In other words, NASA acts as a patron to private groups of specialists, providing them with the necessary technological, financial, and political capital to develop technologies and services for which NASA later becomes a customer.

Large giveaways of public funds to scientific endeavors tend to be the first items eliminated in times of austerity; as such, a US$20 million NASA contest to stimulate private sector lunar exploration capabilities was cut in 2006. Following this, head of the XPrize Foundation, Peter Diamandis, pitched a joint sponsorship to Google cofounders. The US$30 million Google Lunar XPrize emerged out of this context in 2007. Had NASA sponsored it, the prize would have been available only to teams from the United States. When Google and the XPrize Foundation took it over, they made the contest global and added funds for a second place prize (Koman 2007). Dozens of companies formed around billionaire investors, former government scientists, and graduates from engineering and business programs to compete for the prize.

The CEO and cofounder of one such company called Moon Express, software billionaire Naveen Jain, claims that his company will offer more "democratic" access to the Moon: "Now that we're shifting from U.S. government-sponsored space exploration to privately funded expeditions, it's important to look at how the resources of the Moon could benefit everyone" (Jain quoted in Caminiti 2014). How, precisely, private lunar mining ventures will overcome technological constraints, capital requirements, and their own profit-maximizing imperatives to make good on its claims to benefit "everyone" as opposed to "paying customers" is a question met with silence in the media as well as in personal interviews with space entrepreneurs. In that silence, we are forced to acknowledge that the hubbub around mining rare earths on the Moon is a pretext for, or a distraction from, a variety of dissonant agendas broadly concerned with the enclosure of lunar territory.

One outcome of the spectacular growth of inequality over the past several decades has been the creation of greater numbers of billionaire capitalists (Bagchi and Svenjar 2014). Billionaire capitalists need a place to put their money in order to avoid a crisis of accumulation (Harvey 2005).[11] The efforts on the part of venture capitalists to use new space industries to keep their money productive has already been theorized by sociologists Dickens and Ormrod (2007), who elaborated the concept of the "outer spatial fix." This concept captures the process through which capital seeks to overcome periodic crises by drawing new raw materials, territories, and markets into the capitalist system. In the same way that capitalist societies resort to imperialism to alleviate crises of over/under accumulation, an elite group of spectacularly wealthy financiers, industrialists, and entrepreneurs are using private space mining companies as a "sink" for their accelerating accumulation of surplus capital in the postrecession era of austerity and regressive taxation. These new investor networks draw together post-2008 real estate moguls, dot-com billionaires, heirs to private fortunes, idealistic young engineers and seasoned NASA professionals seeking to "ignite a new industrial age" beyond Earth (Dolman 2002). Extreme privilege combined with the cultural capital surrounding space-related endeavors generates a kind of evangelical zeal that manifests in a marked incapacity of space investors to handle deeper questions about their projects. Anyone critical of the "wild west" approach to lunar mining is often dismissed as simply lacking vision, not bold enough, or failing to understand the importance of space exploration.

But one need not forfeit their love of space exploration to pose some urgent and important questions about where all of this is heading, and indeed, whether mining rare earth elements on the Moon has anything to do with making life better for anyone except a select few. After all, what are we to make of an endeavor that insists on propagating fictions of lawlessness and resource exhaustion in order to advance its agenda? It is a strange approach: to predicate the success of space-faring enterprise—which is perhaps the most iconic example of sophisticated human cooperation—on the incapacity of humans to figure out how to live sustainably on Earth. While these endeavors promise fundamental transformations in how resources are produced and consumed, they are betting on the immutable nature of the unsustainable political economic status quo.

One of the current private sector frontrunners in the race to mine the Moon, Moon Express, explains it in this way: "The opportunity is simply driven by technology. What used to require the unlimited budgets of a superpower are now within reach of a private enterprise . . . we are going back often, and spending orders of magnitude less to do it" (Moon Express 2015a). The company proposes to build a "lunar railroad" that "will provide cost-effective exploration and development of Earth's eighth continent" (Moon Express 2013). Global Business

Development Manager for iSpace Technologies echoed this sentiment, arguing for the need to explore the Moon to capitalize on the rare earth resources that "we are going to run out of in the near future."[12] There are several other companies intent on mining lunar resources such as Helium-3 and platinum group metals.[13] However, Moon Express and iSpace Technologies are the only ones that have publicly declared their intention to extract rare earth elements. Representatives from two other firms also conveyed this intention, but spoke on conditions of anonymity.

Among private firms, there are two key ideas about what to do with rare earths that are mined from the Moon. The first is to robotically process ores and bring partially separated elements back to Earth.[14] The second, from a company interviewed on the condition of anonymity, involves a plan to 3-D print processed minerals into a perforated spheroidal shape composed of about seven hundred kilograms of refined mineral to be "stored" in Earth's orbit until it is profitable to introduce them into the global market.

In the first approach, rare earths would be brought back to Earth regardless of market conditions for the purpose of making a profit from lunar resources. Perhaps this explains some companies' continued claims on the importance of mining rare earth elements on the Moon despite the 2015 liberalization of China's exports. From a business standpoint, Moon Express is not currently concerned about needing to make a return on its lunar mining business. In the near term, it expects to profit from providing transportation service to "paying customers" who want to set up their own operations—scientific or commercial—on the Moon.[15]

There is considerable enthusiasm from diverse sectors in developing an integrated Earth-Moon system, although what that system would consist of is subject to debate. In the longer term, according to professor of international law at Western Sydney University, Steven Freeland, the eventual development of an Earth-Moon industrial corridor could benefit earthly environments by alleviating industrial pressures on earthly ecosystems while moving some of the dirtiest mining enterprises off-Earth. These proposals possess a compelling logic. It is endorsed by Vatican planetary scientist Brother G. J. Consalmango, who has lectured extensively on the ethical considerations of expanded space exploration (Lamb 2010). Feasibility aside, Consalmango's concern is that if the immense mineral reserves on the Moon are brought back to Earth, the sudden influx of outer space resources into the global economy could cause commodity prices to crash and create massive unemployment: "On the one hand, it's great. You've now taken all of this dirty industry off the surface of the Earth. On the other hand, you've put a whole lot of people out of work. If you've got a robot doing the mining, why not another robot doing the manufacturing? And now you've just put all of China out of work. What are the ethical implications of this kind of major shift?" (quoted in Lamb 2010). According to Moon Express, introducing lunar resources into the global market is intended to deliberately disrupt the global resource economy as we know it, but for the better. The idea, heavily influenced by regular annual immersions in gift economies such as those at Burning Man,[16] is to innovate our way out of resource scarcity so that a new economy becomes possible: "Once you take away the mind-set of scarcity and replace it with a mind-set of abundance, amazing things can happen here on Earth. . . . The ability to access the resources of the Moon can change the equation dramatically" (Jain quoted in Caminiti 2014). At this point the idea is no more specific than that; it is not the company's responsibility to worry about the "after," it is simply focused on "getting humanity to that point."[17]

The second approach intends to leverage scarcity rather than abundance. It involves a strategy to act as a gatekeeper to the "infinite" resources of outer space by withholding processed minerals from Earth until prices increase enough for lunar goods to be sold on Earth at a profit. One attorney working for an anonymous firm explained that keeping resources in processed form in Earth's orbit would effectively be the same as keeping them locked in a vault on Earth—they could be "kicked" down[18] and retrieved only at the firm's discretion without breaking any current global trade agreements. The world that fits their vision is one in which humanity has laid such waste to Earth's endowments (or stimulated such runaway inflation) that certain rare earth elements will cost in excess of US\$100,000/kg. Even allowing for promised technological improvements that reduce transport costs, a US\$10,000/kg commodity price is hardly something to aspire to, especially when a global peak in the price of some rare earth elements at US\$2,031/kg was sufficient to unleash calls for trade and resource wars.

When questioned on this point, a common rebuttal offered by lunar mining advocates is that the vast majority of the resources will eventually be used only in space: to build solar arrays to beam energy back to Earth, furnish building materials for bases, and provide fuel for transport vehicles to enable humanity to access to the mineral bounty of the solar system (Dudley-Flores and Gangale 2013). Regardless of the nature of this endeavor, a tremendous amount of political capital is still needed to actualize it in such a way that does not lead to international conflict. Thus far the most widely deployed narrative is that of resource scarcity, which has been critically deconstructed in previous chapters. Some scarcity narratives are nefarious, predicting species self-destruction through environmental holocaust or nuclear war (O'Neill 2000), while others revive Malthusian paranoia, warning that we are headed for population-induced disaster if we fail to tap off-Earth resources.[19]

The imagined futures under which lunar mining makes political economic sense leverage potent environmental, military, and elite economic anxieties that a growing human population will eventually lead to resource exhaustion. This framing then sets up a moral imperative to find and acquire those "resources desperately needed on Earth" (Day 2009), lest we meet our demise in overpopulation-induced starvation (Guner 2004) and the accompanying general breakdown of civilization from which we could be saved only by a military takeover to help "those most capably endowed" humans reach colonies off the planet (Dolman 2002). According to this view, which is propagated in popular media and culture, no matter what environmental conservation efforts we make now, we are "only prolonging the ecological endgame for life on Earth" (Autry 2011, 2), so we must look elsewhere.

Another, less common variation of the environmental apocalypse narrative argues for developing the capacity to mine in outer space to be prepared for an eventual planetary "fortress conservation" (Neumann 1998). Moving dirty industry "off-world" will be necessary once environmental crisis reaches such a point that the global economy is "locked down" in the name of environmental protection (Lamb 2010). In both of these framings, investors and private space advocates conflate the infinite expanse of the cosmos with infinite possibilities of accumulation that will be enhanced, rather than threatened, by the obliteration of Earth's biosphere. Accordingly, the only real solution is to formulate an "exit strategy" (Valentine 2012) to offload humans from the planet (O'Neill 2000). Whoever is in place to facilitate this "transition away from Earth" will profit enormously.

These dystopic fantasies would not be worth our attention were it not for the significant minority of scholars, policymakers, and billionaires betting on the inevitability of environmental apocalypse as an impetus for lunar exploitation and militarization. However, the apocalyptic imaginaries required to legitimate lunar exploitation clash with the cornucopian promise ready to be enjoyed with today's technologies. Progressive narratives envisioning human evolution in space do not cohere with the incapacity of humanity to avoid fouling Earth beyond repair. This indicates that the various pushes to mine the Moon do not form a coherent story. But as shown in previous chapters and argued elsewhere, incoherence can be productive for, rather than contradictory to, the expansion of neoliberal logic (see Couldry 2012; Sparke 2009). Conspicuous among such arguments is a marked inability to see beyond the current global political economic status quo, lending weight to Žižek's (2012, 1) observation that, "it seems easier to imagine the 'end of the world' rather than a far more modest change in the mode of production, as if liberal capitalism is the 'real' that will somehow survive even under conditions of ecological catastrophe."

Despite the fanfare, state-funded space programs remain the preponderant force in the race to mine rare earths on the Moon. For example, in October 2014 the Russian Federal Space Agency and the Russian Academy of Sciences and the Sternberg Astronomical Institute unveiled a US$2.5 billion plan to resume lunar exploration in 2016 with the purpose of bringing unprocessed lunar rocks back to Earth for resource extraction as well as extensive scientific analysis (Jamasmie 2014). Vladislav Shevchenko, the head of the Department of Lunar and Planetary Research at the Sternberg Institute, stated: "We think that the lunar surface contains enough rare earth metal. On Earth, reserves of cerium, lanthanum, neodymium, praseodymium and other metals that are primarily used in the manufacture of high-tech products have dwindled. China currently has the monopoly on this market" (quoted in Ter-Ghazaryan 2014). In Russia's case, promoting lunar mining of rare earths is pitched as a strategy to recruit the nascent domestic commercial space sector by assuming the economic risk of developing a manned lunar base (Blosser 2014). In fact, NASA, the European Space Agency, and the Indian Space Research Organization have all announced plans to pursue lunar mining in collaboration with the private sector.

It may be tempting to characterize the arrangement of public and private actors as fluid, and the lines between these sectors as blurred. To some extent, this is accurate: outer space exploration has always been a massive undertaking drawing on the resources and talents of multiple sectors of society. As discussed in the Introduction, the boundary between public and private sector is less a hard truth than a dynamic negotiation of the distribution of benefits and harms as well as the changing limitations to certain forms of democratic accountability. This is equally true for China's national space initiatives, which likewise draw on public and private sector institutions, yet are unequivocally characterized in domestic and international discourse as a purely government endeavor.

The *Jade Rabbit*

Following the successful landing of its lunar rover in 2013, China's National Space Administration (CNSA) became the most advanced actor in the race to mine rare earths on the Moon. Although geochemical analyses of lunar rock and meteorite samples containing rare earth elements were first published in Chinese scientific journals in 1986 (Zhang 1986), China's selenological activities entered a new era with the successful deployment of the *Jade Rabbit* lunar rover.[20] The mission's objective was to achieve China's first soft landing and surface exploration; gather selenological data; and provide a basis, through technological demonstration, on which to develop key technologies for future missions, including a rock sample return mission in 2017 (O'Neill 2013). The *Jade Rabbit* carried twenty kilograms of instruments intended to "achieve lunar shallow structure profiling, long period observation of astronomical variable-source brightness, Earth plasma detection and other scientific tasks" (Sun, Jia, and Zhang 2013, 2708). These included ground-penetrating radar and spectrometers in order to inspect the composition of the soil and the lunar crust.

China's space program shares its origins with the establishment of military-industrial rare earth production in the context of mid-twentieth century Cold War global politics elaborated in chapter 2. The Ministry of Industry and Information Technology, which also oversees China's rare earth sector, is responsible for CNSA.[21] This institutional arrangement logically follows the material applications of rare earths: rare earth alloys are essential to space exploration, satellites, and missile defense technologies. The explicit relationship between these multiple initiatives was concretized in the PRC's founding policy, as well as in the organization of scientific and military institutions. Although it is beyond the scope of this chapter, it is worth noting that the hinterland industrialization, rare earth research, atomic aspirations, and frontier securitization examined in chapter 2 were closely intertwined with the development of China's national space program. This relationship has deepened and matured over time.

China's rare earth news wires reported that over a thousand of the components of the *Jade Rabbit* contained domestic rare earth alloy technology developed specifically for the space program (Shanghai Rare Earth Network 2014). Consisting primarily of new, exclusively Chinese technologies, the *Jade Rabbit* mission was promoted as a bid to advance scientific exploration and human knowledge about a relatively unexplored area of the Moon. Toward that end, CNSA and the European Space Agency signed a mutual support agreement which allows the two agencies to use each other's findings and technological infrastructure (Xiong 2013). The lack of any such agreement between China and the United States is the result of the 2011 Wolfe Amendment, which prohibited NASA from collaborating with China in any capacity, including participation in any international space venture in which China is included (Rosenberg 2013).

The *Jade Rabbit* was launched on December 1, 2013 aboard Long March 3B rockets. It achieved lunar orbit on December 6 and landed on December 14 (Xinhua 2013a). It transmitted images of the lunar surface back to the Beijing command center on December 15 before embarking on a three month natural resources survey, which was cut short due to technical failures (Xinhua 2013b). Two years later, following concerted international efforts, the first of *Jade Rabbit*'s findings were published in international journals (Ling et al. 2015).

Meanwhile, three discourses have since emerged in China concerning rare earths on the Moon. Just as there are debates concerning our future in outer space in Anglophone literature, there are likewise debates in China over whether to pursue scientific lunar exploration under the terms of the OST, to develop mining capability for eventual extraction, or to exploit the Moon as quickly as possible in order to capture geopolitical advantage. In the first discourse, Sinophone academic literature emphasizes the scientific importance of rare earth exploration on the lunar surface because differentiations among certain concentrated arrangements of rare earths tell an as yet poorly understood story of the formation of the Moon (Guo et al. 2014). Furthermore, the leading scientific authority on lunar resources in China, Dr. Ouyang Ziyuan, emphasizes that the purpose of selenological exploration is "not only for the meaningful understanding of the Moon itself, but also to help understand and explore other planets in the solar system, particularly the formation and evolution of the Earth" (Ouyang and Liu 2014, 5). At present, this is the discourse that is translated to other languages and propagated internationally. In an effort to avoid any diplomatic tension over China's advances, the central government has repeatedly framed its pursuit of lunar exploration in the spirit of international cooperation. As one piece in *Xinhua* stated: "Space exploration is the cause of humankind, not just 'the patent' of a certain country. China will share the achievements of its lunar exploration with the whole world and use them to benefit humanity . . . all data . . . will be open to the whole world. China's lunar exploration provides an opportunity for countries dedicated to peaceful use of outer space to advance space technology together" (Xiong 2013).

In the second discourse, domestic state media frames rare earth mining on the Moon as a possibility, reasoning that while China's rare earth magnets industry is growing, eventually the country's efforts to meet increasing global market demand and identify new resources will require China to exploit lunar rare earths if there are no other breakthroughs that would render such a mining expedition unnecessary (China Electrical Intelligence Network 2013). This discourse most closely resembles the more restrained sentiment expressed by planetary scientists in Anglophone literature who are skeptical, but nevertheless committed to "keeping an open mind" regarding the possibility of lunar exploitation (Crawford 2015). The third and most bellicose discourse maintains that mining critical materials on the Moon is a sure way for China to overcome the measures that the United States and Japan have allegedly taken in order to undermine and contain China's rise: "The rich . . . rare earth, uranium, and thorium resources on the Moon can ease China's energy crisis, maintain the status of China as a rare earth power, and facilitate the rapid development of China's aerospace technology. . . . China now has 'first-strike capability' on lunar mineral development" (CM022 2013).

As of this writing, such sentiments remain marginal in Chinese discourse. The importance of maintaining global peace and stability temper expansionist dreams among China's space policymakers. This sensibility has been especially evident in plans for a follow-up mission to the *Jade Rabbit*, which CNSA submitted to the UN Office for Outer Space Affairs in 2014, pursuant to their obligations under the OST.

The Geopolitical Stakes: Fear, Territory, and the "Greater Good"

Because China's central government values its satellites and prizes lunar research as a measure of national progress, it is implausible that China's space program would undertake an activity to provoke an attack on its space assets. Those who have attempted to characterize China's advances in outer space as an asymmetric space warfare strategy, generally in an effort to justify greater military expenditures in the United States or India,[22] have fundamentally misread the situation. The soundest geopolitical analysis of China's lunar activities to date comes from Greg Kulacki, head of the China Program at the Union of Concerned Scientists: "the strategic objective of China's space policy is not to exploit asymmetry between China and the United States, but to end it" (Kulacki 2014, 11).

In general, however, observers in the Anglophone world have interpreted China's lunar exploration in opposite terms, framing the science as secondary. Reflecting broader US anxieties over losing supremacy in the global geopolitical order, past chair of NASA's Lunar Exploration Analysis Group Jeff Plescia commented: "[China's] lunar exploration, while trying to do some science, is more focused on the geopolitical theatre. They are demonstrating that they have the technical capability of doing the most sophisticated deep-space activities. They have a program, and they can keep to the schedule and accomplish mission goals on time [while] the United States has been floundering around for decades, trying to figure out what to do" (quoted in David 2014).

Building on longer-term colonial and postwar conflicts between globalized territorial orders, this is a struggle over who gets to act in the name of the greater good—whose activities are recognized as science and whose activities are construed as trespass. If former Cold War antagonists—the United States and Russia—can represent the paragon of international space cooperation, what accounts for the hostilities between the United States and its largest trading partner, China?

One conspicuous but under-examined contributing factor is US paranoia over the advances made by China's space program in particular, and the transfer of manufacturing, scientific, and technological capacity from the United States to China more generally. The timeline is worth considering: China announced plans to send an unmanned vehicle to the moon in 2005. NASA's lunar innovation prize was cut in 2006. The private sector took on the initiative and announced the Google Lunar XPrize in 2007, but it remained marginal, confined to the domain of space mavens. Fears that China could surpass the United States in outer space was not, itself, sufficient to mobilize investment or political will to support the capitalization of outer space, especially during the collective belt-tightening years following the 2008 US financial crisis. The 2010 rare earth crisis gave shape to Western anxieties over China's rise by providing something concrete and broadly important around which to focus geopolitical insecurities. In April of the following year, the US Congress passed the 2011 Continuing Appropriations Act containing the infamous Wolfe Amendment.[23] These insecurities reached a much sharper pitch following the deployment of the *Jade Rabbit*.

Rather than review the array of paranoid, racist, and saber-rattling discourses that proliferated in Anglophone literature following the deployment of the *Jade Rabbit*, suffice it to say that the fixation on China's achievements in space exploration as threat to the United States or the European Union serves to shift the debate from substantive questions about who should undertake lunar exploration and toward what ends—concerns that are in fact shared across Anglophone and Sinophone discourses—to a narrowly focused one of "how can we beat China?" (Whittington 2013). This generates a "by any means necessary" (Tadjeh 2014) approach that leaves private sector, and in fact illegal, approaches unquestioned. For example, billionaire space entrepreneur Robert Bigelow has successfully convinced counterparts in the federal government that the United States is locked in a territorial race with China, stating: "The big danger here isn't a fear of private enterprise owning and maximizing profitable benefit from the Moon. . . . The big worry is America is asleep and does nothing, while China comes along, lands people on the Moon, and decides, 'We might as well start surveying and laying claim, because who is going to stop us?'" (quoted in Wells 2013). Bigelow's discourse got results. The February 3, 2015 FAA memo declaring US Federal Government support for private claims to the Moon was addressed to him (Chang 2015). Playing on US anxieties over China's ascendancy, Bigelow minimized the potentially destabilizing effects on international law and scientific cooperation of unleashing private sector, profit-maximizing activity.

Instead of leveraging China's advances in lunar exploration in order to reinforce the terms of the OST, Anglophone commentators cried trespass and downplayed the scientific importance of the achievement in order to resurrect a "red menace" and racially code it as Chinese. This flattened the diversity of perspectives within China to conjure an imperialist monolith that simply does not exist. As in other parts of the world, in China the state remains the preponderant force in lunar exploration. In both the Anglophone and Sinophone worlds, those advocating aggressive territorialization and lunar resource exploitation contrary to the OST are likewise an extreme minority. The difference is that in the United States, this minority has organized itself into a politically connected private sector, while in China, this minority is found among military commentators whose sentiments remain contrary to the central government's diplomatic strategies. The most significant difference, of course, is that in the United States this minority has succeeded in codifying its interests into domestic legislation.

The SPACE Act of 2015

On November 24, 2015, US president Barack Obama signed the Spurring Private Aerospace Competitiveness and Entrepreneurship Act, which grants US citizens the legal right to claim outer space resources and to bring civil suits against entities that pose "harmful interference" to the exercise of private property rights in outer space. Chapter 513, section 51303 states: "A United States citizen engaged in commercial recovery of an asteroid resource or a space resource under this chapter shall be entitled to any asteroid resource or space resource obtained, including to possess, own, transport, use, and sell the asteroid resource obtained in accordance with applicable law, including the international obligations of the United States." This legislation, which passed with bipartisan support,[24] is an oblique attack on the reigning res communis regime espoused in the OST and the Moon Treaty. By granting US citizens property rights, primarily over asteroid resources and secondarily over "space" resources, the legislation attempts to present itself as consistent with the very international treaty obligations it undermines.

It is physically impossible to mine rare earths for profit on the Moon or on any other body in outer space in a manner that is consistent with the provisions of the OST. Mining obliterates a given landscape, while profiteering requires exclusive access. This is precisely why mining is so useful for extending territorial control to historically elusive places: because it quite simply, brutally, and unambiguously eliminates the possibility for other uses of the site in question. If it is a US company, rather than a US public venture, that establishes an exclusive mining site in outer space, the geopolitical ambitions of the United States would, in theory, be served either way.

In this case, the private sector can do the dirty work[25] of fulfilling the state's geopolitical agenda while the public sector provides protections and guarantees to the private sector. But in fact, a distinction between the public and private sector obscures more than it clarifies. After all, many of the new space industries were founded by former state space agency personnel, and many of the most effective advocates for the privatization of space have backgrounds in both finance and government. State promotion of the private sector in pursuit of lunar mining closely resembles the cases reviewed in the previous two chapters, wherein the private sector was selectively enlisted to execute the territorial agenda of the state. In this case, the national government provides force and backing to a risky and illegal venture in exchange for anticipated geopolitical advantages.

This is where critical geopolitics helps us see further than conventional geopolitics. Conventional geopolitics would hold that this is simply twenty-first-century statecraft instrumentalizing the private sector to further national interests. For the moment, this particular contrivance of a public-private divide is conceived as enabling US actors on all sides to maximize benefits and dodge international treaty obligations while they territorialize the Moon. The flaw in this reasoning is the assumption that all interests are wedded to the US national interest, so the newly empowered private sector is imagined as acting as an extension of government interests. But there is no such guarantee. Critical geopolitics, by contrast, challenges fixed notions of the state and therefore fixed notions of public and private sector interests. Private sector firms, newly empowered by the US government to sue any entity that damages their private interests in outer space, are free to contract with any paying customer regardless of their national origin or the integrity of their enterprise.

With the case of the Moon, the stakes of the state's investment in private sector mining differ from those discussed in previous chapters. It is not just a matter of pursuing profit and geopolitical control, but of maintaining the status quo of the global political economy. Under the terms of the OST—to which all state actors advancing space mining are party—any mineral extracted from the Moon would have to be distributed in a way that is "to the benefit of all peoples" on Earth. To pursue lunar mining in compliance with the OST would fundamentally change the global political economy of resource production and consumption from profiteering to sharing. There is no having it both ways—the terms of the OST have made it thus. Any state or nonstate entity doing otherwise would clearly be operating with impunity regardless of the verbal gymnastics involved in legislative attempts at the national scale to sidestep these agreements. But by insisting on a false premise of legal ambiguity at best and "chaos" at worst (Whittington 2013), private sector actors can do the dirty work of the state, until such time that international treaties are supplanted or other parties acquiesce to violation as the new norm.

For a particular government to assert the right of its citizenry to mine resources in any particular place, and to secure for that citizenry the right to pursue punitive legal action against any entities who interfere with the exercise of their property rights is, by definition, an assertion of sovereignty over those places, whether they are scattered across multiple celestial bodies or consolidated in one place, such as on the Moon. Such claims directly and unambiguously contradict existing international treaty obligations of the United States. The SPACE Act attempts to evade this by concluding with a Disclaimer of Extraterritorial Sovereignty, elaborated in Section 403: "It is the sense of Congress that by the enactment of this Act, the United States does not thereby assert sovereignty or sovereign or exclusive rights or jurisdiction over, or the ownership of, any celestial body."

The United States need not assert sovereignty over an entire celestial body in order to claim a particular territory therein. After all, that is how the political geography of Earth is organized: no single state controls the entirety of the celestial body we call home, but that does not negate the sovereignty of 192 national governments over their respective territories. The verbal gymnastics of the SPACE Act do not succeed in side-stepping the OST's prohibition of assertions of national sovereignty "by means of use or occupation, or by any other means" (UN 1967, Article II).

None of this is to suggest that a coherent agenda exists between the state and the private sector. Advocates of privatized space exploitation have multiple perspectives on the role of the state. Some denigrate civilian space exploration as too slow (Wingo, Spudis, and Woodcock 2009) and bogged down in bureaucracy, which inhibits the fantastic innovation potential of the private sector (Jones 2013). Others see the state as critical to securing their investments. Of the signing of the SPACE act of 2015, Eric Anderson, cofounder and cochairman of Planetary Resources, Inc. gushed: "This is the single greatest recognition of property rights in history. This legislation establishes the same supportive framework that created the great economies of history, and will encourage the sustained development of space" (quoted in Navarro 2015).

Regardless of their perspective, private sector interlocutors are working toward capturing maximum possible support and minimal regulatory intervention from the public sector. This effectively translates into massive transfers of public wealth to private hands while reducing oversight mechanisms concerning the use of that wealth. This coheres with the extensively theorized relationship between the "retreat of the state" and the "financialization of everything" under contemporary neoliberalism. But as with other cases examined in this book, this is not simply a case of deregulation, but also of reregulation. The proliferation of commercial space agencies represents not a retreat of the state per se, but rather a reconfiguration of state functions to support a program of redistributing public assets into the private sector in the name of beating a bogeyman from the East. Indeed, the most vociferous political, public, and legal opinion holds that the private sector should lead the way, and that "the government should focus on its role as enabler" (Whitehorn 2005). This is overwhelmingly compatible with the US government's approach since the end of the Cold War (United States House of Representatives 1998).

Most fundamentally, the privatization of outer space removes control over the architecture of our future in outer space from deliberative democratic decision-making. Authorizing private sector accumulation on the Moon does democratize competition to a limited extent, but it also undermines democratic control over the terms of such competition as set forth in UN treaties. There is nothing blurry or ambiguous about this shift. The implications for scientific research and discovery alone should give us pause. What happens when technologies used, data collected, and samples returned are no longer open to any individual or organization for scientific purposes, but are instead withheld as intellectual property made available only to the highest bidder? The production of selenological knowledge for the benefit of all humankind was a cornerstone of peaceful cooperation in the use of outer space, and instrumental to collaborative scientific endeavors seeking to understand the origins of our planet and our solar system. Will this knowledge be enclosed and militarized to the detriment of global scientific progress?

Across the globe, proponents of lunar enclosure who insist that legal conventions governing outer space do not apply to them are opening a Pandora's box that the 1967 OST was explicitly designed to keep closed. By interpellating outer space as a new "wild west," they are dismantling the conventions put in place to encourage rapid scientific progress and prevent the weaponization of outer space. Wary of this shift, Russia and China have submitted several drafts of a new UN Treaty on Prevention of the Placement of Weapons in Outer Space and of the Threat or Use of Force Against Outer Space Objects. The United States' UN delegate has voted against it every time, calling it a "diplomatic ploy to gain military advantage" (Hu 2014).

The trade-off between peaceful use of the Moon for the benefit of all humankind and the enclosure of lunar space for private gain is a false choice. None of the premises advanced to justify this radical departure from robust international treaty regimes withstand scrutiny—not environmental eschatology, not capitalist humanitarianism, and most certainly not rare earth scarcity.

Conclusion

The fiction that rare earths are in face rare on earth remains productive in policy, investment, and popular discourse. This chapter reviewed the means deployed by public and private sector actors to conjure a lunar frontier. By discursively framing their activities to negate basic facts of contemporary laws and available quantities of global rare earth resources, they have made notable progress transforming legal regimes, scientific practice, and investment flows around the globe. This history demonstrates the complicated role that the race to mine rare earths on the Moon plays in contemporary struggles over the definition of human activity in outer space. Across the globe, these debates range from scientific and humanistic to paranoid and militaristic.

As with struggles over the subsoils of the northwestern Amazon, the heart of Afghanistan, or the protected areas of Greenland, the race to mine rare earths on the Moon has less to do with actual resource scarcity and much more to do with how rare earth mining in these places might serve frustrated bids for territory, power, and control.

Although doomsday scenarios generated by China's rare earth monopoly have proven useful to generate broader public support for lunar mining, proponents have overlooked the fact that, rather than going to the Moon, we could simply implement proven rare earth recycling techniques here on Earth (Meyer 2011). This would preempt many of the scarcity scenarios lunar mining purports to resolve. If rare earth scarcity were really the issue, a massive untapped frontier is much closer than the Moon: currently, less than 1 percent of all rare earths consumed are subsequently recycled (Molycorp 2013). Surely we can do better: both in finding the reasons to explore the cosmos, and in our everyday production and consumption of rare earth elements.

CONCLUSION

How can we learn from the present to build a better future?

This work has been concerned with the frontier as defined by the mined and to-be-mined: particularly how and why, from among the hundreds of places endowed with rare earth deposits, have the sites examined herein emerged? Chapter 1 defined rare earths and provided a world-historical analysis of their entanglements with global politics following their discoveries. Chapters 2 and 3 showed how rare earths played a crucial role in the development of Inner Mongolia specifically: how embedding rare earth extraction, production, research and development in a comprehensive military-industrial project was essential for laying the foundation on which China's monopoly later emerged in the context of global neoliberalism. At the turn of the millennium, environmental and epidemiological costs of China's rare earth monopoly reached such a point of severity that the central government reoriented national strategy from export dominance to resource conservation. Chapter 4 analyzed the events that precipitated the 2010 crisis and its aftermath. The sudden shock to the global market drew unprecedented attention to the conditions of extractive policy and practice in Baotou and Bayan Obo, which then precipitated dramatic spatial transformations in other parts of the world, such as Afghanistan, Greenland, and the Americas. Chapter 5 unpacked the spatial paradoxes characterizing the Brazilian rare earth frontier, using rare earths as a lens through which to examine the ongoing struggles over the meanings of sovereignty—defined as the right to mine—in northwestern Brazil.

Chapter 6 showed how fictions about rare earth scarcity were productively mobilized as a justification to push the contemporary space race beyond the bounds of existing international treaties, which mandate the peaceful and collectivist use of lunar resources.

This book, though far from an exhaustive catalogue of all established, explored, and prospective mining sites, showed that geopolitical ambitions, territorial struggles, and the politics of sacrifice are as important as geology in determining which places emerge on the global rare earth frontier. The questions taken up here are: What have we learned from looking at three radically different sites? How might we take the best lessons from across history to build a more just and sustainable future?

First, something must be said for the fact that so much fiction surrounds the production of new rare earth frontiers. The myths of absolute scarcity and the persistent idea of "dwindling" rare earths despite abundant evidence to the contrary is an illustration of what contemporary philosopher H. G. Frankfurt (2005) describes as "the most salient feature of our age," that is, bullshit. This is not so much a deliberate lie as a "lack of connection to a concern for the truth" (Frankfurt 2005, 33). Advocates for mining the Moon, Greenland, the Amazon, the ocean floor, and Afghanistan cling to the claim that rare earth elements are in fact rare on earth, and soon we will have used up all available resources. Hence the need to do what it takes to mine rare earths in these forbidding places. This myth works well with another sort, which is the claim that each new site is the largest deposit in the world. These claims are simply not true, but it is not enough to simply name the untruth. Rather, this book examined how misrepresentations, deliberate or not, have been productive for various agendas. These paired myths of global scarcity and local abundance intersect with longer-term territorial and geopolitical anxieties in the face of China's growing global influence.

These fictions are especially potent because they are not entirely divorced from the truth. Rare earth elements are generally difficult to access, and all sites studied here do actually possess minable deposits of rare earths. Although the magnitude of actual deposits may be overstated, they are not entirely false. This requires a different way of looking at the situation, to identify where a lack of connection to a concern for the truth serves political and economic ends. One consistent attribute across all sites examined is that rare earth elements played a small but important part in significant territorial transformations on local and regional scales. This was the case for Baotou; this is what geologists and key figures among the Brazilian federal government desire for Cabeça do Cachorro; and this is what lunar mining advocates in the public and private sector desire: to transform the Moon into Earth's "eighth continent." Thus what is critical about the 2010 rare earth crisis is the way in which it stimulated a radical geographical transformation of the frontiers of prospecting and extraction. That many of the post-2010 discourses are based on false premises is perhaps less significant than the ways in which these imaginaries have transformed resource politics on multiple scales and redefined the scope of just how far some will go in pursuit of wealth and power.

Understanding the Frontier

Each of the three primary sites enriches our notions of "the frontier" as an operative spatial category and potent geographical imaginary. Just as rare earths are embedded in an array of important commodities, an array of important meanings and agendas are embedded in the creation and exploitation of the rare earth frontier. As elaborated in the introduction, frontiers have been theorized as spaces of conflicting regimes of governance, law, and property rights. This is because, as Tsing (2005) notes, frontiers do not exist *a priori* but are conjured into being by extralocal powers. What exists, or had existed at the moments of conjuring were, in the cases of Baotou and São Gabriel, mobile, multiethnic polities. Successive efforts to impose borders ranged from genocidal to integration to developmentalist campaigns, the effectiveness of which has never been absolute. In the case of the Moon, the latest discursive and regulatory offensive seeks to enclose what Cold War-era treaties had designated as global commons. Each of these endeavors are driven by a desire to turn the spaces concerned into something else to serve the geopolitical and accumulationist ends of multiple competing actors.

Invoking a frontier signals an expression of ownership, or an aspiration thereof, while the production of geological knowledge can signal (or be read as) an intention to territorialize. In the case of the Moon, international treaty regimes claimed the "final frontier" for all humanity. Accordingly, these treaty regimes mandated that all materials and research findings be made available to all. This is unique among the cases examined herein. By contrast, following the failure of decades of national integration and infrastructure construction campaigns in Cabeça do Cachorro, President João Figuereido designated portions of the region with considerable geological wealth as Biological Reserves: if massive, state-orchestrated capital could not access the resources, then nor should anybody else. This sentiment is especially apparent in the divisions between the military and large extractive interests on one hand, and indigenous small-scale miners on the other. Actors on both sides agree that mining should be permitted to occur in the region, but they are locked in an intense struggle over the meanings and entitlements according to which extraction should be organized. Under the current legal regime, the visions of both sides are illegal. The territorial orders of late twentieth-century conservationism and state custodialism reign.

But as with any reigning power, political economies at the level of everyday life are much more complex. Ultimately the frontier cannot just circulate as a disembodied idea. It must be enacted in a specific time and place by specific people. Territorial assemblages result from the encounter between the frontier vision and complex local realities. A person's idea about what that means differs according to their positionality. In São Gabriel da Cachoeira, this is especially visible in the draconian penalization of small-scale indigenous miners despite corporate dependence on *garimpeiro* activity, and broad support on the part of certain state actors for small-scale mining. In Baotou, as noted in chapter 3, local officials and police officers play an important role in maintaining small-scale, illegal production while viewing their actions as consistent with national policy mandates to consolidate and rationalize the rare earth industry. In either case, seemingly identical interests—liberalized mining on indigenous land or advancing resource policy mandates—produce radically different outcomes as local actors negotiate different needs, priorities, and identities in relation to the state. This helps explain the "elasticity" of frontier spaces (Weizmann 2007), where local actors maintain an apparently looser relationship to law and order by selectively reinterpreting and incorporating broader political changes into everyday practice.

In the case of the Moon, multiple actors within the state and private sector are actively working to conjure conditions of lawlessness where one of the most effective international treaties to date has held force for sixty years. When examined in comparative world-historical perspective, a consistent trend emerges across all sites examined in this work. Watts (2012) noted how frontier resource exploitation tends to leverage unclear, contradictory, or nonexistent legal regimes. In the cases examined herein, we are actually seeing regulatory offensives to criminalize customary, inclusive, and pacifist resource governance regimes. Where these offensives succeed in criminalizing the customary and litigating against inclusion, they also create the local spaces in which the hazards can be placed. This practice is vividly demonstrated in each of the cases. The Euro-American world externalized its production to colonial frontiers, and later to China. China's central government and the Soviet government externalized rare earth mining and processing to Inner Mongolia Autonomous Region (IMAR). Australia externalized to Malaysia; Mountain Pass to Estonia and China; and China is looking to externalize rare earth mining beyond its borders. Now, multiple high-profile actors are looking to the Moon.

Legal regimes, whatever their qualities, are *produced*: the diverse but immense regulatory offensives undertaken by pro-mining interests across the rare earth frontier shows that where a permissive "wild west" climate does not prevail, certain actors strive to make it so. From the perspective of large-scale mining interests,[1] indigenous and environmental protections act as a barrier to accumulation in the northwestern Amazon, while environmentally motivated consolidation and mining control efforts in Baotou exacerbate overcapacity (and therefore profitability) problems in local industries. The effort to create conditions of lawlessness on the frontier is especially vivid in the case of the Moon: there is one very clearly worded and detailed international treaty which a majority of countries, including all space-faring states, have signed or ratified that explicitly prohibits anything resembling enclosure or privatized gain from lunar resources. This is, of course, anathema to the accumulationist dreams of high-profile new space investors whose insistence that there are no rules—and that China will beat the United States—moved the US government to enact new laws in direct contradiction to international treaties. Although the particular context might be new, the practice of disavowing existing conventions governing land and resource use has been fundamental to colonial, capitalist, and socialist expropriation.

It is important to note that these spaces are not as empty as people sitting in offices would claim. All three spaces are occupied—if not with people, then with transnationally held meanings inhospitable to large-scale mining. But they are also far from "centers of calculation," where political decision-making power and relatively more politically empowered populations tend to reside (Latour 1987). This feature explains the aims to concentrate a destructive and toxic industry away from metropolitan areas despite the logistical challenges involved, while simultaneously serving an important geopolitical purpose by territorializing a region far from centers of power. It is through this spatial relation to centers of power that such spaces come to be described as "marginal," as the extensive ideological and subject-formation campaigns examined in chapters 2 and 5 attest. It is an immense project to convince local residents that the ground beneath their feet is somehow distant, and that their (re)productive activities must be valued and evaluated in terms of their compliance with the interests of a far-off state that maintains an inconsistent presence. Because frontiers are placed at the edges of the known and governed, where the frontier is said to lie is an indication of where centralized power imagines its own limits to lie. There is another side to this dynamic, wherein inhabitants within the frontier region may seek to leverage the imposition of the frontier signifier as a way to gain greater recognition, in the global economy, or as legitimate national citizens, respectively.

The cases of Cabeça do Cachorro and Greenland show that sacrifice zones are not unilaterally imposed from the top-down, but also can be sought after and fought for by local actors who wish to set the terms of the creative destruction characterizing our contemporary economy as it unfolds in their particular place. Local mining proponents feel strongly, if somewhat naively, that the geopolitical and economic spoils of rare earth extraction will outweigh the potential hazards simply because they intend to do mining their own way, on their own terms. They hope for deeper integration into global economies, greater control over local destinies, and broader recognition of local importance won by supplying the world with these vital elements.

The aspirations of people on the new rare earth frontiers are further complicated by the fact that despite their importance, rare earths simply are not gold. Although some post-2010 commentators characterized the global waves of exploration and speculation as a new gold rush (BBC 2011; Gustke 2011; Jeffries 2014), the analogy quickly fell apart in practice. The lucky miner with a gold nugget in hand holds instant wealth, but the same cannot be said for rare earth elements. Rare earth ores, by themselves, are worth very little without undergoing complex and hazardous beneficiation processes. Processing high quality rare earths has proven to be risky business, not just because of the environmental and epidemiological hazards, but also because despite their importance and proliferation, the global market remains decidedly small.

As a result, firms specializing exclusively in rare earth oxides have not fared well. Baotou emerged as the rare earth capital of the world in no small part because of the integration of rare earth mining, processing, and research with regional military, heavy machinery, and high technology industries. With the subcontracting and deindustrialization in the West following Reagan and Thatcher's deregulations and Deng Xiaoping's reforms, Baotou's industrial-scientific architecture became the center of gravity for global rare earth production. In Brazil, CBMM subsidized the development of reclaimed rare earth oxides with its booming niobium monopoly. The Mountain Pass mine in California, which specialized exclusively in rare earth production, reopened in 2012 under the since discredited pretense that it was leading the way in repatriating environmentally superior rare earth production. But prior to declaring bankruptcy in 2015, it relied on the same subcontracting practices that precipitated its demise—shipping minimally processed ore to China and Estonia for further value-added processing. This situation sheds some small measure of light on the seemingly runaway efforts to exploit the Amazon and the Moon when there are abundant resources available from far more accessible and far less controversial sources: to succeed, it appears that the enterprise of rare earth mining must be yoked to other industrial and territorial endeavors.

In this way, Baotou is unique in history because the tremendous investment in building a regional integrated military-industrial base contextualized rare earths and their broader (potential) applications in a very concrete way. Research and development on rare earth applications was integrated with the development goals of nearby munitions, aerospace, energy, heavy machinery, and information technology industries. This was complimented by the multi-mineral extraction approach to the Bayan Obo mine, which in addition to rare earths, is exploited for iron, gold, and niobium. The economic successes of Baotou and the failures of other sites of rare earth mining suggests that large-scale rare earth mining needs to be closely integrated with complementary industries and publicly funded research institutes in order to weather the vicissitudes of global political economy. If this is the case, the glaring absence of strategies to develop regional auxiliary or support industries at new points on the global rare earth frontier leads us to two possible conclusions: first, that mining proponents, investors, firms and policymakers are profoundly unaware of what it takes to build a successful rare earth enterprise, and second, that the quest to open up these new spaces is about something else besides rare earths.

Rare earth elements were not the sole reason that Baotou developed into a hinterland metropolis, but neither were the other interests entirely removed from the industrial and economic realities related to rare earth mining and processing. In some ways, the consolidation of China's rare earth monopoly was one outcome of a much larger set of processes in which rare earths played an important, but by no means exclusive, part. Baotou was built into a military-industrial hinterland to serve the developmental and military needs of the USSR and the People's Republic of China. Rare earths emerged in prominence contemporaneously with the establishment of iron and steel works, aerospace and defense industries, and aligned research institutes. This scientific-industrial base is thoroughly integrated into broader development strategies that evolve over time in response to changing global political economic conditions and domestic needs and aspirations. In many ways, Baotou is a success story of China's nationalist development and Open Up the West Campaign, which is currently a subject of intense interest and ongoing academic interchange between Brazilian and Chinese scholars.

There is a small but growing body of Brazilian scholars who study China's Western development model in order to apply it to the Amazon. The goal is to definitively exercise sovereign control over a region that has provoked territorial anxieties since imperial times. The abundant resources of the Amazon, it is envisioned, could be unlocked to fuel Brazil to a place of global political economic prominence that may one day rival China's.[2] This would require massive infrastructure investment, annihilation of local landscapes and lives, and unprecedented waves of migration and resettlement in order to provide necessary labor power. These scholars look at China's one-party system and echo China's criticisms of democracy as creating chaos. "I would prefer a dictatorship, at least then things got done," is a trope that, unheard in 2010 fieldwork, was repeatedly uttered on long-distance bus rides and in Federal ministries in Brazil in 2014. These comments presaged the political turmoil and rightward shift of national politics in 2016. In Brazil, the idea that economic development and prosperity was more important than anything else, including the integrity of the biosphere and hard-won civil liberties and legal protections, had gained considerable ground against earlier ideas about equitable sustainable development and the need to move Brazil away from the status of primary commodity producer.

This example shows that there are constructive and destructive lessons to be learned from the rise of Baotou and Bayan Obo. When it comes to matters of regional and industrial development, policymakers and planners in the rest of the world would do well to consider the broader industrial, research, and policy support networks that are necessary to sustaining a robust rare earth industry. Studying the rise of Baotou, however, should not be confused with a naïve celebration of authoritarian industrialism. Nor should the eventual rise of China's rare earth monopoly be used to justify the necropolitics that preceded the rise of the military-industrial complex in IMAR. Instead, the best lessons should inspire new thinking on industrial organization, while the particular histories related in this book should serve as a cautionary tale of how nationalist development projects can provide cover for racialized violence. To figure out how to source rare earth elements in a stable, ethical, and sustainable way, we need to understand why they have been sourced in unstable, unethical, and unsustainable ways.

Racist politics complicate the global rare earth frontier in many different ways. The ideas of which landscapes and lives are deemed sacrificable in the name of rare earth mining is often informed by and reflective of existing racial inequalities. Although the abundance of potentially minable deposits identified globally may convey the sense that "everywhere on Earth" has been explored, this should not be misinterpreted to the effect of depoliticizing the practice of geological knowledge production. As the cases examined herein demonstrate, the production of geological knowledge is an act of power, and contests over its meaning have defined struggles between local and extralocal interests over the last long century. Agents of questing European powers, Chinese nationalists, Japanese imperialists, US atomic interests, and Sino-soviet revolutionary communists carried out the surveying and prospecting in Inner Mongolia that led to the identification of the deposits at Bayan Obo. São Gabriel da Cachoeira was surveyed by Imperial Portuguese explorers, the US Army Corps of Engineers during World War II, and later under the military dictatorship before such activities were outlawed in the 1980s and 1990s. It is important to note that localized geological knowledge production continued despite changing legal regimes. As indicated by the ongoing struggles of indigenous garimpeiros on the one hand and the continued conferral of mining permits by Departamento Nacional de Produção Mineral to outside mining interests on the other, criminalizing activities essential to the production of geological knowledge did not stop them. As for the production of selenological knowledge, the anxieties circulating in Anglophone discourse surrounding the *Jade Rabbit* mission exposes the geopolitics of scientific exploration when one party insists on framing the other party's activities as trespass. Although China's lunar program has proceeded in compliance with existing international treaties, in Anglophone discourse their research has been reframed as a violation of the US frontier.

The relationship between existing collections of geological fact and the conditions under which they are collected is critical to informing ongoing struggles over the geography of the global rare earth frontier. In chapter 5, this is pitched between indigenous people and small-scale miners on one hand, and military and corporate mining interests on the other. On the Moon, neoimperial Cold War politics under which the selenological data was gathered are refracted through the private sector mining race, where firms selectively present themselves as proxies for state power when politically expedient. The politics of geological knowledge production in each of the cases reflected longer-term relations of domination. This is something to bear in mind as we consider how to build a more just and sustainable future. Although the production of scientific knowledge cannot be separated from its political and social context, paying careful attention to the social context in which data is gathered and plans are made can help us avoid reinscribing the structural and direct violence of our bloody history.

In a different way, the racialization of toxic rare earth production as a distinctly Chinese problem obscures the common challenge of isolating the dangers of rare earth mining and processing. As detailed in chapters 1 and 3, highly toxic and radioactive elements geologically coincide with rare earth elements, many of which are themselves hazardous to living tissues. Rare earth separation generates tonnes of pollutants, both in the form of industrial acids and the liberation of lead, arsenic, fluoride, uranium, thorium and radon gas in the form of waste products. This is a challenge common to rare earth mining across the globe. Developing and disseminating the best practices for controlling these hazards should be privileged as a point of international cooperation.

What Can We Do?

These cases are particularly illustrative of the dynamics defining the global rare earth frontier: global modernity depends on these resources, yet exploitation is both toxic and expensive. There are compelling reasons for states to develop rare earth production on national soil, but conventional mining and processing methods are too toxic, expensive, or controversial to develop anywhere other than on land that is deemed marginal and sacrificable. Yet, even with the emergence of viable greener alternatives and the elimination of China's export quotas, rare earths remained politically relevant in the push to territorialize lands that had historically eluded centralized state power. In essence, the securitization and crisis narratives surrounding rare earths helped the states involved resolve their own longer-standing frontier problems, the contemporary manifestations of which generated new social and geopolitical meanings for rare earth elements.

The primacy of territorial politics in determining the global geography of the rare earth frontier explains why our global geography of production is so strange, and why the strangeness persists despite better alternatives. But what can we do? There are at least three concrete ways to approach this issue. The first is to make the territorial politics explicit, so we can address the underlying and perhaps unrelated interests that drive rare earth mining to controversial and conflict-prone places. This requires critically interrogating the grandiose claims made by aspiring firms and developmentalist states, while also taking the concerns highlighted by social movements and researchers seriously. This will enable us to more readily focus on substantive and collaborative efforts to build a more sustainable, just, and rational rare earth economy. Programs such as the Extractive Industries Transparency Initiative, and the growing practice of incorporating social and environmental safeguards into major mining and infrastructure development projects provide useful blueprints for rethinking rare earth mining in a way that prioritizes the integrity of local landscapes and lives.

The second is to support the market for more sustainably produced rare earth oxides, such as those extracted from existing waste sites using certified sustainable processes rather than from new holes in the ground. The fact that recaptured rare earth oxides are already being produced by an ISO 14001 certified company shows that the greatest challenge—producing greener rare earths from existing mine wastes—has already been overcome. The reason greener rare earths have not yet gained greater market share is due solely to pricing rather than availability issues. This state of affairs undermines environmental remediation efforts in China and undermines the sustainable development potential of an array of rare earth reliant industries in the rest of the world. A straightforward solution would be to provide tax incentives to rare earth reliant firms engaged in the production of medical equipment, scientific instruments, renewable energy technologies, and energy efficient transportation technologies for a decade or two. This would help create market certainty for greener rare earths and incentivize other enterprises to reprocess rare earths from existing waste sites in a certified environmentally responsible manner. This could also allow time for robust monitoring and certification programs to be developed specific to the rare earth industry, which would further build consumer confidence in the sustainability of rare earth-bearing products. A broader positive outcome would be the reduction of the waste footprint of existing and former mining sites across the globe, and perhaps one day the end of the need to sacrifice new lands to open new rare earth mines.

The third concrete solution is to recycle rare earth elements. In addition to the abundant rare earths present in mine tailings across the globe, less than 1 percent[3] of rare earth elements consumed are currently recycled. This is, in no small part, due to the physical manner in which rare earth elements are used: they are additives, used to "dope" other materials, referred to as "spices" or the "vitamins" of industry. Effective recycling involves energy and chemically intensive processes of separating elements from magnets, alloys, lasers, batteries, hard drives, and other technologies into which they are blended. In her research on the viability of rare earth recycling in the European Union, Verrax (2015) identified the central obstacle to this worthy initiative. The exact composition of each component, as well as the precise quantities of which particular elements are used, varies according to brand and model. In other words, not all laptops, smartphones, or lasers are created equal: the exact composition of each product is confidential. Even laboratory analyses detailing the composition of particular electronics, and how best to extricate rare earths, remain protected under trade secrets.

Furthermore, the feasibility of any rare earth recycling initiative is currently predicated on industry demand, which varies according to element; downstream buyers have been unanimous in their negative response to the prospect of paying premiums for recycled elements, because the quality of recycled rare earths has not yet been demonstrated. There is also a serious social constraint, which is the lack of a large-scale waste collection system for both industrial and individual rare earth technologies: How to systematize the collection of jet propulsion systems on one hand, and broken laptop speakers on the other? And of course, how much more would recycled elements cost compared to those being produced in China?

These are clear obstacles to implementing viable rare earth recycling programs. But clear obstacles also present clear solutions. Just as there are standards governing the material composition of manufactured goods, there can also be standards that require rare earth components to be more readily extricable from potentially recyclable electronics. Likewise, the infrastructure and organizations for collecting multiple forms of waste are already well developed in major consuming economies. Developing a system to collect rare earth bearing products would hardly require starting from scratch. Rather, developing the facilities, training the processing personnel, and educating the public would likely follow similar practices as recycling and composting campaigns that took place in previous decades.

Pursuing any and all of these options would lead to a more just, sustainable, and stable global rare earth production regime. It is entirely possible to live in a world where the hardware of modern life is built on sustainability rather than suffering. It is also possible that our growing demand for rare earth-dependent technologies will continue to be used to justify all manner of brutality in vulnerable landscapes across the globe. The choice is ours.

Areas for Further Research

In addition to the concrete options offered above, the findings presented in this work have several implications for future research. Chief among the epistemological questions to be further researched are those concerned with how to truly think globally in research, theory building, and action from the standpoint that "the global" is dynamic and composed of distinct local instances (Tsing 2005). Foregrounding the local in global studies has had its own perils. For example, it is important to examine global political economy in a way that treats Euro-centrism and Sino-centrism critically without going so far as to propose that the "center" should simply be somewhere else, as proposed by Mignolo (2009), for example. The crucial next step is yoking broad global-scale inquiry to the concept of scattered hegemonies (Grewal 1994) to work beyond colonial and Cold War–era epistemic straits that might incline globally minded researchers to take at face value hemispheric divides, teleologies of development, and particular relationships between the private sector and the state under neoliberalism. On a related note, thinking in a global yet grounded manner also requires that we think of our "globe" in context (Cosgrove 2005). The technologically empowered extension of economic, extractive, military, and political interests to spaces beyond Earth requires that we rethink global epistemologies now that human life is co-constituted with extraglobal technologies, power struggles, and possibilities. It is time to consider outer space as an "area" worthy of the comprehensive study afforded to other places that comprise "area studies" programs.

With respect to Baotou and Bayan Obo, the most obvious need for further research concerns the outcomes of the industrial restructuring and liberalization of rare earth exports at the beginning of 2015. It will be important to evaluate whether removing the export and production quotas had any effect on efforts at environmental remediation and suppressing unauthorized and unregulated rare earth mining. If so, what these were and whether they have been discernible in the everyday life of local inhabitants is of utmost importance if we are to identify possible practices that could inform the production of less devastating future practices of rare earth mining and processing. Of further interest is how industrial restructuring in China's rare earth sector relates to and is informed by the New Silk Road campaign, which involves, among other things, conducting geological prospecting and constructing infrastructure across central Eurasia. This would support a nascent body of scholarship on "global China" (Lee 2014), which seeks to bridge the gap between research on China's overseas activities and scholarship on domestic China.

During fieldwork in São Gabriel da Cachoeira, I learned of an extensive, and entirely unwritten history, of indigenous practices regulating mining activities on their lands, including the establishment of a permitting and tax collection system to ensure that allowing outside small-scale miners also brought benefits to the host communities. This history flies in the face of established narratives of small-scale mining as an entirely unregulated disaster wrought by outsiders on victimized indigenous communities. It is important, therefore, to investigate how certain indigenous groups in the region engage with the practice of mining over time, including their participation, regulation, and control over such practices. This could begin by supporting local efforts to construct an archive of this experience. This would support the expansion of a small but extremely important body of literature pioneered by Graulau (2001) and Lahiri-Dutt (2011) that focuses not just on the place of mining in indigenous livelihoods, but also on the ways in which the importance of this enterprise differs along gender lines. There is important work to be done on the intersection of indigenous and women's agency in mineral extraction and how that contradicts and entangles with visions of masculinized dominion over vertical space.

A further site of considerable ethnographic interest concerns the coproduction of Silicon Valley techtopias, narratives of apocalypse, and west coast utopian experiments. Although utopian experiments such as communal living and festivals of radical self-expression are often cast as the antidote to apocalyptic futures,[4] my findings among nascent private sector space mining firms suggest that, in fact, the utopian experiments inform apocalyptic common sensibilities in a selective convergence of extreme leftist and extreme rightist ideologies. Based within the spectacular accumulation of wealth surrounding technological innovation in Silicon Valley, what remains underexamined are the ways in which promises of technologically enabled futures of convenience and interconnection rely on preserving the current unsustainable political economic status quo. Desires and claims for a utopian future as exercised in the rarified atmosphere of Silicon Valley solidify rather than undermine imaginaries about the inevitability of apocalypse and societal collapse. Given the deepening relationship between technology firms and the US government, this would be a fascinating area for participant observation, ethnographic and archival research.

Finally, because of the contemporary nature of the issues studied, they require ongoing engagement to see where the geography of rare earth extraction settles in the near-term, how practices improve and change over time, and whether the fiction of rare earths as rare will remain operative in agendas to territorialize the places examined in this work.

Appendix
METHODOLOGIES AND APPROACHES

The methodological toolkit for this research consisted of multiscalar interviews and archival research to support a world-historical analysis and critical encompassing comparison among the sites examined herein. As discussed in the introduction, this research started with the contention that much of what drives the production of the global rare earth frontier operates in the epistemological gaps of political economies, political ecologies, and postcolonial theories that drew their ordering logic from the Cold War vision of a world thrice divided. Although Cold War geographies of extractive territoriality cast long shadows over the contemporary production of these particular places, it has long since been identified that Cold War structurings of the world presumed a static division of global space according to which all other life processes were subordinate—while relevant, this could neither describe nor comprise *the* global situation, because in fact there are many situations that vary according to positionality and subjectivity. Rather than presuming a fixed global structure at the outset of this work, I instead followed McMichael's (1990) lead of viewing structures as formed through historical relations. From such a perspective, it is necessary to compare processes across space and across time in order to identify the historical formation of specific phenomena and the world-historical processes with which they are co-constituted. Pomeranz's (2000) work on elucidating *The Great Divergence* between two ends of the Eurasian land mass just before the industrial revolution is a key example of this approach, in which no single case is upheld as the norm against which others are measured. When examined in this way, illustrative differences *and* similarities emerge that allow us to understand the relationship of places presumed to be mutually alien—Inner Mongolia, the Amazon, and the Moon, for example—within a world-historical process.

Methods

In practical terms, this meant that I carried out multiscalar in-depth serial interviews with relevant institutions, stakeholders, and researchers, and advocates in the national and state capitals, nearby cities, and local areas. To complement this approach, I conducted in-depth analyses of several archives within national and local contexts, and made extensive use of library and web-based resources. These interviews and archival research both preceded and followed my environmental analyses of specific sites, where I collected soil and water samples, conducted household interviews, and gathered oral histories from elder members of the communities. Following my visits to specific frontier sites, I then triangulated my interview, archival, and field-based findings. I evaluated institutional and archival discourses from national and international scales in light of my local findings in order to identify gaps, silences, or contradictions. Where possible, I followed up on these matters in subsequent interviews in order to examine the gaps between discourses about the frontier and realities on the ground.

Interviews

Interviewees were selected on the basis of their relevance to, or experience with, frontier (un)making processes identified in each of the three sites. This included individuals involved directly in rare earth policy, economics, and research, as well as those working on broader development, environment, and security issues. I conducted serial interviews with national, state, and local civilian and military officials, members of international policy, research, and investment circles, as well as laborers, activists, village elders and other citizens. I made a point of engaging institutions taking a leading role in shaping local, national, and global rare earth politics in the United States, Brazil, and China, such as those charged with geological survey, national security, regional planning, ethnic affairs, brokering investment deals, and environmental protection. I personally conducted all interviews in local languages and in a semi-structured manner; where appropriate I employed snowball sampling in order to identify other interviewees. The overarching objective of the interviews was to ascertain the ideas and practices involved in shaping the current geography of the global rare earth frontier on local and global scales.

Archives

Such an inquiry requires a fine-grained historical analysis. In addition to interviews, I conducted archival research in the United States, China, Brazil, and

Berlin in order to understand the histories of the key sites before and leading up to their (re)configuration as points along the global rare earth frontier. I consulted the United States Geological Survey records held at the National Archives and Records Administration in Bethesda, Maryland; the Central and East Asian and Latin American maps collections at the Library of Congress in Washington, DC; and made extensive use of the library databases and online collections at the Doe Library and the C.V. Starr East Asian Library, as well as the special collections procured through the Bancroft Library at the University of California, Berkeley. In Berlin, I made use of the loan capabilities of the Wissenschaftszentrum Berlin für Sozialforschung (Berlin Social Science Center) to examine historical survey maps of the Inner Mongolia Autonomous Region.

In China, I consulted the microfiche, online scientific research databases and regional books collections at the National Library in Beijing; the personal and institutional archives of researchers at the China Academy of Social Sciences Institute for Chinese Borderland History and Geography; the local newspapers and periodical collections at the Inner Mongolia Autonomous Region Library in Hohhot for periods between 1947 and 1970; and revolutionary documents collections that archival staff permitted me to see for ninety minutes at a time, after several days' wait,[1] at the Inner Mongolia Autonomous Region Archive in Hohhot. The Baotou Municipal Library contained a wealth of local documents deposited from local bureaus and industries dating back to the early 1970s, and the staff at the local Environmental Protection Bureau selectively opened contemporary archives to the extent they deemed prudent in order to answer my questions about local developments since 2010. The local Ministry of Land and Resources office in Bayan Obo allowed me to consult its annals and past local development plans. Although I did encounter plenty of self-styled gatekeepers, I also benefitted tremendously from the generosity of senior scholars. One such scholar, upon learning that a particular librarian had denied me entrance to a particular archival collection, arranged to have eighty relevant volumes brought into his office on a special forty-eight-hour loan. He then invited me to peruse them, providing me with a scanner and a student assistant during that time.

In Brazil, I examined the historical scholarship on the Brazilian Amazon held at the National Library in Rio de Janeiro. I consulted the Senate Archives as well as the documents collections at the Instituto Socioambiental in Brasília, the historical documents collections at the Public Library of Manaus, and the local records and documents collection at the Instituto Socioambiental and Federal Police post in São Gabriel da Cachoeira. The Public Archives in Manaus were under renovation during my research visits.

Research Considerations

There are several limitations to this study, most of which derive from the choice to forfeit an extensive ethnographic commitment to the sites studied in favor of working out, in practical terms, a transnational research project that intends to show a way out of ossified notions of global order and difference generally segregated between East and West as well as the Global North and Global South. As I pointed out in the introduction and elsewhere (Klinger 2017b), claims of exceptional agency in the making of global history are conspicuous features of both Anglophone and Sinophone literature, while the tendency to folklorize or fossilize Latin American (pre)colonial history is likewise conspicuous in both linguistic canons. In order to carry out this research in a way that did not reinscribe teleological narratives of longue durée global change with fixed arrangements of power and agency across global space, I relied heavily on a critical geographic and transnational feminist approach, both of which demand careful attention to local productions of space through the practices of everyday life. Through this approach, the multiple and particular hegemonies relevant in different places to different actors were brought to light. I adopted this approach in deliberate contradistinction to the prevailing analyses of global rare earth dynamics, which rely on geological determinism and speculative hyperbole at the expense of actual interests and local history. A fine-grained approach with attention to difference reveals that these places are not mutually unintelligible, but rather share important characteristics recognizable as different forms of a frontier process.

By pointing to the shared historical attributes involved in (re)configuring the sites studied into the global rare earth frontier, the aim was to draw careful attention to similarities and differences, in order to demystify how extractive and territorial power work together across global and local scales. By working within the local languages and literatures of each of these sites over a period of five years, as well as cultivating relationships of trust built on a decade of research, life, and engagement with Brazil and China, I endeavored to balance the trade-off between breadth and depth. However, in comparison to my own predoctoral experience conducting extensive rural ethnographies in a single township in southwestern China, the difference between relocating wholesale for a year or more to one place versus structuring my research according to six to twelve-week fieldwork stints unfolding over five years is apparent in three key ways.

First, I eschewed extensive household surveys in favor of targeted multiscalar interviews and repeated escorted site visits with a range of interlocutors. While the strength of this approach lies in capturing multiple narratives of a given place from a range of positionalities, a possible weakness lies in the fact that I did not generate a codifiable body of work (as would be the result of standardized surveys, for example) that could be analyzed with various social science analytics software. This limits in some ways the claims that I can make about the extent of the similarities and differences of the three sites. This would be a compelling project for further research. The question of how different concepts travel across languages and global spaces, and how those traveling concepts intersect with power to generate specific spatial transformations has been an enduring topic of interest for me since acquiring my fluencies in Mandarin and Portuguese over a decade ago. However, I have found that there is little theoretical or empirical work to date examining this question in the context of specific political economic issues. Such questions are for subsequent analysis.

Second, while one of my strengths as a researcher is my linguistic and cultural facility in Brazil and China, the fact that I did not rely on a translator to conduct interviews, particularly in sensitive contexts were interviewees requested not to be recorded, affected the quality of some of my interview notes. The pacing of a translated interview is considerably slower and more modulated than a direct interview, which allows the interviewer time to write more immediate complete notes and to gather thoughts between translations. Working in second and third languages in the context of a direct interview meant that I was often taking notes in shorthand to aid the subsequent write-up process later that day or week. Although I take pride in the diversity of perspectives represented, I cannot avoid a sense of some precious things having gotten lost.

Third, there is a justifiable fatigue among certain policymakers and activists in dealing with international researchers inquiring about social and environmental issues for the purposes of their own scholarship which will not be published in local languages for quite some time, if at all, in addition to an array of (real and perceived) security concerns involved with working in frontier regions. In light of this, I found the practice of multiple returns—as opposed to one long uninterrupted field stint—to be productive for building trust over time as well as for demonstrating sensitivity to security concerns and the workaday constraints shaping the lives of busy people. This allowed me multiple opportunities to approach situations that at first may have been impossible, whether for reasons of political censorship, social turmoil, or suspicions concerning whether I was actually an academic researcher and not an undercover journalist or spy. Local counterparts later expressed appreciation that I did not push certain issues when times were difficult, yet maintained an engagement with them and the issues with which they were working over several years.

In terms of safety and security issues, three points must be made. First, despite the elasticity of law and order that can characterize frontier regions, it is not as though public safety and bodily autonomy are guaranteed in urban metropolises. The consistent, low-level alarmism I encountered in response to my chosen field sites had the effect of misconstruing the San Francisco Bay Area, or Beijing, or Rio de Janeiro as the safe and orderly opposites of Baotou, Bayan Obo, and São Gabriel da Cachoeira. This response was as unexamined as it was predictable. Such attitudes, although largely well intentioned, have the effect of erasing the everyday perils involved in moving through urban spaces as someone gendered female, as though threats of violence and bodily harm only become salient facts of life out on the frontier. This is patently absurd. Although I am privileged by my race in US, Brazilian, and Chinese contexts, because of my gender, my experience is shaped by certain hazards that are consistently present at home and abroad, and must always be navigated in context-specific ways.

Second, the protection of the identity of my contacts and interviewees, especially those in securitized regions, is something that extends far beyond the actual time I spend in the field. By choosing to talk to an outsider who is also a conspicuous US citizen in a time of international furor over invasive US surveillance practices, local contacts risked increased scrutiny and interrogation should someone later decide to view their engagements with me in a suspicious light. I took care to avoid this eventuality by gaining appropriate permissions and letters of introduction from relevant national and local institutions before initiating interviews, and took the added precaution of meeting first with officials and power brokers to mitigate possible misunderstandings on the part of national and local authority figures. I also prepared copies of my research abstract and a list of key questions in local languages, which I circulated in addition to relevant documentation from the institutional review board of the University of California, Berkeley. Because everyone I met received the same paperwork, this helped to establish my credibility and lower any possible stakes that might otherwise be involved in meeting with an outsider. As a further precaution, I assigned a pseudonym to every interviewee cited in the text, except for high-profile public figures. Many more interviewees are left out entirely. As stated in the introduction, information and potentially sensitive documents that were shared with me in confidence are not cited. Following the advice of veteran researchers of contentious topics, I used what was shared with me in confidence to curate and analyze the publicly available information cited herein.

Third, conducting in-depth research into issues deemed to be of critical national significance in the United States, China, and Brazil presented its own unique set of challenges. In many ways, this project was an informal education in how censorship works in multiple contexts. In China, where censorship is openly discussed and practiced as a matter of everyday state functions, I found that it actually became clear quite quickly what information would and would not be available. Censorship tended to be applied to thematic areas of public discourse as opposed to specific issues engaged by specialized audiences. This meant that although topics such as "radioactive waste" were generally considered too sensitive for newspaper articles, specialized academic research into specific aspects of the problem as manifest in particular places was freely accessible in libraries and academic databases. But censorship is also a dynamic practice that changes over time. In the event that a particular article or book was deemed too sensitive to be downloaded at the time I sought it, library databases alerted the researcher with a dialogue box explaining that the article contained secret contents and as a result would not be available for download, with apologies for the inconvenience. The author, title, and publication information of the censored work was still available, thereby allowing me to identify which work was deemed to be secret at a given point in time, and whether any such research on my topic existed in the first place. Since part of my purpose was to review the extent of Sinophone literature on environment and development issues in Baotou related to the rare earth industry, this occasional dead end proved only to be a minor hindrance.

Censorship works differently in different countries. In the United States, where censorship is largely not discussed, it nevertheless shapes knowledge and action in important ways. Whether people fear lawsuits, or whether it is a function of the generalized paranoia precipitated by post-9/11 mass domestic surveillance practices, self-censorship is a conspicuous feature of engagement with industry and private sector actors in the United States. Because this affective posturing forms part of the Anglophone common sense concerning rare earth elements in the United States, it was an instructive practice to encounter and observe. By contrast, in Brazil I encountered a refreshingly open practice with respect to information concerning the issues researched herein. I was invited into back offices of government bureaus, where civil servants literally opened their files for me to peruse; granted multiday tours of industrial sites; and was lent closed archival materials to take back to my hotel to peruse on my own time after hours.

These differences are accounted for by different political and cultural attitudes concerning the power of information. In China, information is explicitly controlled as a matter of state policy. Therefore, in a certain sense, one knows where one stands, although under such conditions ground-level functionaries can sometimes adopt an overzealous approach when dealing with foreign scholars just to err on the side of caution. In the United States, political discourse and culture has a deep affinity with paranoid rhetoric (Hofstadter 2012). Although this has its roots in the "red scares" of the Cold War era, the habit was manifest in the way in which rare earths were discursively cast during and after the 2010 crisis. This affect was conspicuous among industry actors, certain elected representatives, and media commentators. In Brazil, freedom of information is considered a basic citizenship right and a pillar of national security in the post-dictatorship era that was further codified in a 2011 federal law. This reflects a markedly democratic notion of national security, oriented toward preventing domestic abuses of power instead of against perceived foreign military threats. There is an ongoing public education campaign to remind private citizens and public servants of this entitlement, manifest in brightly colored posters declaring "Information Is YOUR Right!" in the lobbies and reception areas of government buildings. I found this ethos to be productively manifest in the process of my research.

Notes

INTRODUCTION

1. For 2014. As compared to iron ore (1,610 million tonnes in 2013), gold (3,923 tonnes in 2014), or potash (56 million tonnes), for example (Basson 2014; PotashCorp 2014; George 2016).

2. Whether the events of 2010 constituted a "crisis" is a matter of considerable debate. The "crisis," real or not, has become parlance for the price increases and attendant geopolitical and economic uncertainty that followed. That the crisis was not what many thought it to be, that China's central government maintained plausible deniability throughout, and that commodity flows may have been only minimally disrupted will be thoroughly examined in chapter 4 and therefore are not points that need to be belabored by using "crisis" in quotation marks hereafter.

3. According to the USGS: "A mineral deposit is a mineral occurrence of sufficient size and grade that might, under the most favorable circumstances, be considered to have economic potential (Cox 1986). Deposits sharing a relatively wide variety and large number of attributes are characterized as a 'type,' and a model representing that type can be developed" (quoted in Berger, Singer, and Orris 2009, 2).

4. Fortunately, the discourse is much more careful and nuanced in academic journals and edited volumes. For example, see Wübbeke 2013 for an analysis of China's rare earth policy narratives. The collection edited by Ryan David Kiggins (2015), titled *The Political Economy of Rare Earths: Rising Powers and Technological Change,* is the first compilation of social science analyses of the contemporary rare earth issue. The chapters in this volume decisively refute the hypothesis that dependence on China constitutes an existential threat to the United States or other countries dependent on China's rare earths.

5. Geographers tend to understand these human-environment relations as taking on three spatial forms: real, perceived, and lived. *Real* space refers to physical space, as in the land and environment, and their physical properties. *Perceived* space refers to people's ideas, cultural norms, and abstractions about space, as in ideas that consider some places more important, dangerous, or strategic than others, for any reason. *Lived* space refers to the actual practice of people and institutions carrying out their daily activities shaped by their physical environments and their ideas about their environments. In practice, lived space combines real and perceived space to produce the world as we know and experience it (Lefebvre 1991).

6. Such inducements are known as a pro-rata nonrenounceable entitlement offer.

7. For a discussion of how artisanal mining practices differ and interact with state-backed large-scale commercial mining enterprises, see Hecht and Cockburn 1990; Hinton, Viega, and Beinhoff 2003; and Lahiri-Dutt and MacIntyre 2011.

8. Senior vice president for investor relations of an anonymous rare earth mining and processing firm, interview by author, June 2014.

9. As discussed in chapter 4, illegal rare earth production and export in China is estimated to exceed official exports by up to 40 percent.

10. See, e.g., Fulp 2012.

11. This has important similarities with obstacles to deep seabed extraction, although deep seabed extraction is demonstrably more feasible than lunar mining.

12. Transportation costs are likewise a barrier to production in São Gabriel da Cachoeira.

13. "If you're going to have a Sputnik moment, how about mining an asteroid for natural resources? . . . There's more natural resources on asteroids than have ever been mined in the history of the earth. So I would say we have to turn space into our backyard" (Tyson 2015).

14. Many of whom came into this wealth by developing and innovating rare earth–dependent technologies.

15. K = potassium, REE = rare earth elements, and P = phosphorus.

16. Further incentivized by career pressures for local officials to increase the local GDP.

17. Closing statements at military geology forum, Manaus, April 26, 2014.

18. China Academy of Sciences delegation to Brazil, interview by author, March 2013.

19. China Academy of Sciences research faculty, interview by author, August 2013.

20. See, e.g., Grasso 2013; Information Office 2012.

21. As discussed in chapters 4 and 5.

22. Such approaches characterize the positivist, Eurocentric approach and all of its pitfalls, which have been roundly and rightly critiqued as a misguided way to study histories of development (McMichael 1990).

1. WHAT ARE RARE EARTH ELEMENTS?

1. At the time, metallic oxides were referred to as earths. For example, magnesia was known as "bitter earths," zirconia as "zirconium earths," and beryllia as "beryllium earths" (Greinacher 1981).

2. "Until 1885, though by that time the scientific interest of the group had been fully demonstrated by the discovery of several new elements, it was supposed that the minerals were almost entirely confined to a few scattered localities in Scandinavia and the Ural mountains. In that year Dr. Auer von Welsbach announced his application of the rare earths to the manufacture of incandescent mantles. Immediately there was a great demand for raw material for the preparation of thoria and ceria. The agents of the Welsbach Company visited all the important mining centers of Europe and America, intent on a search which shortly made it clear that the metals of so-called 'rare earths' are really quite widely distributed in nature" (Levy 1915, 2).

3. Promethium is occasionally excluded from the rare earth group because it is a synthetic radioactive element produced during nuclear fission and is found, on Earth, only in spent nuclear fuel. It is also found in the center of certain stars in the Andromeda galaxy (Cardarelli 2008; Jørgensen 1990).

4. Periodized by Greinacher (1981) as lasting from 1891, when Auer von Welsbach was awarded his patent, to 1930, when the properties of rare earth elements began to be used more widely, but before the launch of various atomic research programs during which the properties of rare earth elements were more systematically discovered.

5. The other 99 percent was thorium.

6. "The development of new uses for the rare earths has been discouraged because the supply has been considered limited and the price, even of foreign ore, has been unstable" (Congress 1952, 23).

7. Pyrophoric: liable to ignite spontaneously upon exposure to air. Google analytics show that use of the word peaked mid-twentieth century, suggesting that Evans et al. (2002) were not alone in watching their lab equipment go up in smoke.

8. Or 800 pounds of neodymium and 130 pounds of dysprosium.

9. Or two tons.

10. MRI machines use over 680 kilograms of magnets each (Molycorp 2012).

11. A few efforts to catalog rare earth applications; environmental, social, and economic impacts; and sites of extraction, enclosure, and pollution are under way. Most notable among these is *The Rare Earth Catalog: Tools for Reckoning with the Anthropocene* currently being organized by Elizabeth Knafo and Jesse Goldstein.

12. "When thorium 232 captures a slow neutron, it converts to thorium 233. The thorium then disintegrates quickly into protactinium 233, which then decomposes, but more slowly, into uranium 233. Uranium 233 is fissionable by slow neutrons and thus potentially a material for sustaining a chain reaction. Thorium, like uranium, occurs widely in the earth's crust, but similarly not often in sufficient concentration to provide economically workable deposits. Before World War II, it was most commonly used in the manufacture of gas mantles" (Jones 1985, 292n1).

13. Brazil-Germany relations during the 1930s suggested that Brazil would support Germany in the event of war. President/dictator Getulio Vargas (1930–45, 1951–54) reportedly enjoyed Adolf Hitler's company and was sympathetic to Nazism/fascism in the 1930s. Germany was Brazil's second largest trading partner until 1940 (Penteado 2006).

14. That is, capturing and enslaving local inhabitants (Helmreich 2014). Headrick (1978), Dumett (1985), Israel (1987), Von Eschen (1997), and Harrison (1998), inter alia, demonstrate how winning the war and the nuclear arms race depended on the exploitation of the colonized world, although the lives lost through forced labor regimes are seldom included in tallies of World War II casualties.

15. These records, based on US military archives, do not cohere with the findings reported by Adam Hochschild (1999) in *King Leopold's Ghost: A Story of Greed, Terror, and Heroism in Colonial Africa*: "With the start of the Second World War, the legal maximum for forced labor in the Congo was increased to 120 days per man per year. More than 80 percent of the uranium in the Hiroshima and Nagasaki bombs came from the heavily guarded Congo mine of Shinkolobwe" (279).

16. "After World War II, owing to the recovery of lanthanide elements in fission products during the reprocessing of spent nuclear waste, the separation of rare earths was greatly improved, and this led to the large commercial-scale solvent-extraction process now widely used to recover lanthanides for industrial applications" (Cardarelli 2008, 423).

17. The chemical and conceptual symbiosis drove advances in rare earth and nuclear research on opposite sides of the globe through the mid-twentieth century. Frank Spedding's discovery of ion exchange for rare earth separation proved crucial to isolating uranium in the 1940s. Xu Guangxian's work on isolating uranium was critical for his discovery of the cascade theory of countercurrent extraction, which revolutionized rare earth production and greatly increased the global rare earth supply in the 1970s.

18. The environmental and epidemiological effects are examined in-depth in chapter 3, with a detailed analysis of Baotou, Bayan Obo, and vicinity.

19. "We have got to take control of our energy future and we can't let that energy industry take root in some other country because they were allowed to break the rules" (Obama quoted in Chapple 2012).

20. Chairman of the Ministry of Land and Resources for Baotou Municipality, interview by author, September 2013.

21. Representative of International Economic Engagement of the Ministry of Foreign Affairs of the People's Republic of China, interview by author, September 2013.

22. Mr. Chen (Chinese Society of Rare Earths), interview by author, September 2013.

23. For a discussion of this dynamic in US, Japanese, and Chinese literatures, see Bruce, Hietbrink, and DuBois 1963; Hirano and Suzuki 1996; and Li, Yang, and Jiang 2012, respectively.

24. A minable concentration is generally defined as a percentage "in the low single-digits"; a spatially defined area of concentrated minerals is called an "occurrence." If it is minable, then it is called a "deposit," "ore deposit," or "mineral deposit" (Zepf 2013).

25. On the black market, elements are immersed in crude acid baths to partially separate them and are then sold to downstream refining facilities via independent traders (Bradsher 2010). This, in part, explains why there is such an extensive delay between discovery and production, and why production is so environmentally devastating.

26. For the sake of simplicity, I am talking about the formation of a bastnäsite Iron-REE-Th deposit here, such as those found in Bayan Obo, Seis Lagos, and Mountain Pass.

27. The field of lunar geology is called selenology.

28. KREEP: K=Potassium, REE=Rare Earth Elements, and P=Phosphorus.

29. Foweraker (1981) has observed that "marginal" and "frontier" tend to overlap.

30. This is meant in the Gramscian sense in that "the moment of hegemony" is revealed when the dominant bloc "also pos[es] the questions around which the struggle rages" (Gramsci, Hoare, and Nowell-Smith 1971, 182 quoted in Goldman 2005, 7).

2. PLACING CHINA IN THE WORLD HISTORY OF DISCOVERY, PRODUCTION, AND USE

1. "Self-sufficiency is a very *Chinese* term. The US committed to an efficient, global system of free trade in order to keep prices low for American companies" (United States government representative, executive branch, interview by author January 2014).

2. In Baotou and Bayan Obo, the legacies of these efforts are present in the rather surprising abundance of protestant churches (Liu 2009). These churches survived the purges of foreign influence in the early years of the PRC by aligning themselves with nationalist campaigns. They were particularly active in the "Three Self-Sufficiencies Patriotic Campaign" of the late 1950s and early 1960s, which consisted of "Self-Governance, Self-Teaching, and Self-Support" (Wang 2010, 15).

3. Reports from the first year of meetings recount several instances of Japanese, American, Russian, and Chinese researchers literally opening their field notebooks for one another (Liu 2009).

4. Also referred to as the Nationalists or the Republicans.

5. A euphemism for purging communists and other "undesirables," which culminated in the Shanghai Massacre of 1927. See Stranahan 1998 and Wakeman 1995, inter alia.

6. This was significant: it included tons of ammonium nitrate (for explosives), heavy artillery, airplanes, and submarines.

7. The production of both of these elements is currently concentrated in China. As of 2013, China produced 80 percent of the global supply of antimony, and was aggressively buying up (and then closing) foreign firms. China currently produces about 85 percent of the global supply of Tungsten (Bromby 2013) while global shortages in the early 2000s have driven the United States and Russia to sell off domestic stockpiles. Both of these elements, like rare earths, are considered critical materials. The fixation on rare earth elements has perhaps hindered a more holistic approach to strategic questions concerning the geography of critical materials production and consumption.

8. Also Romanized as Ho Tzao-lin.

9. Also known in the Sinicized form as Teh Wang.

10. Also Romanized as General Fu Tso-yi.

11. Called Mongokuo, modeled on Japanese advice after Manchukuo in northeastern China. Japan claimed that their imperial forces had nothing to do with this, even stating that the land was "too barren" for them to be interested in incorporating it. This conflicts with reports from missionaries at the time, which corroborated Chinese reports that Mongolian forces were under Japanese command.

12. "The reports said the new nation carved out of northern Chahar province included and area roughly the size of the state of Ohio. Bounded on the north by outer Mongolia, one the east by Jehol province, and on the west by the strongly fortified Chinese province of Suiyuan, its southern border was said to have been placed along the Great Wall, extending at one point with 20 miles north of Kalgan" (Gandhi, Mu, and Honrath 2013).

13. In his remarks on March 5, 1949, Mao emphasized the Suiyuan style of revolutionary victory "as a bloodless method of struggle, but that is not to say it isn't struggle . . . that captures a portion of the Nationalist army and strives for them to stand up on our side of politics" (Bai 1999, i).

14. Also known as the Yalta Agreement.

15. This despite the fact that Article 5 of the previous treaty, the "Agreement on General Principles for the Settlement of Questions between the Republic of China and the USSR" signed on May 31, 1924, contained the following language: "The Government of the USSR recognizes that Outer Mongolia is an integral part of the Republic of China and respects China's sovereignty therein" (Republic of China and USSR 1924). In December 1924, Georgy Chicherin, the commissar for foreign affairs of the USSR, issued the following statement, which was taken to define the "status quo" as used in the Yalta Agreements: "We recognize the Mongolia People's Republic as part of the Chinese Republic, but we recognize also its autonomy in so far-reaching a sense that we regard it not only as independent of China in its internal affairs, but also as capable of pursuing its foreign policy independently" (Associated Press 1937a).

16. In northwestern China as in northwestern Brazil these histories have been extensively studied. See, for example, Figueireido 1967; Bulag 2002; and Tyner 2012.

17. See, e.g., Hsü 1982.

18. These are the Pinceance and Uinta Basins in the United States; the Orinoco Belt in Venezuela, and the Alberta Oil Sands in Canada.

19. IMAR People's Standing Committee Ethnic Affairs Representative, interview by author, April 2013.

20. Or politics and economy, or policy and markets, or power and technology (Clausewitz 1976; Guo 2013; Lauren, Craig, and George 2007). Rich literatures are devoted to defining these couplets and the relationship within them. "War and industry" is decidedly more antiquated than the others; I use it deliberately in order to capture the particular form of the political and economic imperatives that sought to render the southern Mongolian Steppe into a red hinterland for the People's Republic of China and the Soviet Union (Hurst 2010; Jia and Di 2009). For an excellent discussion of the relationship between war and the state, see *Treatise on Nomadology—The War Machine* (Deleuze and Guattari 1987).

21. "There is more to the picture than semiotic systems waging war on one another armed only with their own weapons. *Very specific assemblages of power impose significance and subjectification* as their determinate form of expression, in reciprocal presupposition with new contents: there is no significance without despotic assemblage, no subjectification without an authoritarian assemblage, and no mixture between the two without assemblages of power that act through signifiers and act upon souls and subjects. It is these assemblages, these despotic or authoritarian formations, that give the new semiotic system the means of its imperialism, in other words, the means both to crush the other semiotics and protect itself against any threat from the outside" (Deleuze and Guattari 1987, 418–420; emphasis added).

22. These campaigns have been studied and well documented elsewhere (Chan, Madsen, and Unger 1992; Eberstadt 1980; Hinton 1966; Lin 2990; Pye 1999; Schwartz 1968; Shapiro 2001; Terrill 1999).

23. In Bulletin 16 of the Cold War International History Project, Zhou Enlai reportedly explained to Anastas Mikoyan in February 1949: "We do not have contacts with the

Xinjiang democratic groups. Our former people there were arrested by [one time governor of Xinjiang] Sheng Shicai. Now we are sending a small group of comrades there." As quoted in the "Memorandum of Conversation between Anastas Mikoyan and Zhou Enlai, February 1, 1949 (Evening)" (quoted in Kraus 2010 7n.).

24. Composed of the intelligence organizations of the US Departments of State, the Army, the Navy, the Air Force, and the Joint Chiefs of Staff.

25. See, for example, Xu 1996, 1998; Shih 1998; and Brook and Luong 1999.

26. For further discussion of this, see Bulag 2002 and Klinger 2017a.

27. However, it was not until 2014 that women were deployed to the front lines of national security in Inner Mongolia (Zou 2014).

28. In the popular imagination, women from these places are thought to be exceptionally beautiful.

29. Mr. Li (PLA veteran on the northwestern front), interview by author, April 2013.

30. Representatives of Baotou Municipal Women's Committee, interviews by author, April 2013.

31. Prerevolution name for Baotou, assigned during the Qing dynasty. See Tighe 2005 for an excellent history.

32. Representative of Baotou municipal family planning bureau, interview by author, April 2013.

33. The large circular tents are typical of Mongolian nomads and celebrated as symbols of life, home, and wellness. Natural features that resemble yurts are attributed sacred status in Mongolian spirituality.

34. Traditional Mongolian medicine practitioner in Bayan Obo, interview by author, September 2013.

35. Communist Party youth volunteer Ms. Hu, escorted survey and interviews in Bayan Obo and vicinity, September 2013.

36. Best among the exceptions are Wu 1965; Clark 1973; and Hogan 1999.

37. "During the period 1950–1954 the Chinese Communists, with some Soviet aid, explored a number of areas for uranium resources. In 1955 this quest for uranium, as well as the supporting Soviet aid, was intensified" (Joint Atomic Energy Intelligence Committee 1960, 2).

38. See, e.g., Li et al. 1987.

39. See, for example, Di 1976 and Joint Atomic Energy Intelligence Committee 1960.

40. This is the basis of the solvent extraction methods used today, which utilize the slight variations in the solubility of rare earth compounds between two liquids that do not dissolve into each other (the same principle as oil and water). Countercurrent cascades carry out many extraction steps in a continuous stream that progressively increases the degree of separation until the substance approaches purity (Gray 2009).

3. "WELCOME TO THE HOMETOWN OF RARE EARTHS"

1. The terms globalization and global capitalism are often used interchangeably. This is due to the hegemony of neoliberal ideologies and practices that are understood to be behind both processes.

2. Heterogeneous economic spaces are not peculiar to China, but rather are also found in other "epicenters" of neoliberalism. See, for example, White and Williams 2012.

3. The messiness of the state's involvement in the rare earth sector in the United States, Brazil, and China is a case in point.

4. As Foucault puts it: "The market . . . can only appear if it is produced, and if it is produced by an active governmentality. There will thus be a sort of complete superimposition of market mechanisms, indexed to competition, and governmental policy. *Government must accompany the market economy, from start to finish*" (2010, 121; emphasis added).

5. For an important exception to this analysis, particularly as it pertains to the post-2008 US context, see Martinez 2009.

6. See, e.g., Peck and Ticknell 2002. This paradoxical but nevertheless widespread view conflates the totalizing force and pretensions of neoliberalism with the totality of human life, with the result that even critical appraisals end up "doing the work" of maintaining neoliberal hegemony by discounting practices and spaces that neoliberalism has failed to penetrate (Gibson-Graham 1996).

7. See, e.g., De Angelis 2001.

8. The territorial processes described in the previous chapter could be reread through this framework.

9. See, e.g., Tsing 2005; Scott 2009.

10. This could be read as an example of the pitfalls of environmental concerns framed by what Agnew (1994, 2010) terms the "territorial trap" based on three obsolete geographical assumptions: (1) states are fixed units of sovereign space that are (2) defined by the polarities of domestic versus foreign and internal order versus external anarchy, and (3) function as "containers" of societies organized according to a coherent set of interests.

11. National leadership now views regional environmental issues as commercial and security threats, and addressing them as serving broader political and economic objectives. Consequently, environmental concerns rose to prominence in international discourse in the last five years. The critique that state interventions in China's rare earth sector are driven by more than environmental concerns is valid; this does not negate, however, the severity of environmental and epidemiological harms resulting from rare earth mining and processing in Baotou and Bayan Obo.

12. These first three comprise the "three wastes" mentioned in Chinese scholarship.

13. Because its melting point is second only to tungsten and tantalum carbide, thorium is used in high pressure applications such as petroleum cracking, welding electrodes, carbon-arc lamps, and high temperature laboratory crucibles for melting refractory metals (Cardarelli 2008, 451).

14. Shortly after Marie Curie's breakthroughs at the turn of the twentieth century, drug and cosmetic manufacturers added thorium to everything from toothpaste to laxatives under the assumption that something so energetic as radioactive thorium had to be beneficial (O'Carroll 2011). The mania ended a few decades later following the grotesque deaths of prominent advocates of radioactive tonics (Rowland 1994).

15. For reasons of space, lead is omitted from this discussion. See Klinger (2015) for further reference.

16. Unlike other radioactive elements, radon is gaseous and easily inhaled. It is odorless, colorless, and tasteless. As such, it is generally responsible for the majority of public exposure to ionizing radiation (EPA 1990).

17. The Environmental Monitoring Station measured U-238, Ra-226, Th-232, and K-40.

18. Village resident, male aged fifty-four, author interview, September 2013.

19. Village leader, male aged forty-six, author interview, September 2013.

20. Rural pastoralist, male aged sixty-one, author interview, September 2013.

21. Anonymous local official in Bayan Obo Mining District. Interview by author, September 2013.

4. RUDE AWAKENINGS

1. The KMT is pro-China and prointegration, so they are more likely to echo Beijing's utterances about the island. The Democratic Progressive Party is proindependence and tends to remain silent on the issue in the interest of maintaining smooth relations with what they view as their more developed neighbor (Kao 2014).

2. Anonymous official in Port City A, interview by author, September 2013.

3. Anonymous port workers in Port City A, interviews by author, September 2013.

4. Newspapers as well as interviewees expressed conflicting dates; all fall between October 28 and November 24, 2010.

5. The Middle East has Oil, China has rare earths.

6. For more on this, see chapter 3.

7. For multiple analyses on the relationship between the raw materials WTO suits against China and industrial capacity in developed countries, see Meléndez-Ortiz, Bellmann, and Mendoza 2012; Blanchard and Wei 2013; Fratianni, Savona, and Kirton 2013; and Laïdi 2014.

8. As Chen (2011) has noted, the toxic hazards of production are racially and sexually coded within a global geopolitics of sovereign fear over foreign toxins hailing from distant production frontiers.

9. See, for example, Orris (2002) and Peters (2007).

10. US and international press reported US$1 trillion in 2010; in 2011 and 2012 Afghan president Hamid Karzai claimed that the assets were worth US$3 trillion. In a 2013 meeting with potential Indian investors, Karzai reportedly stated: "Actually it's $30 trillion. The US knocked a zero off to keep our assets a secret" (Mehrotra 2013).

11. See, e.g., Abdullah, Chmyriov, and Dronov 1980.

12. Anonymous US-based economic geologist and mining consultant, interview by author, December 2014.

13. US federal rare earth researcher and lobbyist, interview by author, January 2014.

14. The Molybdenum Corporation of America, or Molycorp operated the mine at Mountain Pass from 1952 to 1977 when it was purchased by Unocal. It belonged to Unocal when it closed in 1999. Chevron acquired it in 2005 and sold the mine back to Molycorp in 2008.

15. These include various forms of tungsten, zinc, tin, niobium, tantalum, vanadium, antimony, phosphorus, pig iron, iron alloys, copper, copper alloys, nickel, and aluminum.

16. Bauxite, coke, fluorspar, magnesium, manganese, silicon carbide, silicon metal, yellow phosphorus, and zinc.

17. CBMM was established in 1955 following the discovery of niobium-bearing pyrochlore in Minas Gerais. Niobium is a soft, ductile metal used to make iron and steel super alloys, which are lighter, stronger, and require less base metal compared to other alloys. The niobium deposit mined in Araxá was formed by an alkaline magmatic intrusion referred to as chimney. These are carbonatite formations similar to those that formed the deposit at Bayan Obo, in which columns of magma pressed against the terrestrial crust in repeated cycles of heating and cooling over millions of years. They are essentially volcanoes that never quite happened. Just as the rare earth mine at Bayan Obo is rich in niobium, the niobium mine in Araxá is rich in rare earth elements. See Klinger 2015a.

18. CBMM's niobium-based technologies are used in nearly every jet engine, automobile body, hybrid fuel cell battery, suspension bridge, and superconductor produced in the last three decades; China, the largest producer and consumer of steel in the world, is entirely dependent on CBMM for its niobium supplies. See Klinger 2015b.

19. Anonymous personnel at CBMM in Brazil, interview by author, March 2014.

20. CBMM earned an ISO 14001 Environmental Management System Certification in 1997 and an OHSA 18001 Health and Safety Management System Certification in 2002.

21. Demonstrated by, among other things, the purchase of a 15 percent retainer of all of CBMM's known deposits in 2011 (Tudor 2011).

5. FROM THE HEARTLAND TO THE HEAD OF THE DOG

1. This is a historically loaded term. "Brazil, Country of the Future" was a phrase coined in 1941 during the presidency of Getúlio Vargas to indicate the potential of the country to become a major world power. Since the end of the dictatorship in 1985, the phrase has been used sardonically to criticize the multiple failed national development campaigns. Rare earths, therefore, are presented as the vehicle to finally deliver the promise of Brazil's unfulfilled greatness.

2. CBMM has yet to publish price or production data for its rare earth products, and reportedly has not permitted major investors from Korea and China to conduct technical due diligence, due to the confidentiality surrounding its separation processes. This suggests that it was unable to produce rare earths at a price that could compete with Molycorp (2015), which reported losses of US$67.2 million for 2013 and US$99.6 million for 2014. Molycorp filed for bankruptcy in 2015.

3. As discussed in this chapter, categorical opposition to mining comes primarily from international advocacy groups and their cosmopolitan counterparts in Brazil.

4. Missionaries, soldiers, federal police officers on temporary assignment, rotating air traffic controllers, and Manaus-based traders, geologists, and politicians used this term. Local Indigenous people and other long-time residents did not.

5. A note on usage: Indigenous interlocutors referred to themselves as *Indios*, which translates as "Indian," and used this term interchangeably with *povos indígenas*, which translates as "Indigenous peoples." This chapter uses both terms.

6. *Garimpeiros*, or small-scale, artisanal, illegal, or clandestine miners, are much more visible in the Brazil case relative to the China case for two primary reasons. First, garimpeiros are politically organized and active in Brazil. Several garimpeiros I interviewed had cultivated allies in state and federal offices; interviewees in government office also referred me to garimpeiros because in their judgment, the garimpeiro perspective needed greater exposure. Second, and by contrast, small-scale or clandestine mining activity was subject to public condemnation and closure campaigns in China. Several officials interviewed in China expressed the sentiment that small-scale mining was irrational and evidence of backwardness. Furthermore, because Baotou and Bayan Obo are securitized areas, there was heightened sensitivity to speaking to foreigners. Even local officials actively supporting the clandestine mining activity stated that there was no possible way for a Westerner to speak to small-scale miners without consequences. Since the consequences would be most acutely felt by those in an already precarious position—the worst I would suffer would perhaps be interrogation and an order to leave—I judged it unethical to pursue this line of inquiry further in China.

7. This event was organized around my visit. A state geologist whom I had interviewed informed the commanding general of CMA that a *gringa* China expert was visiting the region. And furthermore, this gringa was *not* an anthropologist, which was taken to be a positive point by those typically criticized in anthropological enquiry in Amazonian regions. The general invited me to address CMA on the subject of rare earths and to offer my analysis on the environmental security impacts of mining as I had observed them in different parts of the world. Recorded with permission. All translations by author.

8. Fundação Nacional do Indio, the federal organ responsible for defending the rights and interests of Indigenous people.

9. The National Truth Commission determined that at least 8,350 Indigenous people were exterminated as a matter of state policy during the military dictatorship (1964–85), with an incalculably greater number affected, disappeared, or unaccounted for (Ricardo et al. 2014).

10. This exclusion of mineral extraction from "permitted" extractive activity is vividly illustrated in the Integrated Conservation and Development Projects in the Amazon in the

1980s and 1990s. For a description of the failures of extractive reserves in the Amazon, which were intended to achieved conservation and development objectives through the sustainable extraction of nontimber forest products, see Dove 2006.

11. For an examination of the concept of the "ecologically noble savage," see Silva 2012.

12. Geraldo (retired garimpeiro), interview by author, April 2014.

13. To my knowledge, there is no rare earth separation in São Gabriel. It is carried out by downstream buyers in Manaus and Colombia. Local activity consists of extraction and transport.

14. Mr. Santos (Indigenous activist and local government staff, São Gabriel da Cachoeira), interview by author, April 2014. Mr. Santos is a pseudonym.

15. Geologist in Manaus, interview by author, April 2014.

16. FOIRN representative and founding member, interview by author, April 2014.

17. Mr. Domingos (FUNAI staff, São Gabriel da Cachoeira), interview by author, April 2014; FOIRN representative and founding member, interview by author, April 2014.

18. "A atividade visaria lucro, mas não do ponto de vista do mundo capitalista e sim no patamar da sustentabilidade, com uso de técnicas artesanais e concepções indígenas e levando em conta a relação com a natureza. Não queremos a presença de grandes empresas e grandes corporações fazendo o trabalho."

19. On the same days in which I was conducting interviews with Indigenous people in Cabeça do Cachorro, hearing heartfelt accounts such as Mr. Santos' as well as sophisticated legal arguments for the statutory protection for Indigenous mining operations on Indigenous lands, Davi was visiting San Francisco, California, to raise awareness and funds to pressure the Brazilian government to maintain the mining moratorium on Indigenous lands.

20. Mr. Domingos (FUNAI staff, São Gabriel da Cachoeira), interview by author, April 2014.

21. Three anonymous Indigenous veterans of land demarcation struggles, group interview by author, April 2014.

22. Located downstream from São Gabriel da Cachoeira toward Manaus.

23. However, there is no record of any such invasion and the original fort is now in ruins. In the Cabanagem rebellion of 1835–40, it was a political prison for captured rebels (Silva 2012), and during the Federalist Revolution in 1893 in the southernmost state of Rio Grande do Sul, the President Marechal Floriano Peixoto banished leading revolutionaries to Marabitanas (Oliveira 1968; Reis 1942). Because of the enduring concern among the Brazilian military that the Orinoco connection to the Caribbean Atlantic poses a national security threat (Brito 2013), it is now the site of the Fifth Rio Negro Frontier Command and the Fifth Special Border Platoon under the CMA.

24. Two US firms, the Chicle Development Company and the Rubber Development Corporation, established operations in São Gabriel da Cachoeira with the intention of taking over extensive rubber tapping networks to supply the US rubber market via the Rio Negro-Casiquiare-Orinoco waterway. Despite massive investment and resettlement programs, no amount of US public funds could overcome the deleterious effect of South American leaf blight that had driven rubber production to Southeast Asia decades ago, nor were they able to disrupt the established networks of local rubber barons. The extractivist initiative failed at an enormously high cost: in June 1945 an audit of the Brazilian office of the Rubber Development Corporation (RDC) found that costs exceeded income by US$9.1 million during the war, while Brazil's Federal Congress estimated that between seventeen thousand and twenty thousand rubber tappers resettled to work in RDC's groves remained unaccounted for (Dean 1987, 104).

25. The final recommendations stated that "this report and maps be published and distributed to . . . interested officials and agencies of the several participating governments,

for their information and study [and that] copies of the report and maps be placed in the hands of all officials and agencies resident in and adjacent to the region, so that the maps and physical data may be available immediately to residents and students of the region" (US Army Corps of Engineers 1943, 297).

26. RADAM stands for Radar of the Amazon (Radar da Amazônia). It was later expanded to cover all of Brazilian territory and some of neighboring countries. The comprehensive survey is known as RADAM-BRASIL.

27. As characterized by Braudel 1967; Mumford 1934, 1967; and Merchant 1980, inter alia.

28. As characterized by Graulau 2003 and Lahiri-Dutt 2011.

29. Unless otherwise cited, the events described in the next three paragraphs are documented Wright's (2005) ethnographic account of the region, and were repeated to me by FOIRN leadership in 2014.

30. Mr. Domingos (FUNAI staff, São Gabriel da Cachoeira), interview by author, April 2014.

31. FUNAI had five presidents during that contentious year: Nelson Marabuto (September 1984–April 1985), Ayrton Carneiro de Almeida (April 1985), Gerson da Silva Alves (April 1985–September 1985), Álvaro Villas Boas (September 1985–November 1985), and Apoena Meirelles (November 1985–May 1986).

32. An expression akin to "play hot potato" or "pass the buck" in the sense that two or more parties attempt to buy time and avoid taking responsibility for an issue by claiming that responsibility lies with the other.

33. "I don't work with *security*. I support *sovereignty* through the construction of infrastructure. In the amazon there is no private sector to build the things that are needed. So we build them to help develop the region." PCN Director Brigadier Dantas, Ministry of Defense in Brasília, interview by author, March 2014.

34. Many of these were documented in the seven-thousand-page report submitted by public prosecutor Jader de Figuereido Correa in 1967. It is now in the possession of the National Truth Commission. At the time of this writing, it was not available to archival researchers.

35. It became formalized into a federal program in 1996.

36. At the time the US Drug Enforcement Agency documented much higher incidences of coca and marijuana plantations in Acre and just outside of Manaus. (Retired US intelligence official, interview by author, May 2014.) This makes sense for market and logistical reasons. Given this, it is unclear what was being "tested" in the upper reaches of the Rio Negro.

37. This is not surprising, given the reliance of FARC on supply shipments from São Gabriel da Cachoeira (Brasil 2003), the extensive economic ties between Indigenous groups on both sides of the border, and the sophisticated trading relations between Indigenous miners and cocaine producers seeking alternative routes for their goods. The relations are deeper than simply economic, however. Certain Indigenous groups maintain kin relations across the borders. Lack of health services in São Gabriel, aggravated by embezzlement of funds intended to provide services through the Ministry of Heath, provoked a health-related exodus of Indigenous people seeking medical care in Colombia. As part of the fight against FARC, the Colombian Ministry of Sanitation stopped delivering medical supplies and fuel to these frontier clinics in 2001 (Ricardo and Ricardo 2006).

38. With the 2016 Peace Agreement and the demobilization of FARC, it remains to be seen how ongoing militarization of the border will be justified.

39. Four firms supply the technology: Raytheon from the United States, MacDonald Dettwiler from Canada, and the Brazilian firms ATECH and Embraer (Perlo-Freeman 2004). The US Export-Import Bank provided 97 percent of the financing at 8.5 percent interest (Guzmán 2013; L. Martin n.d.).

40. Created via Decree Law 12.836 of 1990.

41. Mining operations at Mountain Pass, for example, were built around rare earth concentrations between 2 percent and 3 percent (Olson et al. 1954).

42. US$99.5 million to US$295 million.

43. In 2003, the primary orientation of the Projeto Calha Norte shifted from sovereignty to developmentalism. The development axis now receives the majority of budget; in March 2014 the head of finances for PCN, Brigadier Dias, reported 1,800 separate development projects, including the construction of six universities in the neighboring province of Roraima. Although PCN does not have an explicit mandate related to natural resource extraction, "the best result of the program is the rapidity with which we are able to execute logistical projects because of our military approach. This will ensure the regularization of economic activity on our frontier."

44. "According to the Amazon Military Command, for two months in mid-2004 the FARC planned assaults on Brazilian Army positions along the 1,600-kilometer border between the two countries. The guerrilla group sought to obtain weapons, ammunition, food, and medical supplies. The military in Querari, a platoon on what was considered the most tense border was reinforced by forty men trained in jungle warfare. It was the last potentially aggressive movement of the guerrillas recorded by the Brazilian military intelligence network in the area of Brazil. At that time, satellite photos from the Ministry of Defense showed the marks of a camp through lower woods in an area customarily used by the guerrillas. Analysts of the Amazon Military Command in Manaus believe that the FARC was gathering a column with about 160 men and women in Jurupari" (Silva 2013, 31–32).

45. Commanding marshal Roberson of federal police post of São Gabriel da Cachoeira, interview by author, April 2014; Antonio, federal police officer, interview by author, April 2014

46. Antonio, federal police officer, interview by author, April 2014.

47. Brigadeiro Dantas, interview by author, March 2014.

6. EXTRAGLOBAL EXTRACTION

1. For example, astronaut Buzz Aldrin, award-winning science fiction author Ursula K. Le Guin, and planetary sciences professor John S. Lewis have written treatises contemplating lunar mining for very different ends.

2. K = potassium, REE = rare earth elements, and P = phosphorus.

3. Second only to orbital space, where satellites are located.

4. Anonymous NewSpace Industries investor, interview by author, August 2012.

5. The principle of the common heritage of all of humanity.

6. Those who take a more moderate position point to the article that vests the "appropriate state party to the treaty" with authority over nongovernmental entities in outer space in order to argue that it is up to national governments to determine which activities are permitted and which are not. The article in question reads: "The activities of nongovernmental entities in outer space, including the Moon and other celestial bodies, shall require authorization and continuing supervision by the appropriate State Party to the Treaty" (UN 1967, Article VI).

7. See, e.g., Grinde and Johansen 1995.

8. "The Moon Treaty was negotiated in the context of the North-South divide marked by the poverty of developing countries that had votes in the UN and the increasing power of multinational corporations to control economic resources. Space advocacy constituencies in the US saw the Moon Treaty as a power grab by poor developing countries to claim space resources through the power of UN bureaucracies that they did not have the technical means to reach on their own" (Beldavs 2013).

9. Space mining entrepreneur, interview by author, November 2015

10. Vice president of space mining firm, interview by author, January 2014.

11. Russian venture capitalist Ilya Glubovich, for instance, reportedly began "chasing down" space startups in Russia in 2013, saying: "I want to give you money. Where are things right now?" (Bort 2014).

12. Global Business Development Manager for iSpace Technologies, lecture delivered at the University New South Wales, Sydney, Australia, November 6, 2015.

13. Planetary Resources intends to mine platinum group metals, principally on asteroids. Shackleton Energy Company plans to mine water and convert it into rocket fuel in the form of hydrogen and oxygen. This would essentially be a lunar gas station. Investor Richard Branson, of Virgin fame, reasons that if humanity is to explore the solar system, it needs to do so independently of Earth's resources. Although these initiatives are very important to the larger picture of the contemporary space race, they are nevertheless very different processes from the proposed rare earth mining on the moon and, as such, are not discussed in this chapter.

14. But first, they must land a rover and traverse a distance of five hundred meters, thereby claiming the Google Lunar XPrize, and then broadcast imagery back to earth in order to "prove the concept" that a company of fifty employees can successfully land on the moon. "Once we accomplish that, then the second or third mission can involve bringing things back from the moon" (Jain quoted in Caminiti 2014).

15. As Moon Express cofounder Bob Richards stated in the San Jose Mercury post: "The goal is to build out a transportation business that we think is profitable by itself for scientific and commercial payloads, but also to really start exploring the moon from an entrepreneurial perspective, which has never been done before. . . . It could be $15 [billion] to $20 billion of infrastructure you'd have to put in place to actually economically liberate that wealth," Richards said. "But those kinds of numbers, although big to a startup like us, are not big to existing mining concerns on Earth. Those are the prices of a typical mine, or even an offshore oil platform. If the resources are there, then the economics are there to liberate them" (quoted in Swift 2012).

16. This is an annual weeklong festival that began as a celebration of radical self-expression and situationist art on Baker Beach in San Francisco in the 1980s and migrated to the Black Rock Desert of Nevada. The ten principles of the gathering are: radical inclusion, gifting, decommodification, radical self-reliance, radical self-expression, communal effort, civic responsibility, leaving no trace, participation, and immediacy (Burningman .org 2015).

17. Anonymous NewSpace Industries investor, interview by author, August 2012.

18. The perforated spheroidal shape described on the previous page was conceived in order to limit the amount of element lost to burn-off upon reentry through Earth's atmosphere to 7–10 percent.

19. See, e.g., Guner 2004.

20. Jointly developed by the Shanghai Aerospace System Engineering Institute and the Beijing Institute of Spacecraft System Engineering, the project began in 2002 and was completed in May 2010 (Xinhua 2008).

21. The array of entities developing China's space program are jointly overseen by CNSA and the state-owned enterprise China Aerospace Corporation, both of which are subordinate to the State Administration for Science, Technology, and Industry for National Defense. The State Council appoints its administrators. The Eighth National People's Congress was established it in its current form in 1993 in order to fulfill the country's mandate to cooperate internationally in scientific research and technological development. CNSA has active agreements with over a dozen countries and several multilateral space exploration organizations.

22. See, e.g., Cheng 2011.

23. For an analysis of this amendment, see Kohler 2015.

24. "Peter Marquez, Vice President of Global Engagement, Planetary Resources, Inc., said, 'Our nation's continued leadership and prosperity in space is enabled by this new law. Planetary Resources is grateful for the leadership shown by Congress in crafting this legislation and for President Obama signing H.R. 2262 into law. We applaud the members of Congress who have led this effort. Marco Rubio (R-FL), Lamar Smith (R-TX), Patty Murray (D-WA), Kevin McCarthy (R-CA), Bill Posey (R-FL) and Derek Kilmer (D-WA) have been unwavering in their support and leadership for the growth of the U.S. economy into the Solar System'" (Planetary Resources 2015).

25. This dynamic has been examined with respect to private defense contractors working in service of US military campaigns in Iraq and Afghanistan (Alexandra, Baker, and Caparini 2008; Gregory 2006; Menkhaus 2003).

CONCLUSION

1. Recall that state-owned enterprises, like other international firms, are driven by profitability concerns.

2. China Academy of Sciences Delegation to Brazil, interview by author, March 2013; China Academy of Sciences public health, ground water, and soil researchers, interview by author, August 2013.

3. What is less than 1 percent composed of? Primarily: rare earth scraps from magnet production, lamp phosphors, and nickel-metal hydride batteries.

4. Indeed, critical social theorists such as Mikhail Bakhtin (1968) and Henri Lefebvre (1991), among others, have rhapsodized the Carnival as containing the seeds for more promising alternative futures.

APPENDIX

1. In one instance in local archives in Inner Mongolia, this entailed the receptionist fabricating an entirely different set of requirements for me to satisfy once he saw that I had complied with all rules posted on the archive's website. These fabricated requirements involved getting letters of permission from a sub-office of a local bureau that—because these rules were made up on the spot—had no experience providing such a service. The put-upon officials in this particular local bureau made a phone call on my behalf, interrogated the receptionist as to why they directed a foreign researcher to their office, and resolved matters.

References

ABC Radio Australia. 2014. "Anti-Lynas Protest in Malaysia as Mining License Renewal Looms." June 23. http://www.radioaustralia.net.au/international/radio/program /asia-pacific/antilynas-protest-in-malaysia-as-mining-licence-renewal-looms /1331426.

Abdullah, S. H., V.M. Chmyriov, and V. I. Dronov. 1980. *Geology and Mineral Resources of Afghanistan Bibliography.* 2 vols. London: British Geological Survey.

Abraham, David S. 2015. *The Elements of Power: Guns, Gadgets, and the Struggle for a Sustainable Future.* New Haven, CT: Yale University Press.

Abraham, Itty. 2011. "Rare Earths: The Cold War in the Annals of Travancore." In *Entangled Geographies: Empire and Technopolitics in the Global Cold War*, edited by Gabrielle Hecht, 101–24. Cambridge, MA: MIT Press.

Acuña, Cristobal de. 1641. *Nuevo Descubrimento del Gran Rio de las Amazonas.* Madrid: Imprenta del Reyno.

Afzali, Ikram. 2014. "Making the Sacrifices in Afghanistan Worthwhile." *The Diplomat.* Accessed December 24, 2014. http://thediplomat.com/2014/12/making-the-sacrifices -in-afghanistan-worthwhile/.

Agence-France Presse. 2010. "Japan's Rare Earth Mineral May Run Out by March." Accessed December 28, 2014. http://phys.org/news/2010-10-japan-rare-earth-minerals -govt.html.

Agence-France Presse. 2015. "Mongol Hangs Himself in China Protest: Group." *China Post.* Accessed April 7, 2015. http://www.msn.com/en-au/news/other/mongol-hangs -himself-in-china-protest-rights-group/ar-AA8CBWP.

Agnew, John. 1994. "The Territorial Trap: The Geographical Assumptions of International Relations Theory." *Review of International Political Economy* 1 (1): 53–80.

Agnew, John. 2010. "Still Trapped in Territory?" *Geopolitics* 15 (4): 779–84.

Alexandra, Andrew, Deane-Peter Baker, and Marina Caparini. 2008. *Private Military and Security Companies: Ethics, Policies and Civil-Military Relations.* London: Routledge.

Ali, Saleem. 2014. "Social and Environmental Impact of the Rare Earth Industries." *Resources* 3 (1): 123–34.

Alliance for Peacebuilding, Afghanistan Watch, Afghan Development Association, British and Irish Agenicies Afghanistan Group, Equal Access, Future Generations Afghanistan, Green Wave, Green Wish for Afghanistan, Global Rights, Global Witness, Heinrich Boll Stiftung, Natural Resource Governance Institute, Open Society Afghanistan, Publish what you pay, Salam Watandar, Sun Development and Environmental Protection Organization, The Liason Office, Transparency International, Transparency International India, Transparency International UK, et al. 2014. "Open Letter." December 2. http://www.globalwitness.org/sites/default /files/library/Letter%20to%20Cameron.pdf

Amah, Edison, Dana Andrews, Ansley Barnard, Christopher Busby, Juliana Carroll, Christopher Ciampi, and Alexander Freeman. 2012. "Defining a Successful Commercial Lunar Mining Program." Paper presented at American Institute of Aerospace and Aeronautics SPACE Conference & Exposition, Pasadena, CA, September 11–13.

Anderson, Benedict. 1982. *Imagined Communities: Reflections on the Origin and Spread of Nationalism.* London: Verso.

Anthony, David. 2010. "China's Stranglehold on World's Rare Earth Supply." *Critical Strategic Minerals,* September 2. http://www.criticalstrategicmetals.com/chinas -stranglehold-on-worlds-rare-earth-supply/.

Archela, Rosely Sampaio, and Edison Archela. 2008. "Síntese Cronológica da Cartografia no Brasil." *Portal de Cartografia das Geociências* 1 (1): 93–110.

Arctic Journal. 2013a. "Kvanefjeld Moving Ahead." December 10. http://arcticjournal.com /oil-minerals/293/kvanefjeld-moving-ahead%60.

Arctic Journal. 2013b. "Rare Earth Minerals Snag Greenland License Approval." September 10. http://arcticjournal.com/oil-minerals/92/rare-earth-minerals-snags -greenland-licence-approval.

Areddy, James T. 2012. "China Coal Sector Has Safety Setback." *Wall Street Journal,* September 3. http://www.wsj.com/articles/SB100008723963904438474045776287223726 80112.

Areddy, James T., David Fickling, and Norihiko Shirouzu. 2010. "China Denies Halting Rare-Earth Exports to Japan." *Wall Street Journal,* September 23. http://online.wsj .com/news/articles/SB10001424052748704062804575509640345070222

Armand, M., and J. M. Tarascon. 2008. "Building Better Batteries." *Nature Geoscience* 451:652–57.

Asher, Michael. 2015. *Rare Earth.* London: Endeavour Press.

Asimov, Isaac. 1982. *Asimov's Biographical Encyclopedia of Science and Technology.* New York: Doubleday.

Associated Press. 1937a. "Missionaries Report Nation Carved from North China by Mongols." *Miami Daily News-Record* (Miami, OK), March 16.

Associated Press. 1937b. "Mongol Tribesmen Carve New Nation out of North China, Called Mongokuo." *Kokomo Tribune* (Indiana), March 16.

Associated Press. 1998. "Contaminated Wastewater: Mining Company Pays Stiff Fine to Settle Charges." *Daily Courier* (Prescott, AZ), July 13.

Associated Press. 2013. "Gold Rush Mine Tailings May Hold Rare Earth Treasure." *San Francisco Chronicle,* July 27.

Atkins, P. W. 1995. *The Periodic Kingdom: A Journey in the Land of the Chemical Elements.* New York: Basic Books.

Autry, Greg. 2011. "Space Policy, Intergenerational Ethics, and the Environment." Paper presented at American Institution of Aeronautics and Astronautics. AIAA SPACE 2011, Long Beach, CA, September 27–29.

Bachman, David. 2007. "Mobilizing for War: China's Limited Ability to Cope with the Soviet Threat." *Issues & Studies* 43 (4): 1–38.

Bagchi, Sutirtha, and Jan Svenjar. 2014. "Does Wealth Inequality Matter for Growth? The Effect of Billionaire Wealth, Income Distribution, and Poverty." Center for Economic Policy Research Discussion Paper no. DP9788, Bonn, Germany.

Bai, G., Z. Yuan, C. Wu, Z. Zhang, and L. Zheng. 1996. *Demonstration on the Geological Features and Genesis of the Bayan Obo Ore Deposit.* Beijing: Geological Publishing House.

Bai, Lina, Sui Wenli, and Lin Zhong. 2004. "The Radiological Impact of the Bayan Obo Rare Earth and Steel Production on the Surrounding Environment." *Rare Earths* 25 (4): 75–77.

Bai, Lina, Licheng Zhang, and Lingxiu Wang. 2001. *Radioactive Environmental Contamination and Prevention Measures in Baotou Rare Earth Production.* Baotou, Inner Mongolia Autonomous Region: Baotou City Radioactivity and Environmental Management Bureau.

Bai, Shewang. 1999. *The 9/19 Suiyuan Uprising.* Baotou Municipality, Inner Mongolia Autonomous Region: Baotou Historical Materials Editorial Committee.

Bakhtian, N. M., A. H. Zorn, and M. P. Maniscalco. 2009. "The Eighth Continent: A Vision for Exploration of the Moon and Beyond." Paper presented at American Institute of Aeronautics and Astronautics Conference and Exposition, Pasadena, CA, September 14–17.

Bakhtin, Mikhail. 1968. *Rabelais and His World*. Cambridge, MA: MIT Press.

Balasubramanian, S. 2011. "Moving the Heaven to Get Some Rare Earth." *Hindu*, June 2. http://www.thehindu.com/todays-paper/tp-features/tp-sci-tech-and-agri/moving -the-heaven-to-get-some-rare-earth/article2069550.ece.

Baletti, Brenda C. 2012. "From Land to Territory: New Geographies of Amazonian Struggle." PhD diss., University of North Carolina, Chapel Hill.

Baltz, Matthew. 2013. "The Limits of State Capacity: US 'Industrial Policy' in the Rare Earth Sector Since 1944." Paper presented at International Studies Association Conference, San Francisco, CA, April.

Bao, Daozu. 2010. "Diaoyu Protests Intensify." *China Daily*, September 11. http://www .chinadaily.com.cn/cndy/2010-09/11/content_11288261.htm.

Baogang Xitu. 2013. "Meeting on the Coordinated Strategic Resource Utilization of Bayan Obo Development and Construction Demonstration Base Held in Beijing." *Industrial Movement* 2013 (6): 8–10.

Bardi, Ugo. 2014. *Extracted: How the Quest for Mineral Wealth Is Plundering the Planet*. White River Junction, VT: Chelsea Green.

Barfield, Thomas J. 1989. *The Perilous Frontier: Nomadic Empires and China*. Cambridge, MA: Basil Blackwell.

Basson, Edwin. 2014. "Raw Materials Outlook." In World Steel Association, *World Steel in Figures 2014*. https://www.worldsteel.org/en/dam/jcr:17354f46-9851-45c2-a1b6 -a896c2e68f37/World+Steel+in+Figures+2014+Final.pdf.

Bauer, Diana. 2010. "DOE's Critical Materials Strategy: United States Department of Energy Office of Policy and International Affairs." PowerPoint presentation in author's possession.

Bayer, Alice. 2014. "'Dalai Lama of the Rainforest' Makes Unique Visit to USA." *Survival International*, February 19. http://www.survivalinternational.org/news /10002.

BBC. 2010. "China Resumes Rare Earth Exports to Japan." Accessed December 28, 2014. http://www.bbc.com/news/business-11826870.

BBC. 2011. "Rare Earths Mining: China's 21st Century Gold Rush." Accessed May 1, 2014. http://www.bbc.com/news/world-asia-pacific-13777439.

Beaudry, B.J., and K. A. Gschneidner. 1974. "Preparation and Basic Properties of Rare Earth Metals." In *Handbook on the Physics and Chemistry of Rare Earths*, 174–233. Amsterdam: Elsevier.

Becker, P. C., N. A. Olsson, and J. R. Simpson. 1999. *Erbium-Doped Fiber Amplifier: Fundamentals and Technology*. San Diego, CA: Academic Press.

Beery, Jason. 2011. "Constellations of Power: States, Capitals, and Natures in the Coproduction of Outer Space." PhD diss., University of Manchester School of Environment and Development.

Beery, Jason. 2012. "State, Capital and Spaceships: A Terrestrial Geography of Space Tourism." *Geoforum* 43 (1): 25–34. doi: http://dx.doi.org/10.1016/j.geoforum.2011.07.013.

Beldavs, Vid. 2013. "How to Form the Lunar Development Corporation to Implement the Moon Treaty." *Space Review*, December.

Belli, P., R. Bernabei, F. Cappella, R. Cerulli, C. J. Dai, F. A. Danevich, A. D'Angelo, A. Incicchitti, V. V. Kobychev, S. S. Nagorny, S. Nisi, F. Nozzoli, D. Prosperi, V. I. Tretyak, and S. S. Yurchenko. 2007. "Search for α Decay of Natural Europium." *Nuclear Physics* 789 (1): 15–29.

Benard, Alexander. 2012. "Obama Must Keep US Military in Afghanistan to Counter China, Russia." *Christian Science Monitor*, January 4.

Bennett, John T. 2010. "Bill Calls for Establishment of First U.S. Rare Earth Minerals Stockpile." *Defense News*, March 18. http://archive.defensenews.com/article/20100318 /DEFSECT04/3180309/Bill-Calls-Establishment-First-U-S-Rare-Earth-Minerals -Stockpile.

Berger, V. I., D. A. Singer, and G. J. Orris. 2009. "Carbonatites of the World, Explored Deposits of Nb and REE—Database and Grade and Tonnage Models." *United States Department of the Interior, United States Geological Survey*. Open-File Report 2009-1139. https://pubs.usgs.gov/of/2009/1139/.

Berle, A. A., and G. C. Means. 1932. *The Modern Corporation and Private Property*. New Brunswick, NJ: Transaction Publishers.

Bernardo e Mello, Officio do Capitão General Manoel. 1763. "Remette o Mappa do Rio Negro, que Mandou Fazer por Filippe Sturm, o qual Explorou a Região, Penetrando nos Dominios Hespanhóes." Limites entre Le Brésil et La Guyane Anglaise, March 12.

Besson, Bernard, and Sophie Weiner. 2016. *The Rare Earth Exchange, Larivière Espionage Thrillers*. New York: Le French Book.

Bi, Rui, ed. 2007. *Baotou War of Resistance*. Baotou, Inner Mongolia Autonomous Region: Baotou Committee of September Third Studies.

Binnemans, K., P. T. Jones, B. Blanpain, T. Van Gerven, Y. Yang, A. Walton, and M. Buchert. 2013. "Recycling of Rare Earths: A Critical Review." *Journal of Cleaner Production* 51:1–22.

Bjerklie, Steve. 2006. "A Batty Business: Anodized Metal Bats Have Revolutionized Baseball. But Are Finishers Losing the Sweet Spot?" *Metal Finishing* 104 (4): 61–62.

Blanchard, Jean-Marc F., and Liang Wei. 2013. "The US, East Asian FTAs, and China." In *Regional Cooperation and Free Trade Agreements in Asia*, edited by Jiaxiang Hu and Matthias Vanhullebusch, 329–48. Leiden, The Netherlands: Koninklijke Brill NV.

Bloomberg News. 2011. "Australia Blocked China Investment on Supply Concerns." *Sydney Morning Herald*. Accessed June 10, 2014. http://www.smh.com.au/business /world-business/australia-blocked-china-investment-on-supply-concerns -20110214-1au8x.html.

Blosser, John. 2014. "Russia to Build Moon Base, Mine for Rare Elements." Newsmax Media. Accessed February 5, 2015. http://www.newsmax.com?Newsfont?lin -Industries-moon-base-rare-earth-elements/2014/12/31/id/6158/.

Boersma, Tim, and Kevin Foley. 2014. *The Greenland Gold Rush: Promise and Pitfalls of Greenland's Energy and Mineral Resources*. Washington, DC: The Brookings Institution, September. http://www.brookings.edu/~/media/research/files/reports /2014/09/24-greenland-energy-mineral-resources-boersma-foley/24-greenland -energy-mineral-resources-boersma-foley-pdf-2.pdf.

Bogard, Paul. 2013. *The End of Night: Searching for Natural Darkness*. New York: Little, Brown.

Bort, Julie. 2014. "This VC Is Helping Space Exploration Become a 'Multi-Billion' Industry." Business Insider. Accessed February 8. http://www.businessinsider.com/vc -space-is-a-multi-billion-industry-2014-10.

Bourzac, Katherine. 2011. "The Rare-Earth Crisis." *MIT Technology Review*. April 19. http:// www.technologyreview.com/featuredstory/423730/the-rare-earth-crisis/.

Bradsher, Keith. 2010a. "Amid Tension, China Blocks Crucial Exports to Japan." *New York Times,* September 24. http://www.nytimes.com/2010/09/24/business/global/24rare .html?pagewanted=all.

Bradsher, Keith. 2010b. "China Is Said to Resume Shipping Rare Earth Minerals." *New York Times*, October 28, B1. http://www.nytimes.com/2010/10/29/business/energy -environment/29rare.html?pagewanted=all.

Bradsher, Keith. 2010c. "China Restarts Rare Earth Shipments to Japan." *New York Times*, November 19. http://www.nytimes.com/2010/11/20/business/global/20rare.html.

Bradsher, Keith. 2010d. "In China, Illegal Rare Earth Mines Face Crackdown." *New York Times*, December 30, B1. http://www.nytimes.com/2010/12/30/business/global/30smuggle.html?pagewanted=2&ref=energy-environment&_r=0.

Bradsher, Keith. 2011a. "China Consolidates Grip on Rare Earths." *New York Times*, September 16, B1.

Bradsher, Keith. 2011b. "The Fear of a Toxic Terun." *New York Times*, June 30, B1.

Bradsher, Keith. 2012. "Specialists in Rare Earths Say a Trade Case against China May Be Too Late." *New York Times*, March 14. http://www.nytimes.com/2012/03/14/business/global/rare-earth-trade-case-against-china-may-be-too-late.html.

Bradsher, Keith, and Hiroko Tabuchi. 2010. "China Is Said to Halt Trade in Rare-Earth Minerals to Japan." *New York Times*, September 24. http://www.nytimes.com/2010/09/25/business/global/25minerals.html?_r=0.

Brasil, Kátia. 2001. "PF Faz Apreensão de Minérios Retirados na TI Alto Rio Negro." *Folha de São Paulo*, April 10.

Brasil, Kátia. 2003. "Brasileiros Abastecem as Farc na Fronteira." *Folha de São Paulo*, June 8. http://www1.folha.uol.com.br/fsp/brasil/fc0806200310.htm.

Braudel, Fernand. 1967. *Capitalism and Material Life, 1400–1800*. Vol. 3, *The Perspective of the World*. Berkeley: University of California Press.

Braudel, Ferdnand. 1985. *The Wheels of Commerce*. London: Fontana Paperbacks.

Braun, Bruce. 2000. "Producing Vertical Territory: Geology and Governmentality in Late Victorian Canada." *Cultural Geographies* 7 (7): 8–46.

"Brazilian Nuclear History." 1947–89. Wilson Center Digital Archive. Washington, DC: Woodrow Wilson International Center for Scholars. Accessed May 8, 2017. http://digitalarchive.wilsoncenter.org/collection/167/brazilian-nuclear-history.

Brennan, Elliot. 2012. "Rare Earths—The Next Oil." *Asia Times*, March 16.

Brennan, Louis, and Alessandra Vecchi. 2011. *The Business of Space: The Next Frontier of International Competition*. New York: Palgrave Macmillan.

Briscoe, Paula. 2013. "Greenland—China's Foothold in Europe?" *Asia Unbound* (blog). Council on Foreign Relations, February 1. http://blogs.cfr.org/asia/2013/02/01/paula-briscoe-greenland-chinas-foothold-in-europe/.

Brito, Adilson. 2013. "Um Rio Negro de Tensões." XXVII Simpósio Nacional de História, Natal, Rio Grande do Norte, July 22–26.

Bromby, Robin. 2013. "First Rare Earths, Now Antimony—It's All about China." Investor Intel. Accessed June 6, 2014. http://investorintel.com/rare-earth-intel/first-rare-earths-now-antimony-%E2%80%95-china-wants-control-now-faces-falling-prices/.

Brook, Timothy, and Hy V. Luong, eds. 1999. *Culture and Economy: The Shaping of Capitalism in Eastern Asia*. Ann Arbor: University of Michigan Press.

Brown, Floyd. 2013. "China's Dangerous Rare Earth Monopoly." Wall Street Daily, LLC. Accessed April 9, 2015. http://www.capitolhilldaily.com/2013/07/china-rare-earth/.

Brown, Kerry. 2007. "The Cultural Revolution in Inner Mongolia, 1967–1969: The Purge of the 'Heirs of Genghis Khan.'" *Asian Affairs* 38 (2): 173–87.

Brownlow, Alec, and Harold Perkins. 2014. "Sacrifice and Security: Geopolitics and Dialectics." Paper session at the Annual Meeting of the Association of American Geographers, Tampa, FL, April 11.

Bruce, D. W., B. E. Heitbrink, and K. P. Dubois. 1963. "The Acute Mammalian Toxicity of Rare Earth Nitrates and Oxides." *Toxicology and Applied Pharmacology* 5 (November): 750–59.

Bruce, Scott Thomas. 2012. "North Korea's Six Trillion Dollar Question." *The Diplomat.* August 30. http://thediplomat.com/2012/08/north-koreas-six-trillion-dollar -question/?allpages=yes.

Bryant, Bunyan, and Paul Mohai, eds. 1992. *Race and the Incidence of Environmental Hazards: A Time for Discourse.* Boulder, CO: Westview Press.

Bulag, Uradyn. 1998. "The Cult of Ulanhu in Inner Mongolia: History, Memory, and the Making of National Heros." *Central Asian Survey* 17 (1): 11–33.

Bulag, Uradyn. 2002. *The Mongols at China's Edge: History and the Politics of National Unity.* New York: Rowman & Littlefield.

Bulag, Uradyn. 2004. "Inner Mongolia: The Dialectics of Colonization and Ethnicity Building." In *Governing China's Multi-Ethinic Frontiers*, edited by Morris Rossabi, 84– 116. Seattle: University of Washington Press.

Bungardt, Walter. 1959. Pyrophoric Flints. United States Patent No. US2908071A, filed May 11, 1954, and issued October 13, 1959.

Bunn, Davis. 2012. *Rare Earth.* Bloomington, MN: Bethany House.

Burnell, James. 2010. "Resource Demands of Alternative Energy Technologies." Paper presented at Geological Society of America Annual Meeting, Denver, CO, October 31.

Burningman.org. 2015. "Welcome Home: A City in the Desert. A Culture of Possibility. A Network of Dreamers and Doers." Accessed February 23, 2015. http://burningman.org/.

Bustamente, Luiz Alberto da Cunha, João Trindade Cavalcante Filho, Márcia Fortuna Biato, and Carlos Jaques Vieira Gomes. 2013. *Análise do Projeto de Lei de Marco Regulatório da Mineração do Brasil.* Brasília, Brazil: Núcleo de Estudos e Pesquisas da Consultoria Legislativa.

Butler, Kiera. 2012. "Your Smartphone's Dirty, Radioactive Secret." *Mother Jones*, n.p.

Butterman, W. C., and J. F. Carlin, Jr. 2004. "Mineral Commodity Profiles: Antimony." Washington, DC: United States Geological Survey. Accessed February 3, 2013. http:// pubs.usgs.gov/of/2003/of03-019/of03-019.pdf.

Cai, Han, Jin Hai, and Sudehualige. 2007. *Republican Inner Mongolia History.* Hohhot: Inner Mongolia Press.

Caijing. 2013. "China's Plan to Create Six Rare Earth Groups Approved by Cabinent." Accessed March 1, 2014. english.caijing.com/cn/2014-01-03/113771290.html.

Caminiti, Susan. 2014. "The Billionaire's Race to Harness the Moon's Resources." CNBC. April 3. http://www.cnbc.com/id/101531789.

Campbell, N. 2000. *The Cultures of the American West.* London: Routledge.

Canadian Security Intelligence Service. 2013. "The Security Dimensions of an Influential China: Highlights from the Conference." Report from the Security Dimensions of an Influential China, Ottawa, Canada, February 28–March 1.

Caramenico, G. 2012. "China's Rare Earth Metals Clampdown Drives New Trade, Mining Ties." World Politics Review. Accessed December 5, 2012. http://www .worldpoliticsreview.com/articles/12517/chinas-rare-earth-metals-clampdown -drives-new-trade-mining-ties.

Cardarelli, François. 2008. *Materials Handbook.* 2nd ed. London: Springer.

Carlin, James F., Jr. 2013. "Antimony." *Mineral Commodity Summaries* (January): 18–19.

Carrington, Anca, ed. 2015. *Money as Emotional Currency: Psychoanalytic Ideas.* London: Karnac Books.

Carswell, Bill. 2002. "The Outer Space and Moon Treaties and the Coming Moon Rush." *Space Daily.* Accessed November 3, 2012. http://www.spacedaily.com/news/oped -02c.html.

Casey, Senator Robert P., Senator Charles E. Schumer, Senator Debbie Stabenow, and Senator Sheldon Whitehouse. 2011. In "Casey: Stop China from Dominating Rare Earth Element Market." Bob Casey: United States Senator for Pennsylvania. Ac-

cessed December 8, 2014. http://www.casey.senate.gov/newsroom/releases/casey
-stop-china-from-dominating-rare-earth-elements-market.

Castilloux, Ryan. 2014. *Rare Earth Market Outlook: Supply, Demand, and Pricing from
2014–2020*. Ontario, Canada: Adamas Intelligence.

Castro, Celso. 2003. "Os Militares e os Outros: Amazônia, Defesa Nacional e Identidades
Sociais." Paper presented at XXVII Encontro Anual da ANPOCS, Caxambu, Minas
Gerais, October 21–25.

Cave, C. J. P. 1944. "Book Review: Clouds and Weather Phenomena." *Popular Astronomy*
52:416.

Central Intelligence Agency. 1952. *Relations between the Chinese Communist Regime and
the USSR: Their Present Character and Probable Future Course*. Edited by Intelligence
Advisory Committee. Washington, DC: Central Intelligence Agency.

Chakhmouradian, Anton R., and Frances Wall. 2012. "Rare Earth Elements: Minerals,
Mines, Magnets (and More)." *Elements* 8 (5): 333–40.

Chan, Anita, Richard Madsen, and Jonathan Unger. 1992. *Chen Village under Mao and
Deng*. Expanded and updated ed. Berkeley: University of California Press.

Chang, Kenneth. 2015. "A Business Plan for Space." *New York Times*, February 10. http://
www.nytimes.com/2015/02/10/science/a-business-plan-for-space.html.

Chao, E. C. T., J. M. Back, J. A. Minkin, M. Tatsumoto, Wang Junewen, J. E. Conrad, E.
H. McKee, Hou Zonglin, Meng Qinrun, and Huang Shengguang. 1997. "The Sed-
imentary Carbonate-Hosted Giant Bayan Obo REE-Fe-Nb Ore Deposit of Inner
Mongolia, China: A Cornerstone Example for Giant Polymetallic Ore Deposits of
Hydrothermal Origin." In *Bulletin 2143*, edited by United State Geological Sur-
vey. Washington DC: Department of the Interior, United States of America.

Chao, Luomeng, ed. 2000. *The Old Revolutionary Base of Inner Mongolia*. Hohhot, Inner
Mongolia Autonomous Region: Inner Mongolia People's Press.

Chapple, Irene. 2012. "Why Minerals Dispute Threatens Electronics Industry." CNN. Accessed December 9, 2013. http://www.cnn.com/2012/03/13/business/rare-earths
-china-u-s-/.

Chen, Mel Y. 2011. "Inanimate Affections." *GLQ: A Journal of Lesbian and Gay Studies* 17
(2–3): 265–86.

Chen, Wanqing, Zhang Siwei, and Zou Xiaonong. 2010. "Evaluation on the Incidence,
Mortality and Tendency of Lung Cancer in China." *Thoracic Cancer* 1 (1): 35–40.

Chen, Zhanheng. 2010. *Outline on the Development and Policies of China's Rare Earth Industry*. Beijing: China Society of Rare Earths.

Chen, Zhanheng. 2011. "Global Rare Earth Resources and Scenarios of Future Rare Earth
Industry." *Journal of Rare Earths* 29 (1): 1–6.

Chen, Zhen, and Luo Yao. 1954. *China's Contemporary Industry and Resources: Imperialism undermined and weakened China's Industry and Mining Affairs*. Beijing: Sanlian
Publishing.

Cheng, Dean. 2011. "China's Space Program: A Growing Factor in U.S. Security Planning."
The Heritage Foundation Accessed February 6. http://www.heritage.org/research
/reports/2011/08/chinas-space-program-a-growing-factor-in-us-security-planning.

Cheng, Jianzhong, and Liping Che. 2010. "Trends in the Exploitation and Development
of China's Rare Earth Resources." *China Rare Earths* 31 (2): 65–69.

Cheng, Yuqi. 1950. "Geology's New Orientations and Responsibilities." *Geology Forum*
15:1–6.

Chengappa, Raj. 2000. *Weapons of Peace*. Noida, India: HarperCollins.

Chernela, Janet. 2014. "Particularizing Univerals/Universalizing Particulars: A Comprehensive Approach to Trafficking in Indigenous Women and Girls in the Northwest
Amazon of Brazil." In *Anthropological Approaches to Gender-Based Violence and*

Human Rights, edited by Sheila Dauer, 34–52. Lansing, MI: Center for Gender in Global Context.

Chi, Shun Wan. 1990. "Economic Aspects of the Sino-Soviet Alliance." MA thesis, University of British Colombia.

"Chicago Man Stakes Claim to Outer Space." 1949. *Science Illustrated* (May): 42–43.

China Daily. 2003. "Activists to Defend Diaoyu Islands." Accessed December 28, 2014. http://www.chinadaily.com.cn/china/target=.

China Daily. 2009. "China Mulls Plan to Curb Rare Earth Smuggling." Accessed December 28, 2014. http://www.chinadaily.com.cn/bizchina/2009-09/14/content_8690293.htm.

China Economic Review. 2014. "Greenland, a Frontier Market Unlike Any Other for China." February. Accessed January 22, 2015. http://www.chinaeconomicreview.com/china-in-the-arctic-greenland-iron-mining.

China Electrical Intelligence Network News Bureau. 2013. "Foreign Media Identifies Jade Rabbit as Lunar Rare Earth Explorer." Accessed February 6, 2014. http://www.chinaelc.cn/ch_hangye/zspl/2013121184055_2.html.

Chossudovsky, Michel. 2005. *America's "War on Terrorism" in the Wake of 9/11*. Pincourt, Québec: Global Research.

Chouinard, Vera. 1994. "Geography, Law and Legal Struggles: Which Ways Ahead?" *Progress in Human Geography* 18 (4): 415–40.

Chowdhury, M. A. I., M. T. Uddin, M. F. Ahmed, M. A. Ali, S. M. A. Rasul, M. A. Hoque, R. Alam, R. Sharmin, S. M. Uddin, and M. S. Islam. 2006. "Collapse of Socio-Economic Base of Bangladesh by Arsenic Contamination in Groundwater." *Pakistan Journal of Biological Sciences* 9:1617–27.

Clancy, John, and Helene Banner. 2012. *EU Challenges China's Rare Earth Export Restrictions*. Brussels: European Commission.

Clark, B. C. 2004. "The Lunar Treasure Chest: Roles of Human Explorers at a Robotic Outpost." Paper presented at 55th International Astronautical Conference, Vancouver, British Colombia, Canada, October 4–8.

Clark, Gardner M. 1973. *Development of China's Steel Industry and Soviet Technical Aid*. Ithaca, NY: Committee on the Economy of China of the Social Science Research Council.

CM022. 2013. "Chinese Expert's 'Slip of the Tongue' on Jade Rabbit Hits CIA Nerve." *China Military Channel*. Accessed May 9, 2017. http://military.china.com/critical3/27/20131204/18193976.html.

CNN. 2000. "Prime Lunar Real Estate for Sale–but Hurry." Accessed November 19, 2013. http://www.cnn.com/2000/TECH/space/11/20/lunar.land/index.html.

Coats, Stan. 2006. "Minerals in Afghanistan." British Geological Survey, Afghanistan Project Office. Accessed April 9, 2015. http://pubs.usgs.gov/fs/2011/3108/fs2011-3108.pdf.

Coêlho, Marilia. 2013. "Atingidos por Mineração São Contra Proposta de Novo Código e Denunciam Impactos." Senado Federal. Accessed January 30, 2014. http://www12.senado.gov.br/noticias/materias/2013/12/04/atingidos-por-mineracao-sao-contra-proposta-de-novo-codigo-e-denunciam-impactos.

Cone, Marla. 1997. "Desert Lands Contaminated by Toxic Spills." *Los Angeles Times*, April 24. http://articles.latimes.com/1997-04-24/news/mn-51903_1_mojave-desert.

Congresso Brasileiro. 1988. *Constituição Da República Federativa Do Brasil De 1988*. Edited by National Congress. Brasília: Casa Civil.

Conley, Heather A. 2013. "Arctic Mineral Resource Exploration." In *Artic Economics in the 21st Century: The Benefits and Costs of Cold*, 19–31. Washington, DC: Center for Strategic and International Studies.

"The Construction of the People's Weapons Maintains Steady Production." 1949. *Inner Mongolia Daily*, 1.

Consumers Association of Penang. 2011. "Chronology of Events in the Bukit Merah Asian Rare Earth Development." Consumers Association of Penang. Accessed April 21, 2014. http://www.consumer.org.my/index.php/health/454-chronology-of-events-in -the-bukit-merah-asian-rare-earth-developments.

Cooper, Keith. 2015. "Earth's Moon May Not Be Critical to Life." *Astrobiology Magazine*, January 27. http://www.astrobio.net/news-exclusive/earths-moon-may-not-critical -life/.

Coppel, E. 2011. "Rare Earth Metals and US National Security." February 1. http:// americansecurityproject.org/wp-content/uploads/2011/02/Rare-Earth-Metals -and-US-Security-FINAL.pdf.

Corbridge, Stuart. 1994. "Bretton Woods Revisited: Hegemony, Stability, Territory." *Environment and Planning A* 26 (12): 1829–59.

Corfield, Richard. 2015. "Mining the Moon Becomes a Serious Prospect." *Physics World* Accessed January 3, 2016. http://www.iop.org/news/15/feb/page_64944.html.

Corrêa, S. L. A., M. L. Costa, and N. P. Oliveira. 1968. "Contribuição Geoquímica à Zona Lateríca do Complexo Carbonatítico de Seis Lagos (Amazonas)." Anais do Congresso Brasileiro de Geologia, Belém, Brazil. 4:1959–68.

Cosgrove, Denis. 2005. "Apollo's Eye: A Cultural Geography of the Globe." In *Hettner Lecture I*. June. Accessed October 3, 2012. http://www.sscnet.ucla.edu/geog/downloads /418/45.pdf.

Couldry, N. 2012. "Universities and the Necessary Counter-Culture against Neoliberalism." *IC-Revista Científica de Información y Comunicación* 9:61–71.

Crawford, Ian A. 2015. "Lunar Resources: A Review." *Progress in Physical Geography* 39 (2): 137–67.

Croat, J. J. 1997. "Current Status and Future Outlook for Bonde Neodymium Permanent Magnets." *Journal of Applied Physics* 81 (8): 4804–9.

Cruz, Fred. 2014. "A situção Política sobre a Mineração na Amazônia." Unpublished manuscript.

Cuadros Justo, L. J. E., and M. M. de Souza. 1986. "Jazida de Nióbio do Morro dos Seis Lagos, Amazonas." In *Principas depósitos minerais do Brazil*, edited by C Schobbenhaus and CE Silva Coelho, 463–68. Brasília: Departamento Nacional da Produção Mineral.

Culver, L., L. Escudero, A. Grindly, M. Hamilton, and J. Sowell. 2007. "Policies, Incentives, and Growth in the NewSpace Industry." Commercial Space Working Paper. Accessed December 1, 2012. http://commercialspace.pbworks.com/f/2007.12.12+Ne wSpace+Policies+Incentives+and+Growth.pdf.

Czarnecki, Ronald. 2010. "China Attacks Sleeping US Defense Industry–Cuts Supply of Rare Earths." Accessed January 3, 2013. http://beforeitsnews.com/story/315/224 /China_Attacks_Sleeping_US_Defense_Industry_Cuts_Supply_of_Rare_Earths .html.

Daly, Brian. 2012. "Man Sues for Ownership of Most of Solar System." IF Press. Accessed February 2, 2015. http://www.lfpress.com/news/weird/2012/03/01/19445886.html.

Danielski, David. 2009. "Expansion in Works for S.B. County Mine with Troubled Environmental Past." The Press-Enterprise, February 9. http://www.pe.com/localnews /environment/stories/PE_News_Local_S_molycorp09.44b8a72.html.

Danielski, David. 2014. "Mountain Pass: Molycorp Fined for Mishandling Toxic Waste." The Press Enterprise, April 21. http://www.pe.com/articles/mine-692838-rare-lead .html.

Davey, Melissa. 2014a. "Malaysian Police Say Australian Activist Faces 'Up to Two Years' Jail." *Guardian*, June 24. https://www.theguardian.com/world/2014/jun/24 /malaysian-police-say-australian-activist-faces-up-to-two-years-jail.

Davey, Melissa. 2014b. "Supporters of Environmental activist jailed in Malaysia Protest in Sydney." *Guardian*, June 27. http://www.theguardian.com/environment/2014/jun/27/environmental-activist-jailed-malaysia-protest-natalie-lowrey.

David, Leonard. 2014. "China Has Big Plans to Explore the Moon and Mars." Space.com. Accessed February 6, 2015. http://www.space.com/27893-china-space-program-moon-mars.html.

David, Leonard. 2015. "Is Moon Mining Economically Feasible?" Fox News Network. Accessed February 2. http://www.foxnews.com/science/2015/01/12/is-moon-mining-economically-feasible/.

Davidson, N. 2007. "Making a Mint out of the Moon." BBC. Accessed April 10, 2015. http://news.bbc.co.uk/2/hi/science/nature/6533169.stm

Davis, Malcolm W. 1926. "Railway Strategy in Manchuria." *Foreign Affairs* 4 (3): 499–502.

Day, Alexander F. 2013. *The Peasant in Postsocialist China: History, Politics, and Capitalism*. Cambridge: Cambridge University Press.

Day, P. J. 2009. "Moon 2.0: Join the Revolution." Google Lunar XPrize video, 8:00. May 1, 2009. http://www.youtube.com/watch?v=SAbgZpaTgvo.

Dean, Warren. 1987. *Brazil and the Struggle for Rubber: A Study in Environmental History*. Cambridge: Cambridge University Press.

De Angelis, Massimo. 2001. "Marx and Primitive Accumulation: The Continuous Character of Capital's 'Enclosures.'" *Commoner* 2 (September): 1–22.

Delegation of the People's Republic of China. 2014. "China—Measures Related to the Exportation of Rare Earths, Tungsten, and Molybdenum: Notification of an Appeal by China Under Article 16.4 and Article 17 of the Understanding on Rules and Procedures Governing the Settlement of Disputes (DSU), and Under Rule 20(1) of the Working Procedures for Appellate Review." Accessed March 7, 2015. https://docs.wto.org/dol2fe/Pages/FE_Search/FE_S_S009-DP.aspx?language=E&CatalogueIdList=124364&CurrentCatalogueIdIndex=0&FullTextSearch=.

Deleuze, Gilles, and Félix Guattari. 1987. *A Thousand Plateaus: Capitalism and Schizophrenia*. Translated by Brian Massumi. Minneapolis: University of Minnesota Press.

Deng, Jinwu. 2009. "A Lifetime of Rare Earth Love: Notes on the China Academy of Science Academician Xu Guangxian." *Science Times*. Accessed August 31, 2014. http://news.sciencenet.cn/htmlnews/2009/1/215299.html.

Deng, Yun; Wang, Keyun. 2007. "Scholarly Achievements of the China-Swiss Northwestern Scientific Expedition." *Science and Technology News: Humanities Editorial* 13:131.

Dent, Peter C. 2012. "Rare Earth Elements and Permanent Magnets (Invited)." *Journal of Applied Physics* 111 (7): 07a721-1–07A721-6.

Desgeorges, Damien. 2013. "Greenland and the Arctic: Still a Role for the EU." *EurActiv: Efficacité et Transparence des Acteurs Européens* 2013 (December 7): n.p.

Di, Zhongheng. 1976. "Chinese Communist Nuclear Forces." *Mingbao Monthly* (August–September), pts. 1–2.

Dickens, P. and Ormrod, J. 2007. *Cosmic Society: Toward a Sociology of the Universe*. London: Routledge.

Digonnet, Michel J. F., ed. 2001. *Rare-Earth-Doped Fiber Lasers and Amplifiers, Revised and Expanded*. 2nd ed. New York: Marcel Dekker.

DiJohn, John. 2010. "Taxation, Resource Mobilization and State Performance." Crisis States Research Centre Working Paper no. 48, London.

Dillon, Robert. 2010. "Murkowski and Bayh Call for Support of Rare Earth Projects in DOE Loan Guarantee Program." United States Senate Committee on Energy and Natural Resources. https://www.energy.senate.gov/public/index.cfm/republican-news?ID=7FA9C642-A42D-4D88-9644-BE679A4E5726.

D'incão, Maria Angela, ed. 1994. *Amazônia e a Crise da Modernização*. Museu Paraense Emelio Goeldi: Editora Isolda Maciel da Silveira.

Ding, Daoheng. 1933. "On the Iron Deposit of Bayan Obo, Suiyuan." Geological Survey of China Report 23. Bejing: Geological Survey of China.

Diniz, Aníbal. 2013. " 'Terras Raras São as Portadoras do Futuro,' Afirma Aníbal Diniz." Partido dos Trabalhadores no Senado. Accessed January 22, 2015. http://anibaldiniz .com.br/terras-raras/index.php/2013-04-04-20-15-47/15-terras-raras-sao-as -portadoras-do-futuro-afirma-anibal-diniz.

Dinwoodie, Ty. 2013. "WTO's Verdict against Chinese Rare Earth Export Quotas Underscores the Criticality of Sustainability." Investor Intel. Accessed January 9, 2015. http://investorintel.com/rare-earth-intel/wtos-verdict-chinese-rare-earths-export -quotas-underscores-criticality-self-sustainability/.

Djenchuraev, Nurlan. 1999. "Current Environmental Issues Associated with Mining Wastes in Kyrgyzstan." MA thesis, Central European University, Budapest.

Dobransky, Steve. 2013. "Rare Earth Elements and US Foreign Policy." American Diplomacy: Foreign Service Despatches and Periodic Reports on U.S. Foreign Policy, October. Accessed May 5, 2017. http://www.unc.edu/depts/diplomat/item/2013/0912 /ca/dobransky_rareearth.html.

Dolman, Everett. 2002. *Astropolitik: Classical Geopolitics in the Space Age*. London: Frank Cass.

Dove, Michael R. 2006. "Indigenous People and Environmental Politics." *Annual Review of Anthropology* 35:191–208.

Duarte, F. 2010. *Tunable Laser Applications*. 2nd ed. Boca Raton, FL: Taylor and Francis.

Dudley-Flores, Marilyn, and Thomas Gangale. 2013. "Forecasting the Political Economy of the Inner Solar System." *Astropolitics: The International Journal of Space Politics and Policy* 10 (3): 183–233.

Dumett, Raymond. 1985. "Africa's Strategic Minerals during the Second World War." *Journal of African History* 26 (4): 381–408.

Easton Cycling Staff. 2009. "Technology Report: Materials/Scandium." Easton Cycling. Accessed February 12, 2013. http://www.eastoncycling.com/bike/wp-content /uploads/2010/04/RD-03-Scandium.pdf.

Eberstadt, Nick. 1980. "Did Mao Fail?" *Wilson Quarterly* 4 (4): 120–31.

Eckert, P., and M. A. Barboza. 2005. "Attracting Non-Space Industry into Space: Catalyst for Lunar Commercialization." International Lunar Conference, Toronto, Canada. September 20. http://sci.esa.int/Conferences/ILC2005/Presentations/EckertP-01 -PPT.pdf.

Edwards, Gordon. 1992. "Uranium: Known Facts and Hidden Dangers." World Uranium Hearings, Salzburg, Austria, September 14.

Eichengreen, Barry, Andrew K. Rose, Charles Wyplosz, Bernard Dumas, and Axel Weber. 1995. "Exchange Market Mayhem: The Antecedents and Aftermath of Speculative Attacks." *Economic Policy* 10 (21): 249–312.

Ekirch, A. Roger. 2005. *At Day's Close: Night in Times Past*. New York: W.W. Norton.

Eliseeva, Svetlana V., and Jean-Claude G. Bünzli. 2011. "Rare Earths: Jewels for Functional Materials of the Future." *New Journal of Chemistry* 35:1165–76.

Ellis, A. 2013. "The Anthropocene: China—The OPEC of Rare Earth Metals?" *St. Andrews Foreign Affairs Review*. Accessed January 3, 2014. http://foreignaffairsreview.co.uk /2013/03/rare-earth/.

Ellis, R. Evan. 2013. "The Strategic Dimnsion of Chinese Engagement with Latin America." Washington, DC: William J. Perry Center for Hemispheric Defense Studies Perry Paper Series I.

Elmquist, Sonja. 2012. "Molycorp CEO Exits Six Months after Neo Material Deal." Thomas Reuters. Accessed December 27, 2014. http://www.bloomberg.com/news/2012-12-11/molycorp-appoints-karayannopoulos-as-ceo-after-mark-smith-exits.html.

Els, Frik. 2011. "On China's Rare Earth Black Market Prices are Falling." Mining.com. Accessed December 28, 2014. http://www.mining.com/on-chinas-rare-earth-black-market-prices-are-falling/.

Emsley, John. 2001. *Nature's Building Blocks: An A-Z Guide to the Elements.* Oxford: Oxford University Press.

Environmental Protection Agency (EPA). 1990. *Toxicological Profile for Radon.* Washington, DC: United States Public Health Service Agency for Toxic Substances and Disease Registry.

Environmental Protection Agency. 2014. "Mining Waste." Accessed October 10, 2014. http://www.epa.gov/osw/nonhaz/industrial/special/mining/.

Epley, Cole. 2014. " 'Huge Economic Opportunity': Mine May Still Come to Southeast Nebraska Mineral Site." *Omaha World Herald*, December 11. http://www.omaha.com/money/huge-economic-opportunity-mine-may-still-come-to-southeast-nebraska/article_af51447f-a590-5b78-9571-39178afc7895.html?mode=jqm.

Epstein, Alex. 2014. *The Moral Case for Fossil Fuels.* New York: Penguin.

Erickson, Luke. 2008. "Land from the Tiller: The Push for Rural Land Privatization in China." *China Left Review.* Accessed June 3, 2013. http://chinaleftreview.org/?p=57.

Espilie, Erin. 2014. *The Lanthanide Series.* Copenhagen, Denmark: Independent Documentary Film.

Evans, D. S., and G. V. Raynor. 1961. "The System Thorium-Ytterbium: Its Alloy Properties." *Journal of the Less Common Metals* 3 (2): 179–80.

Evans, Geoff, James Goodman, and Nina Landsbury, eds. 2002. *Moving Mountains: Communities Confront Mining and Globalization.* London: Zed Books.

Evans, Julie. 2009. "Where Lawlessness Is Law: The Settler-Colonial Frontier as a Legal Space of Violence." *Australian Feminist Law Journal* 30 (1): 3–22.

Evans-Pritchard, Ambrose. 2013. "Japan Breaks China's Stranglehold on Rare Metals with Sea-Mud Bonanza." *Telegraph.* Accessed April 10, 2015. http://www.telegraph.co.uk/finance/comment/ambroseevans_pritchard/9951299/Japan-breaks-Chinas-stranglehold-on-rare-metals-with-sea-mud-bonanza.html.

Exército Brasileiro. 2006. "21st Company of Engineering and Construction." Centro de Comunicação Social do Exército. Accessed April 9, 2015. http://www.exercito.gov.br/03ativid/Amazonia/cancao.htm.

Fackler, Martin, and Ian Johnson. 2010. "Japan Retreats with Release of Chinese Boat Captain." *New York Times*, September 25, A1. http://www.nytimes.com/2010/09/25/world/asia/25chinajapan.html?ref=global&pagewanted=all.

Fan, H. R., F. F. Hu, F. K. Chen, K. F. Yang, and K. Y. Wang. 2006. "Intrusive Age of No. 1 Carbonatite Dyke from Bayan Obo REE-Nb-Fe Deposit, Inner Mongolia: With Answers to Comment of Dr. Le Bas." *Acta Petrologica Sinica* 22 (2): 519–20.

Farias, Elaíze. 2013a. "Upper Rio Negro Indigenous Defends Extractivist Mining." *Amazonia Real*, November 4. http://amazoniareal.com.br/indigena-do-alto-rio-negro-defende-mineracao-extrativista/.

Farias, Elaíze. 2013b. "Rare Earth Research on Indigenous Lands in the Amazon." *Amazonia Real.* Accessed April 10, 2015. http://amazoniareal.com.br/terras-indigenas-da-amazonia-sao-alvos-de-pesquisas-sobre-terras-raras/.

Faris, Stephan. 2014. "Greenland's Prime Minister Looks on Global Warming's Bright Side." Bloomberg Businessweek. Accessed December 20, 2014. http://www.businessweek.com/articles/2014-05-01/greenland-prime-minister-expects-global-warming-mining-riches.

Faust, Jeff. 2011. "Fear of a Chinese Moon." *The Space Review: Essays and Commentary about the Final Frontier*, October 31.

Federici, Silvia. 2004. *Caliban and the Witch: Women, the Body, and Primitive Accumulation*. Brooklyn, NY: Autonomedia.

Ferraz, Rodrigo. 2011. "CETEM Coordena Primeiro Seminário Sobre Terras-Raras no Brasil." Centro de tecnologia mineral. Accessed January 22, 2015. http://www.cetem.gov.br/611-cetem-coordena-primeiro-seminario-sobre-terras-raras-no-brasil.

Figueiredo, Jader de Correia. 1967. *Relatório MI-55-455*. Brazil: Ministry of the Interior.

Filho, Argemiro Procópio, and Alcides Costa Vaz. 1997. "O Brasil no contexto do Narcotráfico Internacional." *Revista Brasileira de Política Internacional* 40 (1): 75–122.

Filho, João Roberto Martins. 2006. "As Forças Armadas Brasileiras e o Plano Colômbia." In *Amazônia e Defesa Nacional*, edited by Celso Castro, 13–30. Rio de Janeiro: Centro de Pesquisa e Documentação de História Contemporânea do Brasil.

Fisher, Travis, and Alex Fitzsimmons. 2013. "Big Wind's Dirty Little Secret: Toxic Lakes and Radioactive Waste." Instutite for Energy Research. Accessed January 9, 2015. http://instituteforenergyresearch.org/analysis/big-winds-dirty-little-secret-rare-earth-minerals/.

Fletcher, James. 2013. "Mining in Greenland—A Country Divided." *News Magazine*, BBC World Service, December 31. http://www.bbc.com/news/magazine-25421967.

Fogel, Joshua A., ed. 2000. *The Nanjing Massacre in History and Historiography*. Los Angeles: University of California Press.

Foreign Affairs Bureau Yunnan Branch Special Committee Archive. 1916. *Reports Concerning Mongolian Opposition to the Imperial System and Announcement of Its Independence*. Edited by Foreign Affairs Bureau Yunnan Branch Special Committee. Jiangsu: Antiquated Texts Publisher of Jiangsu.

Foucault, Michel. 2007. *Security, Territory, Population. Lectures at the Collège de France: 1977–1978*. Edited by Arnold J. Davidson. Translated by Graham Burchell. New York: Picador.

Foucault, Michel. 2010. *The Birth of Biopolitics: Lectures at the Collège de France 1978–1979*. Edited by Michel Senellart. New York: Picador.

Fountain, Lynn M. 2003. "Creating Momentum in Space: Ending the Paralysis Produced by the 'Common Heritage of Mankind' Doctrine." *Connecticut Law Review* 35:1753–88.

Foweraker, Joe. 1981. *The Struggle for Land: A Political Economy of the Pioneer Frontier in Brazil from 1930 to the Present Day*. Cambridge: Cambridge University Press.

Fox, Julia. 1999. "Mountaintop Removal in West Virginia: An Environmental Sacrifice Zone." *Organization and Environment* 12 (June): 163–85.

Frakes, Jennifer. 2003. "The Common Heritage of Mankind Principle and Deep Seabed, Outer Space, and Antarctica: Will Developed and Developing Nations Reach a Compromise?" *Wisconsin Journal of International Law* 21:409–34.

Franken, Al. 2003. *Lies and the Lying Liars Who Tell Them*. New York: Plume.

Frankfurt, H. G. 2005. *On Bullshit*. Princeton: Princeton University Press.

Fratianni, Michelle, Paolo Savona, and John J. Kirton, eds. 2013. *Corporate, Public, and Global Governance: The G8 Contribution*. Burlington, VT: Ashgate.

Fravel, Taylor M. 2008. *Strong Borders, Secure Nation: Cooperation and Conflict in China's Territorial Disputes*. Princeton: Princeton University Press.

Fulp, Mickey. 2012. "Weekly Briefing: What Makes a Critical Metal 'Critical' or a Strategic Element 'Strategic'?" *Mercenary Musings*. Online Newsletter. August 6. http://www.goldgeologist.com/mercenary_musings/musing-120806-What-Makes-a-Critical-Metal-Critical-or-a-Strategic-Element-Strategic.pdf.

Galyen, John. 2011. "Hearing on 'China's Monopoly on Rare Earths: Implications for U.S. Foreign and Security Policy.'" In *Rare Earth Elements: Supply, Trade and Use Dynamics*, edited by J. O. Manino, and E. T. Jones, 73–76. New York: Nova Science.

Gandhi, Amit, Li Mu, and Justin Honrath. 2013. *Tungsten Market Outlook*. New York: CPM Group.

Gao, James Z. 2007. "The Call of the Oases: The 'Peaceful Liberation of Xinjiang, 1949–53." In *Dilemmas of Victory: The Early Years of the People's Republic of China*, edited by Jeremy; Pickowics Brown, Paul, 184–204. Cambridge, MA: Harvard University Press.

Gee, Alastair. 2014. "The Rare-Earths Roller Coaster." *New Yorker*, May 22. http://www.newyorker.com/business/currency/the-rare-earths-roller-coaster.

General Agreements on Tariffs and Trade (GATT). 1994. "Article XX, Measures (b) and (g)." World Trade Organization, Geneva, Switzerland. Accessed November 3, 2012. https://www.wto.org/english/docs_e/legal_e/gatt47_02_e.htm.

Geng, Zhiqiang, ed. 2007. *Annals of Baotou Urban Construction*. Hohhot, Inner Mongolia Autonomous Region: Inner Mongolia University Press.

George, Michael W. 2016. "Gold." United States Geological Survey Mineral Commodity Summaries. https://minerals.usgs.gov/minerals/pubs/commodity/gold/mcs-2016-gold.pdf.

Gibson-Graham, J. K. 1996. *The End of Capitalism (as We Knew It): A Feminist Critique of Political Economy*. Minneapolis: University of Minnesota Press.

Giovannini, Arthur Lemos. 2013. "Contribuição à Geologia e Geoquímica do Carbonatito e da Jazida (Nb, ETR) de Seis Lagos (Amazonas)." MA thesis, Federal University of Rio Grande do Sul.

Gobarev, Viktor M. 1999. "Soviet Policy toward China: Developing Nuclear Weapons 1949–1969." *Journal of Slavic Military Studies* 12 (4): 1–53.

Goldenberg, S. 2010. "Rare Earth Metals Mine Is Key to US Control Over Hi-tech Future." *Guardian*, December 26. http://www.guardian.co.uk/environment/2010/dec/26/rare-earth-metals-us.

Goldman, Joanne Abel. 2014. "The U.S. Rare Earth Industry: Its Growth and Decline." *Journal of Policy History* 26 (2): 139–66.

Goldman, Michael. 2005. *Imperial Nature: The World Bank and Struggles for Social Justice in the Age of Globalization*. New Haven, CT: Yale University Press.

Goldschmidt, Z. B. 1978. "Atomic Properties." In *Handbook on the Physics and Chemistry of Rare Earths*, 1–172. Amsterdam: Elsevier.

Gomes, C. B., E. Ruberti, and L. Morbidelli. 1990. "Carbonatite Complexes from Brazil: A Review." *Journal of South American Sciences* 3 (1)51–63.

Goodhand, Jonathan. 2005. "Frontiers and Qars: The Opium Economy in Afghanistan." *Journal of Agrarian Change* 5 (2): 191–216.

Gordon, Julie, Rob Wilson, and Gunna Dickson. 2012. "Molycorp Buys Neo Material for C$1.3 Billion." Reuters, March 8. http://www.reuters.com/article/2012/03/09/us-molycorp-idUSBRE82800T20120309.

Gozzi, Ricardo. 2011. "Vale Descobre Terras Raras em Salobo, na Amazônia." *Estadão*, October 3. http://economia.estadao.com.br/noticias/negocios,vale-descobre-terras-raras-em-salobo-na-amazonia,86669e.

Grasso, Valerie Bailey. 2013. "Rare Earth Elements in National Defense: Background, Oversight Issues, and Options for Congress." In *Rare Earth Elements: Supply, Trade, and Use Dynamics*, edited by J. O. Manino, and E. T. Jones, 83–112. New York: Nova Science.

Graulau, Jeanette. 2001. "Peasant Mining Production as a Development Strategy: The Case of Women in Gold Mining in the Brazilian Amazon." *Revista Europea de Estudios Latinoamericanos y del Caribe/European Review of Latin American and Caribbean Studies* 71:71–104.

Graulau, Jeanette. 2003. "Gendered Labour in Peripheral Tropical Frontiers: Women, Mining and Capital Accumulation in Post-development Amazonia." In *Women Miners in Developing Countries: Pit Women and Others*, edited by Kuntala Lahiri-Dutt and Martha Macintyre, 289–305. Burlington, VT: Ashgate.

Gravgaard, Anna-Katarina. 2013. "Greenland's Rare Earths Gold Rush." *Foreign Affairs*, October 13, n.p.

Gray, Theodore. 2009. *The Elements: A Visual Exporation of Every Known Atom in the Universe*. New York: Black Dog & Leventhal.

Gregory, Derek. 2006. "The Black Flag: Guantanamo Bay and the Space of Exception." *Geografiska Annaler Series B: Human Geography* 88 (4): 405–27.

Greinacher, E. 1981. "History of Rare Earth Applications, Rare Earth Market Today." In *Industrial Applications of Rare Earth Elements*, edited by K. A. Gschneider, Jr., 3–18. Washington, DC: American Chemical Society.

Greising, David. 2005. "China's Investment Wave: Good or Bad?" *Chicago Tribune*, July 10. http://articles.chicagotribune.com/2005-07-10/business/0507090254_1_cfius -foreign-takeovers-foreign-investment.

Grewal, Inderpal, and Caren Kaplan, eds. 1994. *Scattered Hegemonies: Postmodernity and Transnational Feminist Practices*. Minneapolis: University of Minnesota Press.

Grinde, Donald A., and Bruce Elliot Johansen. 1995. *Ecocide of Native America: Environmental Destruction of Indian Lands and Peoples*. Santa Fe, NM: Clear Light.

Gruner, Brandon C. 2005. "A New Hope for International Space Law: Incorporating Nineteenth Century First Possession Principles into the 1967 Space Treaty for the Colonization of Outer Space in the Twenty-first Century." *Seton Hall Law Review* 35:299–357.

Guenther, Godehard A. 2003. Compact High-Performance Speaker. United States Patent No. US7006653 B2, filed June 27, 2001, and issued August 26, 2003.

Guillamin, Collete. 1995. *Racism, Sexism, Power and Ideology*. New York: Routledge.

Guimarães, Ed Carlos Sousa. 2010. "(In)Justiça e Violência na Amazônia: O Massacre da Fazenda Princesa." *PRACS: Revista Eletrônica de Humanidades do Curso de Ciências Sociais da UNIFAP* 3:109–22.

Gullett, Brian K., William P. Linak, Touati Abderrahmane, Shirley J. Wasson, Staci Gatica, and Charles J. King. 2007. "Characterization of Air Emissions and Residual Ash from Open Burning of Electronic Wastes during Simulated Rudimentary Recycling Operations." *Journal of Material Cycles Waste Management* 9:69–79.

Guner, B. 2004. "A New Hope for International Space Law: Incorporating Nineteenth Century First Possession Principles into the 1967 Space Treaty for the Colonization of Outer Space in the Twenty-First Century." *Seton Hall Law Review* 35:299–357.

Guo, Dijun, Jianzong Liu, Li Zhang, Jinzhu Ju, Jingwen Liu, and Liang Wang. 2014. "The Methods of Lunar Geochronology Study and the Subdivisions of Lunar Geologic History." *Earth Science Frontiers* 21 (6): 45–61.

Guo, P., Z. L. Huang, P. Yu, and K. Li. 2012. "Trends in Cancer Mortality in China: An Update." *Annals of Oncology* 23 (10): 2755–62.

Guo, Shilie; Cao Xinsheng. 2009. "Pollution Prevention Measures of Bayan Obo Associated Radioactive Materials." *Vocational College Journal* 2009 (2): 122–25.

Guo, Sujian. 2013. *Chinese Politics and Government: Power, Ideology and Organization*. New York: Routledge.

Guomindang on Mongolian and Tibetan Affairs Committee. 1941. *Intelligence Report of the Mongolian and Tibetan Affairs Committee on the Secretariat of the Policy Implementation Council for the Mongolian Autonomous Government Governing Structure*. Edited by Committee on Mongolian and Tibetan Affairs. Jiangsu: Antiquated Texts Publisher of Jiangsu.

Gustke, Constance. 2011. "The New Gold Rush: Rare Earths." CNBC. Accessed May 1, 2014. http://www.cnbc.com/id/40913684.

Guzmán, Tracy Devine. 2013. *Native and National in Brazil: Indigeneity after Independence*. Chapel Hill: University of North Carolina Press.

Hall, Stuart. 1992. "The West and the Rest: Discourse and Power." In *Formations of Modernity: Understanding Modern Societies*, edited by Stuart Hall and Bram Geiben, 275–331. Cambridge, UK: Polity Press.

Hammond, C. R. 2000. *The Elements*. Boca Raton, FL: CRC Press.

Han, Jie, Lei Min, and Yuan Junbao. 2010. "China Reiterates the Rare Earth Export Policy, Hopes More Countries Will Cooperate to Strengthen Resource Exploitation." *Xinhua News Network*. Accessed December 1, 2011. http://www.miit.gov.cn/n11293472 /n11293832/n11293907/n11368277/13535078.html.

Hannis, Eric. 2012. "Are We Losing the Race for Rare Earths? U.S. National Security and Economic Health Depend upon the Development of a Domestic Rare Earths Supply Chain." *U.S. News & World Report*. Accessed April 10, 2015. http://www.usnews .com/opinion/blogs/world-report/2012/11/20/the-us-needs-rare-earth -independence-from-china.

Hansen, Kathryn. 2012. "Afghanistan's Mineral Resources Laid Bare." *Earth Magazine*, n.p.

Hao, Yufan, and Weihua Liu. 2011. "Rare Earth Minerals and Commodity Resource Nationalism." In *Asia's Rising Energy and Resource Nationalism: Implications for the United States, China, and the Asia-Pacific Region*, edited by G. B. Collins, 39–52. Washington, DC: National Bureau of Asian Research.

Hao, Yufan, and Jane Nakano. 2011. "Rare Earth Elements, Asia's Resource Nationalism, and Sino-Japanese Relations." National Bureau of Asian Research. Accessed April 11, 2015. http://www.nbr.org/research/activity.aspx?id=137#.Uh4NJz-T5bE.

Harrison, Mark, ed. 1998. *The Economics of World War II: Six Great Powers in International Comparison*. Cambridge: Cambridge University Press.

Hart, Gillian. 2002. *Disabling Globalization: Places of Power in Post-Apartheid South Africa*. Berkeley: University of California.

Harvey, David. 1996. *Justice, Nature, and the Geography of Difference*. London: Blackwell.

Harvey, David. 2005. *A Brief History of Neoliberalism*. Oxford: Oxford University Press.

Hashimoto, K., Z. Kato, N. Kumagai, and K. Izumiya. 2009. "Materials and Technology for Supply of Renewable Energy and Prevention of Global Warming." *Journal of Physics: Conference Series* 144 (1): 1–6.

Hassano, S., J. C. Biondi, and J. H. Javaroni. 1975. *Anomalia Radioativa de Uaupés-Amazonas: Relatório da Viagem*. Edited by Radam/Nuclebrás. Belem: Projeto Radam.

Hatch, Gareth P. 2008. "Going Green: The Growing Role of Permanent Magnets in Renewable Energy Production and Environmental Protection." Magnetics 2008 Conference, Denver, CO. Accessed March 3, 2013. http://www.terramagnetica.com /papers/Hatch-Magnetics-2008.pdf.

Hatch, Gareth P. 2012. "April 2012 Updates to TMR Advanced Rare Earths Project Index." Technology Metals Research. Accessed January 9, 2015. http://www.techmetalsresearch .com/2012/05/april-202-updates-to-the-tmr-advanced-rare-earth-projects-index/.

Hayes, Monte. 1986. "The Fires of Rebellion Still Burning in Some South American Countries." *Santa Cruz Sentinal*, February 9.

Haynes, Gabriella. 2014. "Colouring In: Defining White Australia's Internal Frontiers." *Agora* 49 (2): 4–12.

He, Qiang. 2009. "Study and Implement All Around Scientific Development and Comprehensively Promote the Great Western Development." *Development* 2009 (7): 75.

Headrick, Rita. 1978. "African Soldiers in World War II." *Armed Forces & Society* 4 (3): 501–26.

Heap, Tom. 2010. "Why China Holds 'Rare' Cards in the Race to go Green." BBC. Accessed May 19, 2014. http://news.bbc.co.uk/2/hi/science/nature/8689547.stm.

Hecht, Susanna. 2005. "Soybeans, Development, and Conservation in the Amazon Frontier." *Development and Change* 36:375–404.

Hecht, Susanna, and Alexander Cockburn. 1990. *The Fate of the Forest: Developers, Destroyers, and Defenders of the Amazon.* Chicago: University of Chicago Press.

Hedin, Sven Anders, Folke Bergman, Gerhard Bexell, Birger Bohlin, Gösta Montell, and Donald Burton. 1944. *History of the Expedition in Asia, Part I–IV, Reports from the Scientific Expedition to the Northwestern Provinces of China under the Leadership of Dr. Sven Hedin. The Sino-Swedish Expedition.* Sweden: Göteborg.

Hedrick, James B. 2004. "Rare Earths in Selected US Defence Applications." 40th Forum on the Geology of Industrial Minerals, Bloomington, Indiana, May 2–7.

Heiken, G. H., D. T. Vaniman, and M. French. 1991. *Lunar Sourcebook: A User's Guide to the Moon.* Cambridge: Cambridge University Press.

Heim, Barbara Ellen. 1991. "Exploring the Last Frontiers for Mineral Resources: A Comparison of International Law Regarding the Deep Seabed, Outer Space, and Antarctica." *Vanderbilt Journal of Transnational Law* 23:819–50.

Helmreich, Jonathan E. 1998. *United States Relations with Belgium and the Congo, 1940–1960.* Cranbury, NJ: Associated University Presses.

Helmreich, Jonathan E. 2014. *Gathering Rare Ores: The Diplomacy of Uranium Acquisition, 1943–1954.* Princeton: Princeton University Press.

Helmreich, S. 2009. *Alien Ocean: Anthropological Voyages in Microbial Seas.* Berkeley: University of California Press.

Hemming, John. 1978. *Red Gold: The Conquest of the Brazilian Indians.* Cambridge, MA: Harvard University Press.

Hennigan, W. J. 2011. "MoonEx Aims to Scour Moon for Rare Minerals." *Los Angeles Times,* April 8. http://articles.latimes.com/2011/apr/08/business/la-fi-moon-venture-20110408.

Hickman, Clarence. 1955. Rocket Projectile Having Discrete Flight Initiating and Sustaining Chambers. United States Patent No. US2724237 A, filed March 5, 1946, and issued November 22, 1955.

Hickman, John. 2012. "How Plausible Is Chinese Annexation of Territory on the Moon?" *Astropolitics: The International Journal of Space Politics and Policy* 10 (1): 84–92.

Hilsum, Lindsey. 2009. "Are Rare Earth Minerals Too Costly for the Environment?" MacNiel/Lehrer Productions. News Report Transcript. Accessed December 1, 2012. http://www.pbs.org/newshour/bb/asia/july-dec09/china_12-14.html.

Hinton, Jennifer J, Marcello M. Viega, and Christian Beinhoff. 2003. "Women and Artisanal Mining: Gender Roles and the Road Ahead." In *The Socio-Economic Impacts of Artisanal and Small-Scale Mining in Developing Countries,* edited by G. Hilson. Lisse, 149–88. The Netherlands: Swets & Zeitlinger B.V.

Hinton, William. 1966. *Fanshen.* Berkeley: University of California Press.

Hinton, William. 1990. *The Great Reversal: The Privatization of China, 1978–1989.* New York, New York: Monthly Review Press.

Hirano, S., and K. T. Suzuki. 1996. "Exposure, Metabolism, and Toxicity of Rare Earths and Related Compounds." *Environmental Health Perspectives* March 104 (Suppl. 1): 85–95.

Hochschild, Adam. 1999. *King Leopold's Ghost: A Story of Greed, Terror and Heroism in Colonial Africa.* Boston: Houghton Mifflin.

Hoefle, Scott William. 2013. "Beyond Carbon Colonialism: Frontier Peasant Livelihoods, Spatial Mobility and Deforestation in the Brazlian Amazon." *Critique of Anthropology* 33 (2): 193–213.

Hofstadter, Richard. 2012. *The Paranoid Style in American Politics*. London: Vintage Books.

Hogan, William Thomas. 1999. *The Steel Industry of China: Its Present Status and Future Potential*. Lanham, MD: Lexington Books.

Hoge, James, and James Hoge, Jr. 2010. *China on the World Stage*. New York: Council on Foreign Relations.

Horwitz, Morton J. 1982. "The History of the Public/Private Distinction." *University of Pennsylvania Law Review* 130 (6): 1423–28.

Hotelling, Harold. 1931. "The Economics of Exhaustible Resources." *Journal of Political Economy* 39 (2): 137–75.

Hsü, Kenneth J. 1982. "Geology in and of China: A Tale of Two Maps." *Tectonics* 1 (4): 319–23.

Hu, Jane C. 2014. "The Battle for Space." Slate. Accessed February 8, 2015. http://www .slate.com/articles/health_and_science/space_20/2014/12/space_weapon_law_u_s _china_and_russia_developing_dangerous_dual_use_spacecraft.single.html.

Hu, Jintao. 2003. "People's Republic of China Radioactive Pollution Prevention Law." *Executive Order of the People's Republic of China* 8 (June 28). http://news.xinhuanet .com/zhengfu/2003-06/30/content_944698.htm.

Hui, Zhidan. 2013. "To Stay Is to Wait for Death: Baotou Tailings Cancer Villagers Give Up Hope." *Renmin Ribao*. http://news.renminbao.com/060/2912.htm.

Humphries, Marc. 2013. *Rare Earth Elements: The Global Supply Chain*. Washington, DC: Congressional Research Service.

Hur, Jae. 2010. "China Resumes Rare Earth Exports to China, METI Says." Bloomberg News, November 23. http://www.bloomberg.com/news/2010-11-24/china-resumes -rare-earths-shipments-to-japan-trade-minister-ohata-says.html.

Hurst, Cindy. 2010. *China's Rare Earth Elements Industry: What Can the West Learn?* Washington, DC: Institute for the Analysis of Global Security.

Hyndmann, Jennifer. 2001. "Towards a Feminist Geopolitics." *Canadian Geographer* 45 (2): 210–22.

Information Office of the State Council of the People's Republic of China. 2012. "Situation and Policies of China's Rare Earth Industry." Beijing. Accessed December 30, 2012. http://www.miit.gov.cn/n11293472/n11293832/n12771663/n14676956.files /n14675980.pdf.

Instituto Socioambiental. 2001. "Militares Sequestram e Torturam Índios no Alto Rio Negro (AM)." Insituto Socioambiental. Accessed January 30, 2015. http://site-antigo .socioambiental.org/nsa/detalhe?id=1461.

Instituto Socioambiental. 2013. "Código de Mineração, Urgência Não!" Instituto Socio Ambiental Accessed January 26, 2015. http://www.socioambiental.org/pt-br/noticias -socioambientais/codigo-da-mineracao-urgencia-nao.

International Atomic Energy Agency (IAEA). 2015. "Report of the International Post-Review Mission on the Radiation Safety Aspects of the Operation of a Rare Earth Processing Facility and Assessment of the Implementation of the 2011 Mission Recommendations." Accessed June 3, 2016. https://www.iaea.org/sites/default/files /lynas-report-20052015.pdf.

International Business Pubications of the United States. 2012. *China Mineral and Mining Investment Sector Investment and Business Guide*. Vol. I, *Strategic information and important regulations*. Washington, DC: Global Investment Center.

International Seabed Authority. 2013. *Law of the Sea Compendium*. Accessed March 3, 2015. https://www.isa.org.jm/files/documents/EN/Pubs/LOS/index.html.

Israel, Adrienne. 1987. "Measuring the War Experience: Ghanaian Soldiers in World War II." *Journal of Modern African Studies* 25 (1): 159–68.

Ives, M. 2013. "Boom in Mining Rare Earths Poses Mounting Toxic Risks." *Yale Environment* 360. Accessed March 4, 2014. http://e360.yale.edu/feature/boom_in_mining_rare_earths_poses_mounting_toxic_risks/2614/.

Jacobi, P. 2009. "Seis Lagos the Largest Niobium Reserve in the World Is Still Waiting to Be Developed." Geologo.com. Accessed February 2, 2015. http://www.geologo.com.br/seislagos.asp.

Jacobs, Andrew. 2011. "Anger over Protesters' Deaths Leads to Intensified Demonstrations by Mongolians." *New York Times*, May 31, A8.

Jaggard, Victoria. 2009. "Who Owns the Moon? The Galactic Government vs. the UN." National Geographic News. Accessed February 2, 2015. http://news.nationalgeographic.com/news/2009/07/090717-who-owns-moon-real-estate.html.

Jamasmie, Cecilia. 2014. "Russia Pushes Forward Plans to Mine the Moon." Mining.com. Accessed February 5, 2015. http://www.mining.com/russia-pushes-forward-plans-to-mine-the-moon-13769/.

Jasper, William F. 2011. "U.S. Hostage to China for Rare Earth Metals." *New American*. Accessed December 4, 2012. http://www.thenewamerican.com/usnews/politics/item/3629-us-hostage-to-china-for-rare-earth-minerals.

Jeandel, C., H. Delattre, M. Grenier, C. Pradoux, and F. Lacan. 2013. "Rare Earth Element Concentrations and Nd Isotopes in the Southeast Pacific Ocean." *Geochemistry, Geophysics, Geosystems* 14 (2): 328–41.

Jefferson, Thomas. 1814. *The Federalist Papers*. National Archives and Records Administration Online. Accessed May 5, 2017. https://founders.archives.gov/documents/Jefferson/03-07-02-0167.

Jeffries, Duncan. 2014. "The Search for Transparency in a Global Gold Rush for Rare Earths." *Guardian*, February 3, https://www.theguardian.com/sustainable-business/global-mining-rare-earth-elements-transparency.

Jia, Heping, and Lihui Di. 2009. "Xu Guangxian: A Chemical Life." *Chemistry World*. Royal Society of Chemistry. Accessed April 14, 2014. https://www.chemistryworld.com/news/xu-guangxian-a-chemical-life/3004349.article.

Jiang, Zefeng. 2013. Nickel-Based Alloy Material with Good Comprehensive Performance. Beijing: State Intellectual Property Office of the People's Republic of China. International Patent Number C22C30/00.

Johnson, D. L., and L. A. Lewis. 2007. *Land Degradation: Creation and Destruction*. Lanham, MD: Rowman & Littlefield.

Johnson, Jay, Gary Pecquet, and Leon Taylor. 2007. "Potential Gains from Trade in Dirty Industries: Revisiting Lawrence Summers' Memo." *Cato Journal* 27 (3): 397–410.

Johnston, Alastair Iain. 2013. "How New and Assertive is China's New Assertiveness?" *International Security* 37 (4): 7–48.

Joint Atomic Energy Intelligence Committee. 1960. *The Chinese Communist Atomic Energy Program*. Washington, DC: Central Intelligence Agency.

Jones, Alisdair, Marina Borrell Falco, Mathilde Merenguetti, and Caroline Stern, eds. 2011. *A Mineração Brasileira*. Engineering and Mining Journal Global Business Reports, 1–45.

Jones, Nicola. 2013. "A Scarcity of Rare Metals is Hindering Green Technologies." *Yale Environment 360* (November 18). https://e360.yale.edu/features/a_scarcity_of_rare_metals_is_hindering_green_technologies.

Jones, Richard. 2010. "Inside China's Secret Toxic Unobtainium Mine." *Daily Mail*, January 10. http://www.dailymail.co.uk/news/article-1241872/EXCLUSIVE-Inside-Chinas-secret-toxic-unobtainium-mine.html.

Jones, T. D. 2013. "Mission." Accessed November 19, 2013. http://www.planetaryresources.com/mission/.

Jones, Vincent. 1985. *Manhattan, the Army, and the Atomic Bomb*. Washington, DC: US Army Center of Military History.

Jørgensen, Christian K. 1990. "Heavy Elements Synthesized in Supernovae and Detected in Peculiar A-type Stars." *Noble Gas and High Temperature Chemistry Structure and Bonding* 73:199–226.

Jowett, Phillip. 2005. *Rays of the Rising Sun*. Vol. 1, *Japan's Asian Allies 1931-45, China and Manchukuo*. Tulsa, Oklahoma: Helion.

Jucá, Senador Romero. 1996. "Projeto Lei 1610: Exploração de Recursos nas Terras Indigenas. edited by Chamber of Deputies of the Federal Legislature of Brazil." Chamber of Deputies. Accessed March 1, 2015. http://www2.camara.leg.br/atividade-legislativa/comissoes/comissoes-temporarias/especiais/54a-legislatura/pl-1610-96-exploracao-recursos-terras-indigenas.

Justo, L. J. E. C., and M. M. Souza. 1986. "Jazida de nióbio do Morro dos Seis Lagos, Amazonas." *Principais Depósitos Minerais Brasileiros—Ferro e Metais da Indústria do Aço* 2:463–68.

Kao, Shih-yang. 2014. "Views on Diaoyu/Senkaku Islands Dispute in Taiwan." Correspondence with author, December 4.

Kato, Yasuhiro, Koichiro Fujinaga, Kentaro Nakamura, Yutaro Takaya, Kenichi Kitamura, Junichiro Ohta, Ryuichi Toda, Takuya Nakashima, and Hikaru Iwamori. 2011. "Deep-Sea Mud in the Pacific Ocean as a Potential Resource for Rare-Earth Elements." *Nature Geoscience* 4 (8): 535–39. doi: 10.1038/ngeo1185.

Katzenbach, E. L., and G. Z. Hanrahan. 1955. "The Revolutionary Strategy of Mao Tse-Tung." *Political Science Quarterly* 70 (3): 321–40.

Kellner, Douglas. 2007. "Bushspeak and the Politics of Lying: Presidential Rhetoric in the War on Terror." *Presidential Studies Quarterly* 37 (4): 622–45.

Kennedy, Charles. 2013. "Pentagon Says Rare Earth Elements Less at Risk." CNBC. Accessed December 27, 2014. http://oilprice.com/Finance/investing-and-trading-reports/Pentagon-Says-Rare-Earth-Elements-Less-at-Risk.html.

Kent, Paula. 1953. Cerium Base Alloy. United States Patent No. US2642358 A, filed September 20, 1949, issued June 16, 1953.

Khrushchev, Nikita. 1971. *Khrushchev Remembers*. Translated by Strobe Talbott. New York: Bantam Books.

Kidman, R., B. Lei, W. Orton, M. Park, and J. Yan. 2012. "Securing Access to Strategic Minerals and Resources: Rare Earth Elements." Defense Logistics Agency White Paper, Washington, DC.

Kiggins, Ryan David, ed. 2015. *The Political Economy of Rare Earth Elements: Rising Powers and Technological Change*. London: Palgrave Macmillan.

King, Amy, and Shiro Armstrong. 2013. "Did China Really Ban Rare Earth Metals Exports to Japan?" East Asia Forum. Accessed December 28, 2014. http://www.eastasiaforum.org/2013/08/18/did-china-really-ban-rare-earth-metals-exports-to-japan/.

Kipnis, Andrew. 2007. "Neoliberalism Reified: Suzhi Discourse and Tropes of Neoliberalism in the People's Republic of China." *Journal of the Royal Anthropological Institute* 13 (2): 383–400.

Kirby, William C. 1984. *Germany and Republican China*. Stanford, CA: Stanford University Press.

Kjeldgaard, Dannie. 2003. "Youth Identities in Global Cultural Economy: Central and Peripheral Consumer Culture in Denmark and Greenland." *European Journal of Cultural Studies* 6 (3): 285–304.

Klare, Michael. 2008. *Rising Powers, Shrinking Planet.* New York: Henry Hold.

Klare, Michael. 2013. *The Race for What's Left: The Global Scramble for the World's Last Resources.* New York: Picador.

Klinger, Julie Michelle. 2011. Author's unpublished fieldnotes from visit to Inner Mongolia Autonomous Region Museum.

Klinger, Julie Michelle. 2013a. Author's unpublished fieldnotes from Inner Mongolia Autonomous Region Museum.

Klinger, Julie Michelle. 2013b. "Rare Earths: Lessons for Latin America." *Berkeley Review of Latin America Studies* (Fall):39–42.

Klinger, Julie Michelle. 2013c. Author's unpublished fieldnotes from Bayan Obo, Baotou, Inner Mongolia Autonomous Region, China.

Klinger, Julie Michelle. 2015a. "On the Rare Earth Frontier." PhD diss., University of California, Berkeley.

Klinger, Julie Michelle. 2015b. "Rescaling China-Brazil Investment Relations in the Strategic Minerals Sector." *Journal of Chinese Political Science* 20 (3): 227–42.

Klinger, Julie Michelle. 2017a. "Frontier Love Stories: Gendered Affect and Nation-Building in Northern China." *Gender, Place, and Culture* (under review).

Klinger, Julie Michelle. 2017b. "The Uneasy Relationship between 'China' and 'Globalization' in Post–Cold War Scholarship." In *Beyond Neoliberalism: Social Analysis after 1989,* edited by Marian Burchardt and Gal Kirn, 215–33. New York: Palgrave Macmillan.

Klotz, Irene. 2015. "Exclusive—The FAA: Regulating Business on the Moon." Reuters. Accessed February 4, 2015. http://www.reuters.com/article/2015/02/03/us-usa-moon-business-idUSKBN0L715F20150203.

Knafo, Elizabeth, and Jesse Goldstein. 2014. "The Rare Earth Catalog: Tools for Reckoning with the Anthropocene." Unpublished manuscript.

Knapp, Freyja L. 2016. "The Birth of the Flexible Mine: Changing Geographies of Mining and the E-Waste Commodity Frontier." *Environment and Planning A*:1–21.

Koerth-Baker, M. 2012. "Rare Earth Elements That Will Only Get More Important." *Popular Mechanics.* Accessed January 7, 2013. http://www.popularmechanics.com/technology/engineering/news/important-rare-earth-elements.

Kohler, Hannah. 2015. "The Eagle and the Hare: US-Chinese Relations, The Wolf Amendment, and the Future of International Cooperation in Space." *Georgetown Law Journal* 103 (4): 1135–62.

Kojevnikov, Alexei B. 2004. *Stalin's Great Science: The Times and Adventures of Soviet Physicists.* London: Imperial College Press.

Koman, Richard. 2007. "Moon 2.0: Google Funds $30 Million Lunar X Prize." NewsFactor, Accessed February 2, 2014.

Kopenawa, Davi, and Bruce Albert. 2013. *The Falling Sky: Words of a Yanomami Shaman.* Cambridge, MA: Harvard University Press.

Korinek, Jane, and Jeonghoi Kim. 2010. "Export Restrictions on Strategic Raw Materials and Their Impact on Trade and Global Policy." In *OECD Trade Policy Studies,* 103–30. Paris: OECD Publishing.

Korporaal, Glenda. 2016. "Lacaze Puts Struggling Lynas Back on Track." *The Australian,* May 16. http://www.theaustralian.com.au/business/companies/amanda-lacaze-puts-struggling-lynas-back-on-track/news-story/3e9e6a2ef4b1fb0635e9f55937a9b38f.

Koslofsky, Craig. 2011. *Evening's Empire: A History of the Night in Early Modern Europe.* Cambridge: Cambridge University Press.

Kosynkin, V. D., S. D. Moiseev, C. H. Peterson, and B. V. Nikipelov. 1993. "Rare Earths Industry of Today in the Commonwealth of Independent States." *Journal of Alloys and Compounds* 192:118–20.

Kraemer, Susan. 2010. "California to Mine Rare Earth Materials Again." *Clean Technica*, December 28. http://cleantechnica.com/2010/12/28/california-to-mine-rare-earth-materials-again/.

Kraus, Charles. 2010. "Creating a Soviet Semi-Colony? Sino-Soviet Cooperation and Its Demise in Xinjiang, 1949–1955." *Chinese Historical Review* 17 (2): 129–65.

Krebs, Robert E. 2006. *The History and Use of Our Earth's Chemical Elements: A Reference Guide*. 2nd ed. Westport, CT: Greenwood Press.

Krieckhaus, Jonathan Tabor. 2006. *Dictating Development: How Europe Shaped the Global Periphery*. Pittsburgh, PA: University of Pittsburgh Press.

Krishnamurthy, Nagaiyar, and C. K. Gupta. 2005. *Extractive Metallurgy of Rare Earths*. Boca Raton, FL: CRC Press.

Krugman, Paul. 2010. "Rare and Foolish." *New York Times*, October 18, Opinion.

Kudlow, Larry. 2011. "How to Combat an Arrogant China?" *National Review*, January 18. http://www.nationalreview.com/node/257420/print.

Kulacki, Gregory. 2014. *An Authoritative Source on China's Military Space Strategy*. Cambridge, MA: Union of Concerned Scientists.

Kurlantzick, Joshua. 2008. *Charm Offensive: How China's Soft Power Is Transforming the World*. New York: New Republic Books.

Lahiri-Dutt, Kuntala, and Martha Macintyre, eds. 2011. *Women Miners in Developing Countries: Pit Women and Others*. Aldershot, UK: Ashgate.

Laïdi, Zaki. 2014. "Towards a Post-Gegemonic World Order: The Multipolar Threat to the Multilateral Order." *International Politics* 51 (3): 350–65.

Lakatos, Christine. 2014. " 'Cleantech Crash': CBS Lesley Stahl 'Exhausted'after Listing Nine Obama-Backed Green Energy Failures, Try 32 and Countiing." Accessed January 7, 2015. http://greencorruption.blogspot.com/2014/01/cleantech-crash-cbs-lesley-stahl.html - .VK68IydaYag.

Lamb, R. 2010. "The Ethics of Planetary Exploitation and Colonization." Discovery News. Accessed February 2. http://news.discovery.com/space/astronomy/the-ethics-of-planetary-exploration-and-colonization.htm.

Landry, Benjamin David. 2013. "A Tragedy of the Anticommons: The Economic Inefficiencies of Space Law." *Brooklyn Journal of International Law* 30:523–78.

Lang, Ian. 2015. "China's Lunar Plan Revealed—Strip Mine the Moon for Rare Minerals." National Monitor. Accessed February 2, 2015. http://thediplomat.com/2015/01/china-leads-race-to-the-moon/.

Lapido-Loureiro, Francisco Eduardo. 2013. *O Brasil e a Reglobalização da Indústria das Terras Raras*. Rio de Janeiro: Ministerio de Comércio, Tecnologia e Inovação.

Latour, Bruno. 1987. *Science in Action: How to Follow Scientists and Engineers through Society*. Cambridge, MA: Harvard University Press.

Lattimore, Owen. 1962. *Studies in Frontier History. Collected Papers: 1928–1958*. London: Oxford University Press.

Lauren, Paul, Gordon A. Craig, and Alexander L. George. 2007. *Force and Statecraft: Diplomatic Problems of Our Time*. 4th ed. New York: Oxford University Press.

Lave, R., M. W. Wilson, E. S. Barron, C. Biermann, M. A. Carey, C. S. Duvall, L. Johnson, K. M. Lane, N. McClintock, D. Munroe, R. Pain, J. Proctor, B. L. Rhoads, M. M. Robertson, J. Rossi, N. F. Sayre, G. Simon, M. Tadaki, and C. Van Dyke. 2014. "Intervention: Critical Physical Geography." *The Canadian Geographer/Le Géographe canadien* 58:1–10.

Laver, Michael. 1986. "Public, Private and Common in Outer Space: Res Extra Commercium or Res Communis Humanitatis Beyond the High Frontier?" *Political Studies* 34 (3): 359–73.

Lawn, Philip, and Matthew Clarke, eds. 2008. *Sustainable Welfare in the Asia-Pacific.* Northampton, MA: Edward Elgar.

Lazaro, Christine. 2014. "NASA Invites Private Companies to Apply for Mining on the Moon." *International Business Times,* February 12.

Lazenby, Henry. 2013. "Alternative Rare Earths Remain Elusive Three Years After Curbs." *Mining Weekly,* May 17.

Le Bas, M. J. 2006. "Reinterpretation of Zircon Date in a Cabonatite Dyke at the Bayan Obo Giant REE-Fe-Nb Deposit, China." *Acta Petrologica Sinica* 22 (2): 517–18.

Le Bas, M. J., J. Kellerre, T. Kejie, F. Wall, C. T. William, and P. S. Zhang. 1992. "Carbonatite Dykes at Bayan Obo, Inner Mongolia, China." *Minerology and Petrology* 46 (3): 195–228.

Le Billon, Phillipe. 2004. "The Geopolitical Economy of 'Resource Wars.'" *Geopolitics* 9 (1): 1–28.

Lee, Ching Kwan. 2014. "The Specter of Global China: Contesting the Power and Peril of Chinese State Capital in Zambia." Global China Colloquium I, University of California, Berkeley, September 12.

Lee, Yoolim. 2011. "Malaysia Rare Earths in Largest Would-Be Refinery Incite Protest." Bloomberg. Accessed June 5, 2014. http://www.bloomberg.com/news/articles/2011 -05-31/malaysia-rare-earths-in-largest-would-be-refinery-incite-protest.

Lefebvre, Henri. 1991. *The Production of Space.* Oxford: Blackwell.

Lefebvre, Henri. 2009. *State, Space, World.* Edited by Neil Brenner and Stuart Elden. Minneapolis: University of Minnesota Press.

Leifert, Harvey. 2010. "Restarting U.S. Rare Earth Production?" American Geosciences Union. Accessed January 12, 2015. http://www.earthmagazine.org/article/restarting -us-rare-earth-production.

Lemon, John J. 1972. "Edward Nixon Explains His Work in Campaign." *Spokane Daily Chronicle,* 4. Accessed 12 April 2015. http://news.google.com/newspapers?nid =1338&dat=19720904&id=SqhYAAAAIBAJ&sjid=svgDAAAAIBAJ&pg =7305,601818.

Levy, Stanley Isaac. 1915. *The Rare Earths: Their Occurence, Chemistry, and Technology.* London: E. Arnold.

Lewis, John. 1996. *Mining the Sky: Untold Riches from the Asteroids, Comets, and Planets.:* New York: Perseus Books.

Lewis, John Wilson, and Litai Xue. 1988. *China Builds the Bomb.* Stanford, CA: Stanford University Press.

Li, Chengdong, and Peng Zhang. 2012. "Fully Promote Baotou Environmental Optimization Economic Development." *Environmental Protection* 2012 (12): 58–60.

Li, Danhui. 1999. "Historical Investigations on the Origins of the 1962 Xinjiang Yita Incident." *Research Data of the Times* 5:1–22.

Li, Deqian, Yong Zuo, and Shulan Meng. 2004. "Separation of Thorium (IV) and Extracting Rare Earths from Sulfuric and Phosphoric Acid Solutions by Solvent Extraction Method." *Journal of Alloys and Compounds* 374 (1–2): 431–33. doi: http://dx.doi.org /10.1016/j.jallcom.2003.11.055.

Li, Guangshou. 2010. "Baotou Rare Earth Mining Causes Ecological Crisis; Crop Failures and Groundwater Contamination." Sina Finance. Accessed November 28, 2014. http://finance.sina.com.cn/roll/20101202/03429042460.shtml.

Li, Hongyan 1987. "Rare Earth Industrial Foundation: Baotou's Environmental Radioactivity Level." *Rare Earths* 1987 (6): 6.

Li, Jiagu. 1998. "Remarks on the Impact of the 'Soviet-Japan-China Accord' on China's War of Resistance." *Japanese Resistance War Research* 1:54–73.

Li, Jue, Rongtian Lei, Yi Li, and Yingxiang Li, eds. 1987. *Contemporary China's Nuclear Industry*. Beijing: China Academy of Social Sciences Press.

Li, Kuo Chin. 1955. *Tungsten: Its History, Geology, Ore-Dressing, Metallurgy, Chemistry, Analysis, Applications, and Economics*. London: Chapman and Hall.

Li, Lin. 1990. "Proceedings of the Second Annual Meeting of China Rare Earth Research Society." Baotou, Inner Mongolia.

Li, Rong. 2005. *History of the Peoples of China War of Resistance against Japan*. Beijing: Central Literature Publishing House

Li, Rose Maria. 1989. "Migration to China's Northern Frontier, 1953–82." *Population and Development Review* 15 (3): 503–38.

Li, Shenxue, Liu Zi, and Zhao Yiqing. 2012. "The Problems and Policy Responses to My Country's Rare Earth Production and Development." *China Population, Resources, and Environment* 22:196–201.

Li, Xirong. 1996. "Changes in Geology's Development Focus and Ideology." *Geology of Inner Mongolia:* 1–10.

Li, Y., J. L. Yang, and Y. Jiang. 2012. "Trace Rare Earth Element Detection in Food and Agriculture Products Based on Flow Injection Walnut Shell Packed Microcolumn Preconcentration Coupled with Inductively Coupled Plasma Mass Spectrometry." *Journal of Agriculture and Food Chemistry* 60 (12): 3033–41.

Li, Yunping, Xiangyi Quan, De Chen, Liang Lin, and Zhaojun Wang. 2003. "The Impact of Endemic Flourosis of Children's Intellectual Development in Baotou." *China Public Health Management* 19 (4): 337–38.

Liang, Jun. 2010. "'No Ban' on Exports of Rare Earths to Japan." *People's Daily*, September 25. http://en.people.cn/90001/90776/90883/7149176.html.

Lima, Júnior. 2011. "Minérios Sob o Solo do Amazonas São Avaliados em R$ 4,3 trilhões." *Diário do Amazonas*, December 11. http://new.d24am.com/noticias /economia/minerios-sob-o-solo-do-amazonas-sao-avaliados-em-r-43-trilhes /44177.

Lima, Paulo César Ribeiro. 2012. *Rare Earths: Strategic Elements for Brazil*. Brasília: Legislative Consultancy for Area XII: Mineral, Hydro, and Energy Resources.

Lin, Chun. 2009. "The Socialist Market Economy: Step Forward or Step Nackward for China?" *Science & Society* 73 (2): 228–35.

Lin, Justin Yifu. 1990. "Collectivization and China's Agricultural Crisis: 1959–1961." *Journal of Political Economy* 98 (6): 1228–52.

Lin, Justin Yifu. 2011. *Demystifying the Chinese Economy*. Cambridge: Cambridge University Press.

Ling, Ming-Xing, Yu-Long Liu, Ian S. Williams, Fang-Zhen Teng, Xiao-Yong Yang, Xing Ding, Gang-Jian Wei, Lu-Hua Xie, Wen-Feng Deng, and Wei-Dong Sun. 2013. "Formation of the World's Largest REE Deposit Through Protracted Fluxing of Carbonatite by Subduction-Derived Fluids." *Scientific Reports* 3 (1776): 1–8.

Ling, Zongcheng, Bradley L. Jolliff, Alian Wang, Chunlai Li, Jianzhong Liu, Jiang Zhang, Bo Li, Lingzhi Sun, Jian Chen, Long Xiao, Jianjun Liu, Xin Ren, Wenxi Peng, Huanyu Wang, Xingzhu Cui, Zhiping He, and Jianyu Wang. 2015. "Correlated Compositional and Mineralogical Investigations at the Chang'e-3 Landing Site." *Nature Communications* 6 (8880): 1–9. doi: 10.1038/ncomms9880.

Lissauer, J., J. Barnes, and J. Chambers. 2011. "Obliquity Variations of a Moonless Earth." *Icarus* 217:77–87.

Listner, Michael. 2011. "A First Look at Austria's New Domestic Space Law." *Space Review* (December): n.p.

Listner, Michael. 2012. "The Moon Treaty: It Isn't Dead Yet." *Space Review* (March): n.p.

Liu, Chunzi. 2009. "Study on the Development of Christianity after the Founding of Inner Mongolia Autonomous Region: Hohhot and Baotou Churches as the Primary Objects of Study." *Mongolia Studies Periodical* 2:1–5.

Liu, Hongzhong. 1996. "A Review of Radioactivity in Rare Earth Production and Its Effects on Human Health." *Rare Earths* 17 (4): 59–64.

Liu, Lican. 2013. *Strong Country, Cancer Village*. Hong Kong: Mingpao.

Liu, Q. B., W. Y. Wang, X. H. Liu, H. L. Li, S. H. Li, F. J. Feng, G. Wang, L. Z. Wang, Z. J. Wang, Q. Bao, D. S. Ying, M. R. Zhou, B. T. Guo, X. K. Zhang, and J. Y. Pan. 2005. "Investigation on Combined Toxicosis of Fluorine and Aluminum Associated in Chen Barag Qi, Inner Monglia." *Chinese Journal of Endemiology* 24 (1): 50–52.

Liu, S. H. 1978. "Electronic Structure of Rare Earth Metals." In *Handbook on the Physics and Chemistry of Rare Earths*, 234–336. Amsterdam: Elsevier.

Liu, Xiaoyuan. 2006. *Reins of Liberation: An Entangled History of Mongolian Independence, Chinese Territoriality, and Great Power Hegemony, 1911–1950*. Washington, DC: Woodrow Wilson Center.

Loewenstein, Antony. 2014. "Australian Uranium Mining Company in Greenland is Tearing the Country in Half." *The Ecologist*. Accessed December 20, 2014. http://www .theecologist.org/News/news_analysis/2399055/australian_uranium_mining_in _greenland_is_tearing_the_country_in_half.html.

Lucey, P. G, G. J Taylor, and E. Malaret. 1995. "Abundance and Distribution of Iron on the Moon." *Science* 268:1150–53.

Lucius, Casey J. 2014. "China's Critical Resources Strategy." In *China and International Security: History, Strategy, and 21st-Century Policy*, edited by C. Donovan and Thomas M. Kane Chau, 169–80. Santa Barbara, CA: ABC-CLIO.

Lujan, N., and J. W. Armbruster. 2011. "The Guiana Shield." In *Historical Biogeography of Neotropical Freshwater Fishes*, edited by N. Lujan and J. W. Armbruster, 211–24. Berkeley: University of California Press.

Luo, Guihuan. 2007. "An Introduction to the Participants of the Northwestern Scientific Expedition Team (1927–1933)." *Studies in the History of Natural Sciences* 27 (Suppl.): 71–82.

Luo, Z. D. 1993. "Epidemiological Survey on Chronic Arsenic Poisoning in Inner Mongolia." *Journal of Endemic Disease of Inner Mongolia* 18:4–6.

Luxemburg, Rosa. [1913] 2004. *The Accumulation of Capital*. New York: Routledge.

Lynas Corporation, Ltd. 2009. *Quarterly Report for the Period Ending 30 September 2009*. Sydney, Australia: Lynas Corporation Ltd. Corporate Office.

Lynas Corporation, Ltd. 2016. "Our Commitments and Responsibilities." Accessed June 10, 2016. https://www.lynascorp.com/Pages/Environment.aspx.

Ma, Caimei. 1995. "The Position and Use of Inner Mongolian Anti-Japanese Struggles in China's War of Resistance." *Yin Shan Academic Journal* 3: 8–13.

Ma, Pengqi, Yongsheng Gao, and Laizi Yu. 2009. "Complexified Use and Environmental Protection of Baotou Bayan Obo Resources." *Decision-Making and Consultation Newsletter* 2009 (2): 88–92.

Macalister, Terry. 2013. "Greenland Government Falls as Voters Send Warning to Mining Companies." *Guardian*, March 15. http://www.theguardian.com/world/2013/mar /15/greenland-government-oil-mining-resources.

Macdonald, Fraser. 2007. "Anti-Astropolitik: Outer Space and the Orbit of Geography." *Progress in Human Geography* 31 (5): 592–615.

Machacek, Erika, and Niels Fold. 2014. "Alternative Value Chains for Rare Earths: The Anglo-Deposit Developers." *Resources Policy* 42 (December): 53–64.

MacMillan, Gordon. 1995. *At the End of the Rainbow? Gold, Land, and People in the Brazilian Amazon*. New York: Columbia University Press.

Maginnis, Robert. 2011. "America Brought China's Rare-Earth Elements Monopoly on Itself." Accessed October 3, 2012. http://www.humanevents.com/2011/10/13/america-brought-chinas-rareearth-elements-monopoly-on-itself/.

Maize, K. 2012. "End Game for Rare Earth Dispute?" *Managing Power.* Accessed December 3, 2013. http://www.managingpowermag.com/supply_chains/End-Game-for-Rare-Earth-Dispute_379_p2.html.

Malthus, Thomas. 1798. *An Essay on the Principle of Population.* London: J. Johnson.

Mancheri, Nabeel, Lalitha Sundarasan, and S. Chandrashekar. 2013. *Dominating the World: China and the Rare Earth Industry.* Bangalore, India: National Institute of Advanced Studies.

Mao, Guangyun, Guo Xiaojuan, Kang Ruiying, Ren Chunsheng, Yang Zuopeng, Sun Yuansheng, Zhang Chuanwu, Zhang Xiaojing, Zhang Haitao, and Wei Yang. 2010. "Prevalence of Disability in an Arsenic Exposure Area in Inner Mongolia, China." *Chemosphere* 80:978–81.

Mao, Zedong. 1967. *Selected Works.* Beijing: Foreign Languages Press.

Margonelli, Lisa. 2009. "Clean Energy's Dirty Little Secret." *Atlantic,* May 1.

Marques, Adriana Aparecida. 2007. "Amazônia: Pensamento e Presença Militar." PhD diss., University of São Paulo.

Martin, Geoffrey, ed. 1915. *The Rare Earth Industry: Including the Manufacture of Incandescent Mantles, Pyrophoric Alloys, and Electrical Glow Lamps, Together with a Chapter on The Industry of Radioactive Substances.* London: Crosby Lockwood and Son.

Martin, Lockheed. n.d. "Lockheed Marting Awareded Major Contract for Brazil's SIVAM Project." PRNewswire. Accessed January 26, 2015. http://www.prnewswire.com/news-releases/lockheed-martin-awarded-major-contract-for-brazils-sivam-project-75021222.html.

Martinez, Mark Anthony. 2009. *The Myth of the Free Market: The Role of the State in a Capitalist Economy.* Sterling, VA: Kumarian Press.

Martyn, Paul. 2012. "Rare Earth Minerals: An End to China's Monopoly is in Sight." *Forbes,* June 8. http://www.forbes.com/sites/ciocentral/2012/06/08/rare-earth-minerals-an-end-to-chinas-monopoly-is-in-sight/.

Mason, Paul. 2012. *Rare Earth.* New York: OR Books.

Matich, Teresa. 2015. "Rare Earths Outlook 2015: Another Tough Year Ahead." Shanghai Metals Market. Accessed April 12, 2015. http://www.metal.com/newscontent/68515_rare-earths-outlook-2015-another-tough-year-ahead.

Mazurkewich, Karen, and Norma Greenaway. 2010. "Motherlode in Afghanistan? U.S. Identifies US$ 1-Trillion in Mineral Deposits." *National Post,* 1.

Mbembe, J. A., and L. Meintjes. 2003. "Necropolitics." *Public Culture* 15 (1): 11–40.

McGwin, Kevin. 2014. "Uranium Ban Overturned." *Arctic Journal,* October 24. http://arcticjournal.com/oil-minerals/211/uranium-ban-overturned.

McMahon, Robert J. 1994. *The Cold War on the Periphery: The United States, India, and Pakistan.* New York: Columbia University Press.

McMichael, Philip. 1990. "Incorporating Comparison within a World-Historical Perspective: An Alternative Comparative Method." *American Sociological Review* 55 (3): 385–97.

McSween, Jr., Y. Harry, and Gary R. Huss. 2010. *Cosmochemistry.* Cambridge: Cambridge University Press.

Mehmood, Amber. 2009. "Chinese Coal Mines: The Industrial Death Trap." *Journal of Pakistan Medical Association* 59 (9): 649–50.

Mehrotra, Kartikay. 2013. "Karzai Woos India Inc. as Delay on U.S. Pact Deters Billions." Bloomberg News. Accessed December 24, 2014. http://www.bloomberg.com/news

/2013-12-15/karzai-tells-investors-u-s-will-meet-his-security-pact-demands .html.

Meléndez-Ortiz, Ricardo, Christophe Bellmann, and Miguel Rodriguez Mendoza, eds. 2012. *The Future and the WTO: Confronting the Challenges*. Geneva, Switzerland: International Center for Trade and Sustainable Development.

Mendelejew, Dmitri. 1869. "Über die Beziehungen der Eigenschaften zu den Atomgewichten der Elemente." *Zeitschrift für Chemie* 12:405–6.

Meng, Qirun. 1982. "The Genesis of the Host Rock Dolomite of Bayan Obo Iron Deposits and the Analysis of Its Sedimentary Environment." *Geological Review* 1982 (5): 481–89.

Menkhaus, Ken. 2003. "Quasi-States, Nation-Building, and Terrorist Safe Havens." *Journal of Conflict Studies* 23 (2): n.p.

Merchant, Carolyn. 1980. *The Death of Nature: Women, Ecology, and the Scientific Revolution*. San Francisco: Harper Collins.

Merchant, Carolyn. 2003. "Shades of Darkness: Race and Environmental History." *Environmental History* 8 (3): 380–94.

Merrill, Dennis. 1990. *Bread and the Ballot: The United States and India's Economic Development*. Chapel Hill: University of North Carolina Press.

Mertie, John B., Jr. 1953. *Monazite Deposits of the Southeastern Atlantic States*. Washington, DC: United States Geological Survey.

Messias da Costa, Wanderley. 2014 "Projeção do Brasil no Atlântico Sul: Geopolítica e Estratégia." *Confins: Revista Franco-Brasileira de Geografia*. Accessed December 4, 2014. http://confins.revues.org/9839.

Meyer, L. 2011. "Rare Earth Metal Recycling." Paper presented at IEEE International Symposium on Sustainable Systems and Technology, Chicago, IL, May 16–18.

Mignolo, Walter D. 2009. "Epistemic Disobedience, Independent Thought, and Decolonial Freedom." *Theory Culture Society* 26:159–81.

Ministry of Commerce of the People's Republic of China. 2012. "Catalogue for the Guidance of Foreign Investment Industries (Amended 2011)." Accessed October 10, 2014. http://www.mofcom.gov.cn/article/bh/201204/20120408086699.shtml.

Ministry of Commerce of the People's Republic of China. 2013a. "Ministry of Commerce Issues Rare Earth Quotas for the Second Half of 2013." Accessed December 5, 2013. http://www.mofcom.gov.cn/article/h/redht/201307/20130700181591.shtml.

Ministry of Commerce of the People's Republic of China. 2013b. "The Second Round of Rare Earth Export Quotas is 390 Tonnes Greater than the First Round." Accessed November 26, 2014. http://pep.mofcom.gov.cn/article/j/cb/201407/20140700665428.shtml.

Miranda, Evaristo Eduardo de. 2007. *Quando o Amazonas Corria para o Pacífico*. Petrópolis: Editora Vozes.

MNO. 1958. "Recorded Conversation of PRC Chairman Mao Zedong with the Czechoslovak Military Delegation." July 17. Central Military Archive, Prague. Box 25/50/4/2.

Molycorp. 2012. "Health Care." Accessed February 9 2013. http://www.molycorp.com /products/rare-earths-many-uses/health-care/.

Molycorp. 2013. "Rare Earth Recycling." Accessed November 19 2013. http://www .molycorp.com/technology/rare-earth-recycling/.

Molycorp. 2015. "Molycorp Reports Fourth Quarter and Fully Year 2014 Financial Results." Accessed May 2, 2015. http://www.molycorp.com/molycorp-reports-fourth-quarter -full-year-2015-financial-results/.

Momsen Jr., Richard P. 1979. "Projeto Radam: A Better Look at the Brazilian Tropics." *GeoJournal* 3 (1): 3–14.

Monahan, Ken. 2012. *China's Rare-Earth Policies Erode Technology Profits*. Washington, DC: Bloomberg Government Report.

Moon Express. 2013. "Return." Accessed November 19, 2013. http://www.moonexpress
.com/missions.html - future.

Moon Express. 2015a. "Company." Accessed February 2, 2015. http://www.moonexpress
.com/company.

Moon Express. 2015b. "Resource." Accessed January 3, 2016. http://www.moonexpress
.com/resource.

Morena, Simon J. 1956. Rare Earth Master Alloys. Edited by United States Patent Office.
Washington, DC: Beryllium Corp.

Mortier, Jan, and Benjamin Finnis. 2015. "China Leads Race to the Moon." Diplomat. Accessed February 2, 2015. http://thediplomat.com/2015/01/china-leads-race-to-the
-moon/.

Moskowitz, Clara. 2011. "China Will Own the Moon, Space Entrepreneur Worries." Space
.com. Accessed June 10, 2014. http://www.space.com/13331-china-space-race
-moon-ownership-bigelow-ispcs.html.

Mote, F. W. 1999. Imperial China: 900–1800. Cambridge, MA: Harvard University Press.

Muldavin, Joshua. 1993. Untitled. Unpublished manuscript.

Muldavin, Joshua. 2000. "The Paradoxes of Environmental Policy and Resource Management in Reform-Era China." Economic Geography 76 (3): 244–71.

Muldavin, Joshua. 2003. Untitled. Unpublished manuscript.

Mumford, Lewis. 1934. Techniques and Civilization. New York: Harcourt Brace.

Mumford, Lewis. 1967. The Myth of the Machine. Vol. 2, The Pentagon of Power. New York:
Harcourt Brace Jovanovich.

Nabuco, Joaquim, Francisco Xavier Ribeiro de Sampaio, and Francisco José Rodriguez. 1903.
Annexes du Premier Mémoire du Brésil. Vol. 3, Documents D'origine Portugaise. Paris.

Nachukdorji, Sh. 1955. Nationalism and Revolution in Mongolia. Translated by Owen Lattimore. New York: Oxford University Press.

Najem, G. R., and L. K. Voyce. 1990. "Health Effects of a Thorium Waste Disposal Site."
American Journal of Public Health 80 (4): 478–80.

Nascimento, Messias Luiz do, and Alcindo José de Sá. 2008. "Quinta Pelotão Especial da
Fronteira: Territorialidades e Temporalidades na 'Cabeça do Cachorro-AM.'" Revista de Geografia 23 (2): 34–54.

National Development and Reform Commission. 2005. "Ten Major Issues in China's Rare
Earths in 2004." National Development and Reform Commission. Accessed November 30, 2013. http://www.ndrc.gov.cn/gyfz/gyfz/t20050707_27869.htm.

National People's Congress. 1955. First Five-Year Plan for Development of the National Economy of the People's Republic of China in 1953–1957. Edited by State Council of the
People's Republic of China. Beijing: Foreign Languages Press.

National Public Radio. 2013. "China Shoots 'Jade Rabbit' Rover to the Moon." Accessed
January 10, 2015. http://www.npr.org/2013/12/06/249261283/china-shoots-jade
-rabbit-rover-to-the-moon.

Navarro, Alyssa. 2015. "President Obama Signs Pro-Asteroid Mining Bill into Law." Tech
Times. Accessed January 5, 2015. http://www.techtimes.com/articles/110935
/20151127/president-obama-signs-pro-asteroid-mining-bill-into-law.htm.

"Negotiations between the United States and China Respecting Joint Efforts in the Explo-
ration of China for Minerals of Importance in the Atomic Energy Programs of the
Two Governments." United States Ambassador (Stuart) in China to the United State
Secretary of State 1948. Foreign Relations of the United States 8:1018–29.

Neumann, Roderick P. 1998. Imposing Wilderness: Struggles Over Livelihood and Nature
Preservation in Africa. Berkeley: University of California Press.

Neves, Eduardo Góes. 1998. "Paths in the Dark Waters: Archaeology as Indigenous History in the Upper Rio Regro Basin, Northwest Amazon." PhD, Indiana University.

Nevins, J. 2004. "Contesting the Boundaries of International Justice: State Countermapping and Offshore Resource Struggles between East Timor and Australia." *Economic Geography* 80 (1): 1–22.

Niinistö, Lauri. 1987. "Industrial Applications of Rare earths, an Overview." *Inorganica Chimica Acta* 140:339–43.

Nonini, Donald M. 2008. "Is China Becoming Neoliberal?" *Critique of Anthropology* 28 (2): 145–76.

Nutman, A., B. F. Windley, and W. J. Xiao. 2007. "Structural and Tectonic Correlation Across the Central Asian Orogenic Collage: Implications for Continental Growth and Intracontinental Deformation." In *Guidebook, IGCP–480: Geological Excursion to Inner Mongolia, China, to Study the Accretionary Evolution of the Southern Margin of the Central Asian Orogenic Belt*, edited by P. Jian, YR Shi, FQ Zhang, LC Miao, LQ Zhang, A. Kroner, 50–72. Beijing: China Academy of Science.

Nuttall, Mark. 2012. "Imagining and Governing the Greenlandic Resource Frontier." *Polar Journal* 2 (Special Issue: Politics, Science and Environment in the Polar Regions): 113–24.

"Obituário: Mateus Marcili dos Santos Silva." 2011. *Zero Hora*, July 9. http://zh.clicrbs.com .br/rs/obituario/mateus-marcili-dos-santos-silva-24344.html.

O'Carroll, Eoin. 2011. "Marie Curie: Why Her Papers Are Still Radioactive." *Christian Science Monitor*, November 7. http://www.csmonitor.com/Innovation/Horizons /2011/1107/Marie-Curie-Why-her-papers-are-still-radioactive.

Office of the United States Trade Representative. 2014a. "United States Wins Victory in Rare Earths Dispute with China: WTO Report Finds China's Export Restraints Breach WTO Rules." In *U.S. Wins Case against China, Helps American Manufacturers Compete*. Online: Office of the United States Trade Representative. Accessed January 5, 2015. http://www.ustr.gov/about-us/press-office/press-releases/2014 /March/US-wins-victory-in-rare-earths-dispute-with-China.

Office of the United States Trade Representative. 2014b. "U.S. Trade Representative Michael Froman Announces U.S. Victory in Challenge to China's Rare Earth Export Restraints." In *U.S. Wins Case against China, Helps American Manufacturers Compete*. Online: Executive Office of the President. Accessed January 5, 2015. http://www .ustr.gov/about-us/press-office/press-releases/2014/August/USTR-Froman-US -Victory-in-Challenge-to-China-Rare-Earth-Export-Restraints.

Oldroyd, David Roger. 1996. *Thinking about the Earth: A History of Ideas in Geology*. London: Althlone Press.

Oliveira, José Carlos. 2013. Código de Mineração: Um Raio-x do Setor e a Necessidade de Reformas. In *Rádio da Câmara*. Brasilia, Brazil. Accessed March 13, 2014. http:// www2.camara.leg.br/camaranoticias/radio/materias/REPORTAGEM-ESPECIAL /457860-CODIGO-DE-MINERACAO-UM-RAIO-X-DO-SETOR-E-A -NECESSIDADE-DE-REFORMAS.html.

Oliveira, José Lopes de. 1968. *Grande Enciclopédia da Amazônia*, edited by Carlos Rocque. Vol. 6, *Fortificações da Amazônia*. Belém do Pará: Amazônia.

Olson, J. C., D. R. Shawe, L. C. Pray, and W. N. Sharp. 1954. "Rare-Earth Mineral Deposits of the Mountain Pass District, San Bernardino County, California." *Geological Survey Professional Paper 261*. Washington, DC: United States Government Printing Office.

Olsvig, Sara. 2013. "Greenland's Decision-Making on Uranium: Towards a Democratic Failure." *Arctic Journal*, October 18. http://arcticjournal.com/opinion/greenlands -decision-making-uranium-towards-democratic-failure.

O'Neill, Gerard K. 2000. *The High Frontier: Human Colonies in Space*. 3rd ed. Burlington, Ontario, Canada: Apogee Books.

O'Neill, Ian. 2013. "China's Rover Rolls! Yutu Gegins Moon Mission." Discovery News, December 14. http://news.discovery.com/space/chinas-rover-rolls-yutu-begins -moon-mission-131214.htm.

Ong, Aihwa. 2007. "Neoliberalism as a Mobile Technology." *Transactions of the Institute of British Geographers* 32 (1): 3–8.

Orris, G. J., and J. D. Bliss. 2002. "Mines and Mineral Occurrences in Afghanistan." United States Geological Survey. Accessed February 20, 2015. http://geopubs.wr.usgs.gov /open-file/of02-110/.

Orris, Greta J., and Richard I. Grauch. 2013. "Rare Earth Elements Mines, Deposits, and Occurrences." *Mineral Resources Online Spatial Data Catalog.* Washington, DC: United States Geological Survey, Department of the Interior.

Osborne, Hannah. 2015. "Mining the Moon: US Firm Plans 'Gas Station in Space' as China Searches for Rare Earth Elements." International Business Times. Accessed February 2, 2015. http://www.ibtimes.co.uk/mining-moon-us-firm-plans-gas-station -space-china-searches-rare-earth-elements-1486289.

Oskin, Becky. 2013. "Radioactive Mountain Is Key in US Rare Earth Woes." *Live Science,* June 11. http://www.livescience.com/37356-heavy-rare-earth-mining-america.html.

Ostini, Marco. 2011. "Return to the Moon." In *Big Ideas: The Smartest Stuff on TV, Radio & Online.* Australian Broadcasting Corporation. Accessed April 14, 2015. http:// www.abc.net.au/tv/bigideas/stories/2011/03/08/3155115.htm.

Ouyang, Ziyuan, and Jianzong Liu. 2014. "Compiled Research on the Formation and Evolution of the Moon with Lunar Geological Map." *Earth Science Frontiers* 21 (6): 1–6.

Oyunbilig. 1997. "Inner Mongolia was Never Part of China." Accessed July 17, 2011. http:// www.innermongolia.org/english/index.html.

Pandey, Gyanendra. 2001. *Remembering Partition: Violence, Nationalism and History in India.* Cambridge: Cambridge University Press.

Parry, Simon, and Ed Douglas. 2011. "In China, the True Cost of Britain's Clean, Green Wind Power Experiment: Pollution on a Disasterous Scale." Daily Mail, January 26. http://www.dailymail.co.uk/home/moslive/article-1350811/In-China-true-cost -Britains-clean-green-wind-power-experiment-Pollution-disastrous-scale.html.

Parthemore, Christine. 2011. "Statement of Christine Parthemore, Fellow, Center for a New American Security. Hearing on 'China's Monopoly on Rare Earths: Implications for U.S. Foreign and Security Policy.'" In *Rare Earth Elements: Supply, Trade, and Use Dynamics,* edited by J. O. Manino and E. T. Jones, 77–82. New York: Nova Science.

Paul, Sonali. 2015. "China's Rare Earth Quotas Go, Possible New Moves Stoke Supply Doubts." Reuters, January 7. http://www.reuters.com/article/2015/01/07/china -rareearths-producers-idUSL3N0UL65220150107.

Pearson, Sophia. 2012. "Molycorp Directors, Majority Shareholders Sued over Stock 'Price Bubble.'" Bloomberg Businessweek. Accessed December 27, 2014. http://www .bloomberg.com/news/2012-02-27/molycorp-directors-majority-shareholders -sued-over-stock-price-bubble-.html.

Peck, Jamie, and Adam Ticknell. 2002. "Neoliberalizing Space." *Antipode* 34 (3): 380–404.

Pecora, W. T., M. R. Klepper, D. M. Larrabee, A. L. M. Barbosa, and Resk Frayha. 1950. Mica Deposits in Minas Gerais Brazil. In *Geologic Investigations in the American Republics, 1949.* Washington, DC: United States Government Printing Office.

Peek, Andrew L. 2014. "The New Space Race, and Why Nothing Else Matters." *Fiscal Times,* July 11. http://www.thefiscaltimes.com/Columns/2014/07/11/New-Space-Race-and -Why-Nothing-Else-Matters.

Penteado, Carlos José Asumpção. 2006. "The Brazilian Participation in World War II." MA thesis, United States Army Command and General Staff College.

Perkowski, Jack. 2012. "Behind China's Rare Earth Controversy." Forbes. Accessed January 12, 2015. http://www.forbes.com/sites/jackperkowski/2012/06/21/behind-chinas-rare-earth-controversy/.

Perlo-Freeman, Sam. 2004. "Offsets and the Development of the Brazilian Arms Industry." In *Arms Trade and Economic Development: Theory, Policy, and Cases in Arms Trade Offsets*, edited by Jurgen Brauer and J. Paul Dunne, chap. 13. New York: Routledge.

Peters, Stephen G. 2007. *Preliminary Assessment of Non-Fuel Mineral Resources of Afghanistan, 2007*. Reston, VA: United States Geological Survey.

Phillipine News Agency. 2011. "Pentagon, USGS Launch New Initiative to Explore Mineral Resources in Afghanistan." Phillipines News Agency, November 30. http://search.proquest.com/docview/906784424?accountid=14496.

Pinheiro, Colonel Alvaro de Souza. 1995. "Guerrilla in the Brazilian Amazon." *Foreign Military Studies Office*. Accessed April 14, 2015. http://fmso.leavenworth.army.mil/documents/amazon/amazon.htm.

Planetary Resources. 2012. "Asteroid Mining Plans Revealed by Planetary Resources, Inc." Accessed April 14, 2015. http://www.planetaryresources.com/2012/04/asteroid-mining-plans-revealed-by-planetary-resources-inc/.

Planetary Resources. 2015. "President Obama Signs Bill Recognizing Asteroid Resource Property Rights into Law." Accessed December 15, 2015. http://www.planetaryresources.com/2015/11/president-obama-signs-bill-recognizing-asteroid-resource-property-rights-into-law/.

Plumer, Brad. 2011. "How to Free the World from China's Rare-Earth Stranglehold." *Washington Post*, September 16. http://www.washingtonpost.com/blogs/wonkblog/post/how-to-free-the-world-from-chinas-rare-earth-chokehold/2011/09/16/gIQA0Zg1XK_blog.html.

PMSA-WIL. 2013. "Port of Oakland: Charting a New Course." Port of Oakland. Accessed April 15, 2015. http://www.portofoakland.com/pdf/about/PMSA-WIL_2013.pdf.

Pomeranz, Kenneth. 1993. *The Making of Hinterland: State, Society, and Economy in Inland North China, 1853–1937*. Oakland: University of California Press.

Pomeranz, Kenneth. 2000. *The Great Divergence: China, Europe, and the Making of the Modern World Economy*. Princeton: Princeton University Press.

Portales, Isabel M. Estrada. 2011. "Terras Raras são Cruciais para Defesa do Brasil." *Diálogo*. Accessed April 14, 2015. http://dialogo-americas.com/pt/articles/rmisa/features/regional_news/2011/12/27/aa-brazil-rare-earth.

Portela, Fernando, and José Genoino Neto Neto. 2002. *Guerra de Guerrilhas no Brasil: A Saga do Araguaia*. São Paulo: Editora Terceiro Nome.

PotashCorp. 2014. *Food Matters: 2013 Online Integrated Report*. http://potashcorp.s3.amazonaws.com/2013_PotashCorp_Annual_Integrated_Report.pdf.

Primack, Dan. 2014. "Vinod Khosla Hits Back at 60 Minutes for Cleantech 'Errors.'" *Fortune Magazine*. Accessed 7 January 2015. http://fortune.com/2014/01/14/vinod-khosla-hits-back-at-60-minutes-for-cleantech-errors/?source=yahoo_quote&utm_source=hootsuite&utm_campaign=hootsuite.

Proctor, James D. 1998. "Geography, Paradox and Environmental Ethics." *Progress in Human Geography* 22 (2): 234–55.

Pulido, Laura. 1996. "A Critical Review of the Methodology of Environmental Racism Research." *Antipode* 28 (2): 142–59.

Pye, Lucian W. 1999. "An Overview of 50 Years of the People's Republic of China: Some Progress, but Big Problems Remain." *China Quarterly* 159:569–79.

Qi, Shuwen. 2010. "China Denies Banning Rare Earth Exports to Japan." *People's Daily*, September 25.

Qiu, Jane. 2007. "China's Green Accounting System on Shaky Ground." *Nature Geoscience* 448:518–19.

Raclin, Grier C. 1986. "From Ice to Ether: The Adoption of a Regime to Govern Resource Exploitation in Outer Space." *Northwestern Journal of International Law & Business* 7:727–61.

Radambrasil, Projeto. 1976. *Folha 19: Pico da Neblina*. Brasília, Brazil: Departamento Nacional da Produção Mineral.

Randall, David, and Jonathan Owen. 2012. "To the Chinese and Indians, the Spoils of a Terrible War." *Independent*, March 18.

Redclift, Michael, and Colin Sage. 1998. "Global Environmental Change and Global Inequality in North/South Perspectives." *International Sociology* 13 (4): 499–516.

Reddy, D. Raja. 2009. "Neurology of Endemic Skeletal Fluorosis." *Neurology India* 57 (1): 7–12.

Redford, Kent, and Allyn Mclean Stearman. 1993. "Forest-Dwelling Native Amazonians and the Conservation of Biodiversity: Interests in Common or in Collision?" *Conservation Biology* 7 (2): 248–55.

Reilly, James F., II. 2013. "Avoiding Extraterrestrial Claim Jumping: Economic Development Policy for Space Exploration and Exploitation." In *Energy Resources for Human Settlement in the Solar System and Earth's Future in Space*, edited by W. A. Ambrose, J. F. Reilly, II, and D. C. Peters, 141–50. Tulsa, OK: American Association of Petroleum Geologists.

Reinstein, E. J. 1999. "Owning Outer Space." *Northwestern Journal of International Law & Business* 20:59–98.

Reis, Artur César Ferreira. 1942. "Roteiro Histórico das Fortificações no Amazonia." *Revista do Serviço do Patrimônio Histórico e Artístico Nacional* 6:118–68.

Ren, Huibin. 2013. "Residents Seriously Ill in Villages Contaminated by Baogang Tailings Dam." *Economic Survey News*, April 26.

Ren, Y., Y. Zhan, and Z. Zhang. 1994. "Study of Heat Events of Ore-Forming Bayan Obo Deposit." *Acta Geoscientifica Sinica* 30 (31): 95–101.

Reporter, Staff. 2012. "Jiangxi's Illegal Rare Earth Mining Too Profitable to Quit." *Want China Times*, April 19. http://www.wantchinatimes.com/news-subclass-cnt.aspx?id=20120419000013&cid=1503.

Republic of China and USSR. 1924. "Agreement on General Principles for the Settlement of Questions between the Republic of China and the Union of Soviet Socialist Republics." *American Journal of International Law* 19 (2): 53–56.

Reuters. 2011. "Truck Kills Herder in China Inner Mongolia Protest: Group." Accessed December 12, 2014. http://www.reuters.com/article/2011/10/24/us-china-innermongolia-idUSTRE79N14L20111024.

Ricardo, Beto, Cleber Buzzato, Clovis Brighenti, Daniel Pierri, Egon Heck, Egydio Schwade, Fany Ricardo, Gilberto Azanha, Ian Packer, Iara Ferraz, Inimá Simões, Isabel Harari, Laura Faerman, Levi Marques Pereira, Luis Francisco de Carvalho Dias, Luiz Henrique Eloy Amado, Manuela Carneiro da Cunha, Marcelo de Souza Romão, Marcelo Zelic, Marco Antonio Delfino de Almeida, Maria Inês Ladeira, Neimar Machado de Sousa, Orlando Calheiros, Patrícia de Mendonça Rodrigues, Porfirio Carvalho, Rafael Pacheco Marinho, Rogerio Duarte do Pateo, Spensy Pimentel, Tatiane Klein, and Vincent Carelli. 2014. "Violações de Direitos Humanos dos Povos Indígenas." In *Comissão Nacional da Verdade: Relatório*, edited by Comissão Nacional da Verdade, 198–256. Brasília: Biblioteca da Comissão Nacional da Verdade.

Ricardo, Beto, and Fany Ricardo, eds. 1990. *Povos Indígenas no Brasil*. São Paulo: Centro Ecumênico de Documentação e Informação.

Ricardo, Beto, and Fany Ricardo, eds. 2006. *Povos Indígenas no Brasil—2001/2005*. Brasília: Instituto Socioambiental.

Ricardo, Fany, and Alicia Rolla, eds. 2013. *Mineração em Terras Indígenas na Amazônia Brasileira*. Brasília: Instituto Socioambiental.

Ricci, Thiago. 2014. "Ex-coronel Admite ter Recebido Ordens para Bombardear Praça com 10 mil Pessoas em 1964." *O Globo*, September 30. http://oglobo.globo.com/brasil /ex-coronel-admite-ter-recebido-ordens-para-bombardear-praca-com-10-mil -pessoas-em-1964-14098944.

Richardson, Ben, and Mark Williams. 2010. "China Denies Japan Rare-Earth Ban amid Diplomatic Row." Bloomberg News. Accessed December 28, 2014. http://www .bloomberg.com/news/2010-09-23/china-denies-japan-rare-earth-ban-amid -diplomatic-row-update1-.html.

Riederer, Rachel. 2014. "Whose Moon Is It Anyway?" *Dissent* 61 (4): 6–10.

Rincon, Paul. 2013. "China Space: 'Jade Rabbit' Lunar Mission Blasts Off." BBC News. Accessed February 2, 2015. http://www.bbc.com/news/science-environment-25178299.

Risen, James. 2010. "U.S. Identifies Vast Mineral Riches in Afghanistan." *New York Times*, June 14.

Robinson, William I. 2004. *A Theory of Global Capitalism: Production, Class, and State in a Transnational World*. Baltimore: Johns Hopkins University Press.

Rodhan, N. R. F. al-. 2012. *Meta-Politics of Outer Space*. Hampshire, UK: Palgrave Macmillan.

Rofel, Lisa. 2007. *Desiring China: Experiments in Neoliberlism, Sexuality, and Public Culture*. Durham, NC: Duke University Press.

Rosenberg, Zach. 2013. "This Congressman Kent the US and China From Exploring Space Together." *Foreign Policy*, December 17.

Rossabi, Morris. 2004. *Governing China's Multiethnic Frontiers*. Seattle: University of Washington Press.

Rossini, Marco Bimkowski. 2012. "Identificação de Estruturas Geológicas e Anomalias de Contraste Espectral na Região do Carbonatito Morro dos Seis Lagos, NW do Amazonas." BA thesis, Federal University of Rio Grande do Sul.

Rostow, W. W. 1960. *The Stages of Economic Growth: A Non-Communist Manifesto*. Cambridge: Cambridge University Press.

Rowland, R. E. 1994. *Radium in Humans: A Review of U.S. Studies*. Argonne, IL: United States Department of Energy, Office of Energy Research, Office of Health and Environment Research, and Assistant Secretary for Environment, Safety, and Health, Office of Epidemiology and Health Surveillance.

Rowlatt, Justin. 2014. "Rare Earths: Neither Rare, Nor Earths." *BBC News Magazine*. Accessed January 5, 2015. http://www.bbc.com/news/magazine-26687605.

Rowley, Emma. 2013. "Rare Earths: West Bids to Challenge China's Monopoly, But Is It Too Late?" *Telegraph*, August 19. http://www.telegraph.co.uk/finance/commodities /10253189/Rare-earths-West-bids-to-challenge-Chinas-monopoly-but-is-it-too -late.html.

Rupen, Robert A. 1955. "Notes on Outer Mongolia Since 1945." *Pacific Affairs* 28 (1): 71–79.

Sablotne, Jayne. 2012. "Space Pioneers LLC Sues Better Business Bureau over Property Rights in Space." PR Newswire. Accessed November 26, 2014. http://www .prnewswire.com/news-releases/space-pioneers-llc-makes-a-ripple-in-time-suing -bbb-over-private-property-rights-in-space-141880803.html.

Sadeh, E., ed. 2011. *The Politics of Space: A Survey*. London: Routledge.

Salter, Andrew W., and Peter T. Leeson. 2014. "Celestial Anarchy: A Threat to Outer Space Commerce?" *Cato Journal* 34 (3): 581–96.

Salvador, Frei Vicente. 1627. *História do Brasil*. Rio de Janeiro: Biblioteca Nacional.

Santos, Fabiano Vilaça dos. 2008. "O Governo das Conquistas do Norte: Trajetórias Administrativas no Estado do Grão-Pará e Maranhão." PhD diss., University of São Paulo.

Santos, J. O. S. 2003. "Geotectônica dos Escudos das Guianas e Brasil Central." In *Geologia, Tectônica e Recursos Minerais do Brasil*, edited by L. A. Brizzi, R. M. Schobbenhaus, R. M. Vidotti, and J. H. Gonçalves, 169–226. Brasília: Companhia de Pesquisa do Recursos Minerais.

Santos, L. M. J. 2013. "Greenland Won't Give EU Preference over China on Rare Earth Access." Santos Republic. Accessed November 24, 2014. http://thesantosrepublic .com/2013/01/28/greenland-wont-give-eu-preference-over-china-on-rare-earth -access/.

Scerri, Eric R. 2007. *The Periodic Table: Its Story and Its Significance*. New York: Oxford University Press.

Schearf, Daniel. 2014. "North Korea's Rare Earths Could be Game Changer." Voice of America. Accessed January 5, 2015. http://www.voanews.com/content/north-korea -rare-earths-game-changer/1832018.html.

Schroeder, Richard A. 2000. "Beyond Distributive Justice: Resource Extraction and Environmental Justice in the Tropics." In *People, Plants, and Justice: The Politics of Nature Conservation*, edited by Charles Zerner, 52–65. New York: Columbia University Press.

Schwartz, B. I. 1968. "The Reign of Virture: Some Broad Perspectives on Leader and Party in the Cultural Revolution." *China Quarterly* 35:1–17.

Scott, James. 2009. *The Art of Not Being Governed: An Anarchist History of Upland Southeast Asia*. New Haven: Yale University Press.

Scott, Peter Dale. 2010. *American War Machine: Deep Politics, the CIA Global Drug Connection, and the Long Road to Afghanistan*. Lanham, MD: Rowman & Littlefield.

Şebnem, Önder, Eda Biçer Ayşe, and Selen Denemeç Işıl. 2013. "Are Certain Minerals Still Under State Monopoly?" *Mining Turkey* (September).

Sellers, L. J. 2016. *Point of Control*. Seattle, WA: Thomas & Mercer.

Senado Federal. 2013. "Subcomissão Temporária de Elaboração de Marco Regulatório da Mineração em Terras Raras." Accessed May 8, 2017. https://legis.senado.leg.br /comissoes/comissao?1&codcol=1663.

Shaiken, Harley. 1986. *Work Transformed: Automation and Labor in the Computer Age*. New York: Holt, Rinehart & Winston.

Shanghai Rare Earth Network. 2014. "Shanghai Developments of New Metal Alloys Help the Jade Rabbit Trek the Moon." Accessed February 7, 2015. http://www.sh-re.com /ZDSYS/ZDSYS_show.aspx?id=8788.

Shapiro, Judith. 2001. *Mao's War against Nature: Politics and the Environment in Revolutionary China*. New York: Cambridge University Press.

Sharma, Rohit, Pradeep Kumar, Neha Bhargava, Amit Kumar Sharma, Shalabh Srivastava, Shweta Jain, and Vijay Agrawal. 2013. "Dental and Skeletal Fluorosis: A Review." *Medico-Legal Update* 13 (2): 151–55.

Shedd, Kim B. 2014. "Tungsten." *Mineral Commodity Summaries* (February):174–75.

Shefa, Xinhua. 2014. "The Jade Rabbit Lunar Rover Successfully Completed its Inaugural Scientific Mission." *Asia Pacific Daily*. December 18. http://www.apdnews.com/news /61661.html.

Shen, Grace Yen. 2014. *Unearthing the Nation: Modern Geology and Nationalism in Republican China*. Chicago: University of Chicago Press.

Shervais, J. W., and James J. McGee. 1999. "KREEP Cumulates in the Western Lunar Highlands: Ion and Electron Microprobe Study of Alkali-suite Anorthosites and Norites from Apollo 12 and 14." *American Mineralogist* 84:806–20.

Shi, Jizhong. 2012. "Ding Daoheng Discovered Bayan Obo." *Ancient Guizhou* (October):58.

Shih, Chih-yu. 1998. "A Postcolonial Reading of the State Question in China." *Journal of Contemporary China* 7 (17): 125–39.

Sie, S.T. 1994. "Past, Present and Future Role of Microporous Catalysts in the Petroleum Industry." *Studies in Surface Science and Catalysis* 85:587–631.

Silva, Carlos Luis Del Cairo. 2012. "Environmentalizing Indigeneity: A Comparative Ethnography on Multiculturalism, Ethnic Hierarchies, and Political Ecology in the Colombian Amazon." PhD diss., University of Arizona, Phoenix.

Silva, Major Gerson Rolim da. 2013. "FARC's Influence in Brazil." PhD diss., United States Army Command and General Staff College, Fort Leavenworth, KS.

Simões, Janaína. 2011. "Brasil tem uma das Maiores Reservas de Terras Raras do Planeta." *Inovação Technológica*: n.p.

Sims, Jim. 2011. "Molycorp Acquires Controlling Stake in AS Silmet, Expands Operations to Europe, Doubles Near-Term Rare Earth Oxide Production Capacity." Accessed February 5, 2014. http://www.molycorp.com/molycorp-acquires-controlling-stake -in-as-silmet-expands-operations-to-europe-doubles-near-term-rare-earth-oxide -production-capacity-2/.

Slater, Candace. 1994. "'All That Glitters': Contemporary Amazonian Gold Miners' Tales." *Comparative Studies in Society and History* 36 (4): 720–30.

Smil, Vaclav. 1993. *China's Environmental Crisis: An Inquiry into the Limits of National Development*. Armonk, NY: M. E. Sharpe.

Smith & Wesson. 2014. "Model 340PD." Accessed February 5, 2014. https://www.smith -wesson.com/firearms/model-340-pd.

Smith, M. P., L. S. Campbell, and J. Kynicky. 2014. "A Review of the Genesis of the World Class Bayan Obo Fe-REE-Nb Deposits, Inner Mongolia, China: Multistage Processes and Outstanding Questions." *Ore Geology Reviews* 64:459–76.

So, Alvin. 2007. "Peasant Conflict and the Local Predatory State in the Chinese Countryside." *Journal of Peasant Studies* 34 (3–4): 560–81.

Souza, Leonam, and Ari Cavedon. 1984. "RADAM Project and the Mapping of the Natural Resources in the Amazon Region." Paper presented at Symposium on the Humid Tropics, Belém, Brazil, November 2–17.

Souza, Oswaldo Braga de. 2014. "Deputado Pede Afastamento do Relator do Código da Mineração." Instituto Socioambiental. Accessed January 26, 2015. http://www .socioambiental.org/pt-br/noticias-socioambientais/deputado-pede-afastamento -do-relator-do-codigo-da-mineracao.

Sparke, Matthew. 2009. "Globalization." In *Dictionary of Human Geography*, edited by Derek Gregory, Ron Johnston, and Geraldine Pratt, 5th ed., 308–11. Hoboken, NJ: Wiley-Blackwell.

Spedding, F. H. 1961. "Introductory Address." Rare Earth Research Conference, University of Denver, September 24–27.

Stahl, Leslie. 2014. "The Cleantech Crash." 60 Minutes. Accessed January 5, 2015. http:// www.cbsnews.com/news/cleantech-crash-60-minutes/.

Sta Maria, Stefanie. 2012. "Australia Says It Again: No to Lynas Waste." Free Malaysia Today, February 21. http://www.freemalaysiatoday.com/category/nation/2012/02 /21/australia-says-it-again-no-to-lynas-waste/.

Stanway, David. 2011. "Will China's Rare Earth Crackdown Work?" Reuters. Accessed December 28, 2014. http://www.reuters.com/article/2011/08/30/china-rareearth -idUSL3E7JG1LA20110830.

State Bureau for Letters and Calls. 2014. "Duties of the State Bureau for Letters and Calls." Accessed November 28, 2014. http://www.gjxfj.gov.cn/2009-11/23/c_133327698 .htm.

St. Clair, Jeffrey. 2003. "Outsourcing US Guided Missile Technology: US Workers Charge Treason Outsourcing US Missile Technology to China." *Counterpunch*, October 25–27.

Steinberg, Jim. 2014. "EPA Fines San Bernardino County's Molycorp $27, 300." *Sun*, April 21. http://www.sbsun.com/environment-and-nature/20140421/epa-fines-san -bernardino-countys-molycorp-27300.

Steinberg, P. E. 2013. "Of Other Seas: Metaphors and Materialities in Maritime Regions." *Atlantic Studies* 10 (2): 156–69.

Stone, B. 2007. "I'm Going to the Moon. Who's with Me?" TED Conferences. Accessed November 24, 2014. http://www.ted.com/talks/bill_stone_explores_the_earth_and _space.html?quote=172.

Stone, Richard. 2009. "As China's Rare Earth R&D Becomes Ever More Rarefied, Others Tremble." *Science* 325 (5946): 1336–37.

Stone, Richard. 2014. "Mother of All Lodes." *Science* (August 15): 725–27.

Stop Lynas Save Malaysia. 2014. "An Australian Company Exporting a Toxic Legacy." Accessed June 10. http://stoplynas.org.

Stover, Dawn. 2011. "The Myth of Renewable Energy." *Bulletin of the Atomic Scientists*, November 22.

Stoyer, June. 2013. "Jack Lifton Explains Why Recycling Rare Earth Metals Matters." The Clean Energy View Radio Show. Accessed April 14, 2015. http://hosts.blogtalkradio .com/theorganicview/2013/10/08/jack-lifton-explains-why-recycling-rare-earth -metals-matters#.

Stranahan, Patricia. 1998. *Underground: The Shanghai Communist Party and the Politics of Survival, 1927–1937*. New York: Rowman & Littlefield.

Strange, Susan. 1996. *The Retreat of the State: The Diffusion of Power in the World Economy*. Cambridge: Cambridge University Press.

Su, Rui 2004. "Third Baogang Testimonial: Decision-Making and Planning." *North Country Stories*, aired December 3, 2004. Inner Mongolia: Inner Mongolia Television Network.

Sun, Zezhou, Yang Jia, and He Zhang. 2013. "Technological Advancements and Promotion Roles of Chang'e-3 Lunar Probe Mission." *Science China: Technological Sciences* 56 (11): 2702–8.

Swift, Mike. 2012. "Silicon Valley Entrepreneurs Shoot for the Moon." *San José Mercury News*, January 1. http://www.mercurynews.com/ci_19641873.

Syndicated Press. 1937. "Nations Fear Stalemate in Peace Effort: Chinese Military Status under Study." *Middletown Times Herald*, November 8, 1–2.

Tabuchi, Hiroko. 2010. "The Hunt for Rare Earths." *New York Times*, November 25, B3.

Tadjeh, Yasmin. 2014. "New Chinese Threats to U.S. Space Programs Worry Officials." *National Defense Magazine*, July 7.

Taffler, Richard, and David Tuckett. 2007. "Emotional Finance: Understanding What Drives Investors." *Professional Investor* (Autumn): 26–27.

Tamer, Mehmet Numan, Banu Kale Köroğlu, Çağatay Arslan, Mehmet Akdoğan, Mert Köroğlu, Hakan Çam, and Mustafa Yildiz. 2007. "Osteosclerosis Due to Endemic Fluorosis." *Science of the Total Environment* 373 (1): 43–48.

Tanton, Tom. 2012. *Dig It! Rare Earth and Uranium Mining Potential in the States*. Washington, DC: American Legislative Exchange Council.

Taylor, L. A., and G. L. Kulcinski. 1999. "Helium-3 on the Moon for Fusion Energy: The Persian Gulf of the 21st Century." *Solar System Research* 33:338.

Taylor, S. R., and P. Jakes. 1974. "The Geochemical Evolution of the Moon." *Proceedings of the Fifth Lunar and Planetary Science Conference* (Suppl. 5). *Geochimica et Cosmochimica Acta* (2): 1287–1305

Ter-Ghazaryan, Aram. 2014. "Moon Exploration Will Reduce the Shortage of Rare Earth Minerals." *Russia beyond the Headlines*, October 26. http://rbth.com/science_and

_tech/2014/10/26/moon_exploration_will_reduce_the_shortage_of_rare_earth_meta_40887.html.

Terrill, Ross. 1999. *Mao: A Biography*. Palo Alto, CA: Stanford University Press.

Tighe, Justin. 2005. *Constructing Suiyuan: The Politics of Northwestern Territory and Development in Early Twentieth-Century China*. Leiden: E. J. Brill.

Ting, Ming Hwa, and John Seaman. 2013. "Rare Earths: Future Elements of Conflict in Asia?" *Asian Studies Review* 37 (2): 234–52.

Tipper, G. H. 1930. "Recent Mineral Developments in India." *Journal of the Royal Society of Arts* 78 (4039): 616n35.

Topf, Andrew. 2013. "Prices, Quotas, and Smugglers: Making Sense of Rare Earth Smugglers in China." Rare Earth Investing News. Accessed December 28, 2014. http://rareearthinvestingnews.com/15702-prices-quotas-and-smugglers-making-sense-of-rare-earths-in-china.html.

Townley, Helen E. 2013. "Applications of the Rare Earth Elements in Cancer Imaging and Therapy." *Current Nanoscience* 9 (5): 686–91.

Triggs, G., and D. Bialek. 2002. "The New Timor Sea Treaty and Interim Arrangements for Joint Development of Petroleum Resources of the Timor Gap." *Melbourne Journal of International Law* 3:322–64.

Tse, Pui-Kwan. 2011. "China's Rare Earth Industry." *United States Geological Survey Open-File Report 2011–1042*. Reston, VA: Department of the Interior of the United States.

Tsing, Anna Lowenhaupt. 2005. *Friction: An Ethnography of Global Connection*. Princeton: Princeton University Press.

Tu, Xinquan. 2012. "China Responds to Rare Earth WTO Complaint." Xinhua. Accessed April 15, 2015. http://news.xinhuanet.com/english/china/2012-03/13/c_122830662.htm.

Tucker, Robert D. 2014. "Afghanistan: Rare Earth Elements Could Beat the Taliban." Scientific American. Accessed December 24, 2014. http://www.scientificamerican.com/slideshow/afghanistan-rare-earth-elements-could-beat-taliban/.

Tudor, Alison. 2011. "Chinese Consortium Buying Stake in Brazilian Miner." *Wall Street Journal*, September 2. http://search.proquest.com/docview/887044922?accountid=14496.

Turner, Wallace. 1973. "3 Nixon Brothers: Close but Hardly Equal." *San Bernardino County Sun*, Monday, April 30, 6A.

Tyner, James A. 2012. *Genocide and the Geographical Imagination: Life and Death in Germany, China, and Cambodia*. Plymouth, UK: Rowman & Littlefield.

Underwood, James R. and Peter Guth, eds. 1998. *Military Geology in War and Peace*. Boulder, CO: Geological Society of America.

UN (United Nations). 1967. "Treaty on Principles Governing the Activities of States in the Exploration and Use of Outer Space, Including the Moon and Other Celestial Bodies." *Resolution 2222*. Accessed November 30, 2014. http://www.unoosa.org/oosa/en/ourwork/spacelaw/treaties/introouterspacetreaty.html

UN. 1979. "Agreement Governing the Activities of States on the Moon and Other Celestial Bodies." New York City.

United Nations Development Programme. 2012. *Extractive Industries and Conflict*. New York: United Nations Interagency Framework Team for Preventative Action.

United Nations Office for Outer Space Affairs. 1984. "Agreement Governing the Activities of States on the Moon and Other Celestial Bodies." *Resolution 34/68*. Accessed November 26, 2014. http://www.unoosa.org/oosa/SpaceLaw/moon.html.

United Nations Office for Outer Space Affairs. 2012. "Accession to the Agreement Governing the Activities of States on the Moon and Other Celestial Bodies." New York.

United States Army Corps of Engineers. 1943. "Report on Orinoco-Casiquiare-Negro Waterway, Venezuela-Colombia-Brazil." Washington, DC.

United States Congress. 2011. 112th Congress. H.R. 618, Rare Earths and Critical Materials Revitalization Act of 2011. https://www.congress.gov/bill/112th-congress/house-bill/618/text.

United States Department of the Army. 1952. *TM 5-545 Geology and Its Military Applications*. Washington, DC: United States Government Printing Office.

United States Department of Defense. 2011. "Airborne Geophysical Exploration Program Launched in Afghanistan." Accessed November 26, 2014. http://www.defense.gov/releases/release.aspx?releaseid=14942.

United States Department of Defense Undersecretary of Defense for Acquisition, Technology, and Logistics. 2013. Annual Industrial Capabilities Report to Congress, October. Washington, DC.

United States Department of Energy. 2011. Critical Materials Strategy. Washington, D.C.

United States Geological Survey (USGS). 1953. *Geological Survey Circular 237*. Washington, DC: United States Department of the Interior.

United States Geological Survey. 2013. 2012 Minerals Yearbook: China [Advance Release]. United States Department of the Interior and United States Geological Survey. Accessed November 25, 2014. http://minerals.usgs.gov/minerals/pubs/country/2012/myb3-2012-ch.pdf.

United States Geological Survey. 2014. "Boron." *Mineral Commodity Summaries* (February): 32–33.

United States Geological Survey, Department of the Interior of the United States. "1942–1947. 57.4.3 Records of the Foreign Geology Section." In *Report on Strategic Minerals in Brazil, 1942–47; Records Concerning the Mica and Quartz Programs, 1943–1945*. Bethesda, MD: National Archives and Records Administration.

United States House of Representatives. 1952. 82nd Congress, 2nd Session, House Document 527, v. 4. Washington, DC.

United States House of Representatives. 1998. Committee on Commerce, Science, and Transportation. Hearing before the Subcommittee on Science, Technology, and Space. September 23. Washington, DC.

United States House of Representatives. 2011a. Subcommittee on Energy of the Committee on Energy and Natural Resources of the United States Senate Critical Minerals and Materials Legislation. June 9. Washington, DC.

United States House of Representatives. 2011b. 111th Congress. Public Law 111-383: Ike Skelton National Defense Authorization Act for Fiscal Year 2011. Washington, DC.

United States House of Representatives. 2011c. 112th Congress. To Reestablish a Competitive Domestic Rare Earths Minerals Production Industry; A Domestic Rare Earth Processing, Refining, Purification, and Metals Production Industry; a Domestic Rare Earth Metals Alloying Industry; and a Domestic Rare-Earth-Based Magnet Production Industry and Supply Chain in the Defense Logistics Agency of the Department of Defense. Washington, DC.

US China Business Council. 2011. "China Business Environment Survey Results: Market Growth Continues, Companies Expand, But Full Access Elusive for Many." New York and Beijing.

Utley, William C. 1937. "Will Japan Ever Conquer China? Nipponese Invaders Face Different Problem Than in the Past; Her Vast Neighbor Today Presents Unified Front." *Indiana Weekly Messenger*, September 16.

Valentine, D. 2012. "Exit Strategy: Profit, Cosmology, and the Future of Humans in Space." *Anthropological Quarterly* 85 (4): 1045–67.

Various. 1948–58. *Inner Mongolia Daily*, Grassland.

Vecchi, Roberto 2014. "O Passado Subtraído do Desaparição Forçada: Araguaia como Palimpsesto." *Estudos de Literatura Brasileira Contemporânea* 43:133–49.

Veronese, Keith. 2015. *Rare: The High-Stakes Race to Satisfy Our Need for the Scarcest Metals on Earth.* Amherst, NY: Prometheus Books.

Verrax, Fanny. 2015. "Recycling toward Rare Earths Security." In *The Political Economy of Rare Earth Elements: Rising Powers and Technological Change*, edited by Ryan David Kiggins, 156–77. New York: Palgrave Macmillan.

Voiçu, G., M. Bardoux, and R. Stevenson. 2001. "Lithostratigraphy, Geochronology and Gold Metallogeny in the Northern Guiana Shield, South America: A Review." *Ore Geology Reviews* 18 (3–4): 211–36.

Von Clausewitz, Carl. 1976. *On War.* Princeton: Princeton University Press.

Von Eschen, Penny M. 1997. *Race against Empire: Black Americans and Anticolonialism, 1937–1957.* Ithaca, NY: Cornell University Press.

Wakeman, Frederic. 1995. *Policing Shanghai, 1927–1937.* Berkeley: University of California Press.

Wakita, H., P. Rey, and R. A. Schmitt. 1971. "Abundances of the 14 Rare-Earth Elements and 12 Other Trace Elements in Apollo 12 Samples: Five Igneous and One Breccia Rocks and Four Soils." *Proceedings of the Lunar and Planetary Science Conference* 2:1319–29.

Wang, Fang 2012. "The Cause and Effects of Moving Westward in the Inner Mongolian Midwest." *Bulleting of the Baotou Textile Arts Academy* 1:14–16.

Wang, Guozhen. 2007. "Some Suggestions Regarding the Management of Radioactive Waste and the 'Three Wastes' Resulting from Rare Earth Smelting." *Sichuan Rare Earths* 2007 (3): 2–5.

Wang, Junzhi. 2010. *China's Rare Earth Protection War.* Beijing: China Economic Publishing House.

Wang, Lili, Xing Wanping, and Ji Yulang. 2002. "A Case-Control Study of Lung Cancer among Workers in Baotou Iron and Steel Company." *Journal of Baotou Medical College* 2002 (4): 10–14.

Wang, Min. 2013. "Enter the "Jade Rabbit" Decoding My Country's First Lunar Rover." Xinhua, December 13. http://news.xinhuanet.com/world/2013-12/13/c_118552981.htm.

Wang, Peng. 2006. "'One in Seven,' Under the Shadow of the Tailings Lake." *Eastern Broadcasting Network*, August 10. http://smgtv.eastday.com/eastday/node21/node148/node13584/userobject1ai161799.html.

Wang, Suping, ed. 2000. *Fifty Years of the Women's Mobilization in Baotou.* Baotou City, Inner Mongolia Autonomous Region: Inner Mongolia Autonomous Region Baotou City Women's Committee Press.

Wang, X. C., K. Kawahara, and X. J. Guo. 1999. "Fluoride Contamination of Groundwater and Its Impacts on Human Health in Inner Mongolia." *Journal of Water Services Research and Technology AQUA* 48:146–53.

Wang, Xiang, Jiali Xie, and Lan Chen. 2007. "Research of Extraction Procedure Toxicity and Hg-Th-Pb Leaching Characteristic in Rare Earth Mine Tailing." *Journal of Chinese Rare Earth Society* 25 (Suppl.): 9–14.

Wang, Xiaoqiang, Guo Chengxiang, Bai Lina, Cai Yingmao, Liu Yingxia, and Tao De. 2009. "Radioactive Thorium Contamination of Soil Surrounding the Baogang Tailings Pond: Trends and Prevention Measures." *Radiation Protection* 29 (4): 270–75.

Wang, Zhe, and Baocheng Dai. 2011. "Analysis of Ammonia Pollution in the Yellow River in the Baotou, Inner Mongolia Section of the Water Body." *Anhui Agricultural Science* 39 (13): 8044–45.

Wardle, Brian. 2009. *Principles and Applications of Photochemistry.* Hoboken, NJ: Wiley.

Watts, M. J. 1999. "Petro-Violence: Some Thoughts on Community, Extraction, and Political Ecology." Berkeley Workshop on Environmental Politics, Institute of International Studies, University of California, Berkeley, September 24.

Watts, Michael. 2012. "A Tale of Two Gulfs: Life, Death, and Dispossession along Two Oil Frontiers." *American Quarterly* 64 (3): 437–67.

Wayne, P. A. 2012. "Molycorp, Inc MCP Securities Stock Fraud." Online Legal Media. Accessed November 26, 2014. http://www.lawyersandsettlements.com/lawsuit /molycorp-inc-mcp-securities.html.

Weber, Robert J., and David J. Reisman. 2012. "Rare Earth Elements: A Review of Production, Processing, Recycling, and Associated Environmental Issues." United States Environmental Protection Agency. http://nepis.epa.gov/Exe/ZyPURL.cgi?Dockey =P100EUBC.txt.

Wegmann, Philip. 2014. "Conservatives Derail Democrat's Bill on Rare Earth Minerals." The Heritage Foundation, July 23. http://dailysignal.com/2014/07/23/conservatives -derail-democrats-bill-rare-earth-minerals/.

Weizmann, Eyal. 2007. *Hollow Land: Israel's Architecture of Occupation.* London: Verso.

Wells, Jane. 2013. "US Must Beat China Back to the Moon: Entrepreneur." CNBC. Accessed February 6, 2015. http://www.cnbc.com/id/101195299.

Welsbach, Carl Auer, von. 1889. Incandescent Device. United States Patent No. US399174A, filed March 8, 1887, issued March 5, 1889.

Wen, Dongguang, Zhang Fucun, Zhang Eryong, Cheng Wang, Han Shuangbao, and Yan Zheng. 2013. "Arsenic, Fluoride and Iodine in Groundwater of China." *Journal of Geochemical Exploration* 135:1–21.

Wen, Tiejun. 2004a. "Civil Primitive Accumulation and Government Conduct: Conclusions from the Wenzhou Investigation." In *What Is It That We Want?*, edited by Tiejun Wen, 59–64. Beijing: Huaxia Publishers.

Wen, Tiejun. 2004b. *Deconstructing Modernity, Talks by Wen Tiejun.* Guangzhou: Guangdong People's Publishing.

Westad, Odd Arne. 2003. *Decisive Encounters: The Chinese Civil War, 1945–1950.* Stanford, CA: Stanford University Press.

White, Jon. 2013. "Can We Make a National Heritage Site on the Moon?" *New Scientist* 2931:27.

White, Richard J., and Colin C. Williams. 2012. "The Pervasive Nature of Heterodox Economic Spaces at a Time of Neoliberal Crisis: Towards a 'Postneoliberal' Anarchist Future." *Antipode* 44 (5): 1625–44.

White, W. W., and L. A. Kimble. 1979. Process for Purifying Rare-Earth Compositions Using Fractional Sulfate Precipitation. United States Patent No. 4265862, filed January 25, 1979, issued May 5, 1981.

Whitehorn, Will. 2005. Statement of Will Whitehorn at House Space and Aeronautics Subcommittee Hearings on the Future Market for Commercial Space. House Space and Subcommittee Hearings. April 20.

Whittington, Mark. 2013. "U.S. Can still Beat China Back to Moon." *USA Today*, December 28. http://www.usatoday.com/story/opinion/2013/12/28/mark-whittington -china-space/4209223/.

Williams, Dee Mack. 2002. *Beyond Great Walls: Environment, Identity, and Development on the Chinese Grasslands of Inner Mongolia.* Stanford, CA: Stanford University Press.

Wilson, Nigel. 2015. "Why China Scrapped Quotas on Rare Earth Metal Exports." *International Business Times*, January 6. http://www.ibtimes.co.uk/why-china-scrapped -quotas-rare-earth-metal-exports-1482171.

Winchester, Simon. 2009. *The Map That Changed the World: William Smith and the Birth of Modern Geology.* New York: HarperPerennial.

Wingo, D., P. Spudis, and G. Woodcock. 2009. "Going beyond the Status Quo in Space." SpaceRef, Last Modified June, 28, 2009, accessed November 19, 2013.

Winichakul, Thongchai. 1988. *Siam Mapped: A History of the Geo-Body of the Nation.* Honolulu: University of Hawai'i Press.

Woodward, Bob. 2006. *State of Denial.* New York: Simon and Schuster.

World Trade Organization (WTO). 2013. "China—Measures Related to the Exportation of Various Raw Materials." Geneva, Switzerland.

World Trade Organization. 2014. "China—Measures Related to the Exportation of Rare Earths, Tungsten and Molybdenum." Geneva, Switzerland.

Wright, Robin M. 2005. *História Indígena e do Indigenismo no Alto Rio Negro.* São Paulo: Mercado de Letras.

Wu, Chengyu. 2007. "Bayan Obo Controversy: Carbonatites versus Iron Oxide-Cu-Au-(REE-U)." *Resource Geology* 58 (4): 348–54.

Wu, Fengshi. 2009. "Environmental Politics in China: An Issue Area in Review." *Journal of Chinese Political Science* 14 (4): 383–406.

Wu, Fulong. 2010a. "How Neoliberal is China's Reform? The Origins of Change during Transition." *Eurasian Geography and Economics* 51 (5): 619–31.

Wu, Lanfu 1999. *Collected Essays of Wulanfu.* Vol. 1. Beijing: Central Culture Press.

Wu, Shellen Xiao. 2010b. "Underground Empires: German Imperialism and the Introduction of Geology in China, 1860–1919." PhD diss., Princeton University.

Wu, Yuan-Li. 1965. *The Steel Industry in Communist China.* New York: Frederick A. Praeger.

Wübbeke, Jost. 2013. "Rare Earth Elements in China: Policies and Narratives of Reinventing an Industry." *Resources Policy* 38 (3): 384–94.

Xia, Y., and J. Liu. 2004. "An Overview on Chronic Arsenism via Drinking Water in PR China." *Toxicology* 198 (1–3): 25–29. doi: 10.1016/j.tox.2004.01.016.

Xie, Tao, and Benjamin Page. 2010. "Americans and the Rise of China as a World Power." *Contemporary China* 19 (65): 479–501.

Xin, Ling. 2009. "Xu Guangxian: Father of Chinese Rare Earths Chemistry." *Bulletin of the China Academy of Science* 23 (2): 98–102.

Xin, Min. 2006. "Investigation of the Pollution of the Baogang Tailings Dam: Radioactive Contaminated Water Seeping Toward Yellow River." *Yangtze Evening Post,* September 21.

Xing, Yulin, and Shitian Lin. 1992. "Brief on the Organization of the Northwestern Scientific Expedition Team." *China Border History and Geography Research* 3:22–29.

Xinhua. 2008. "China's First Lunar Landing Vehicle Engineering Prototype Vehicles." Accessed April 15, 2015. http://news.xinhuanet.com/tech/2008-04/24/content _8038623.htm.

Xinhua. 2012. "Rare Earths Defended." Accessed March 16, 2015. http://english.cri.cn/6909 /2012/03/16/3141s687424.htm.

Xinhua. 2013a. "China's Moon Rover 'Jade Rabbit' Separates from Lander." Accessed February 6, 2015. http://news.xinhuanet.com/english/photo/2013-12/15/c_132968376.htm.

Xinhua. 2013b. "China's Moon Rover Continues Lunar Survey after Photographing Lander." Accessed December 22, 2015. http://news.xinhuanet.com/english/china /2013-12/22/c_132987209.htm.

Xiong, Ping. 2013. "Commentary: Chang'e-3's Soft Landing Marks China's Hard Success." Accessed February 6, 2015. http://news.xinhuanet.com/english/china/2013-12/14/c _132968313.htm.

XPrize Foundation. 2015. "Google Lunar X Prize." Accessed February 2, 2015. http://lunar .xprize.org/.

Xu, Ben 1996. *Toward Post-Modernity and Post-Colonialism.* Beijing: China Social Sciences Press.

Xu, Ben. 1998. "From Modernity to Chineseness: The Rise of Postcolonial Theory in Post-1989 China." *Positions* 6 (1): 203–37.

Xu, Chenggang. 2011. "The Fundamental Institutions of China's Reforms and Development." *Journal of Economic Literature* 49 (4): 1076–51.

Xu, Jodi. 2014. "Molycorp Bonds Drop to All-time Low after Quarterly Loss." Bloomberg News. Accessed December 27, 2014. http://www.bloomberg.com/news/2014-11-06/molycorp-bonds-drop-to-all-time-low-as-quarterly-loss-widens.html.

Yan, Fangfang, Jia Jingtao, Liao Chunsheng, Wu Sheng, and Xu Guangxian. 2006. "Rare Earth Separation in China." *Tsinghua Science and Technology* 11 (2): 241–47.

Yang, Guobin. 2005. "Environmental NGOs and Institutional Dynamics in China." *China Quarterly* 181 (1): 44–66.

Yang, Zhanfeng. 2013. History of the Baotou Institute of Rare Earths. *Baotou Research Institute of Rare Earths.* Accessed November 26, 2014. http://www.brire.com/english/ezhjs/ebyjs.htm.

Yu, G. J., E. C. Huo, H. M. Zhang, H. Q. Wang, B. Chen, and H. P. Zhang. 2005. "Surveillance of Endemic Fluorosis in Inner Mongolia in 2002 and 2003." *Chinese Journal of Endemiology* 24 (6): 649–50.

Yu, Guangjun, Chenyu Le, and Shaojiang Li. 2008. "Research on the Regional Differences of Economic Development in Inner Mongolia." *Inner Mongolia University Journal of Philosophy and Social Sciences* 6:4–10.

Yuasa, Shino. 2010. "Traders: China Resumes Rare Earth Exports to Japan." *Washington Times*, September 29.

Zak, Anatoly. 2013. *Russia in Space: The Past Explained and the Future Explored.* Charlotte, NC: Griffin Media.

Zepf, Volker. 2013. "Rare Earth Elements: A New Approach to the Nexus of Supply, Demand, and Use Exemplified along the Use of Neodymium Permanent Magnets." PhD diss., Augsburg University.

Zhang, Licheng, Wang Li, Liu Yi, Zhang Lihong, and Zhang Yixiang. 2001. "Inquest of Residents in Baotou Bayan Obo Mining District 1995–1997." *Journal of Baotou Medical College* 17 (1): 7–9.

Zhang, Peishan, Kueming Yang, and Kejie Tao, eds. 1995. *Mineralogy and Geology of Rare Earths in China.* Beijing: Science Press.

Zhang, Zhefu. 1986. "Rare Earth Geochemistry of Meteorites and the Moon." *Mineralology, Petrology, and Geochemistry Bulletin* 10 (1): 26–28.

Zhao, Guangli. 1994. "Proceedings of the Third Annual Meeting of the Chinese Rare Earth Academic Research." Baotou, Inner Mongolia, China.

Zhao, Jianmei, ed. 2010. *Annals of the Bayan Obo Mining District (1994–2009).* Hulunbeier: Inner Mongolia Culture Press.

Zhao, L. J., C. Wang, Y. H. Gao, and D. J. Sun. 2013. "National Annual Monitoring Report of Drinking-water-borne Endemic Fluorosis in 2010 and 2011." *Chinese Journal of Endemiology* 32 (2): 177–82.

Zhao, Lancai, Liu Zhihe, Qiao Dongliang, Zhang Lanping, Cheng Jie, Li Fusheng, Chen Yingmin, Chen Yao, Sun Huaiyu, Hu Shiliang, Cui Shanxiu, and Fu Zhangchun. 1994. "Survey of Natural Radionuclide Levels in Yellow River Delta Environment and Food." *China Radiological Health* 2:24–33.

Zhao, Zhenbei. 1999. *Handbook on the Prevention of Fluorosis among Livestock in Shadegesumu.* Wulate Front Banner, Inner Mongolia: Animal Husbandry Office.

Zhou, Jianbo, Zheng Yongfei, Yang Shaoyong, Shu Yong, Wei Chunsheng, and Xie Zhi. 2002. "Paleoplate Tectonics and Regional Geology at Bayan Obo in Northern Inner Mongolia." *Geological Journal of China Universities* 8 (1): 46–62.

Zielinski, Sarah. 2010. "Rare Earth Elements Not Rare, Just Playing Hard to Get." Smith-
    sonian Institution, November 18. Accessed November 26, 2014. http://www
    .smithsonianmag.com/science-nature/rare-earth-elements-not-rare-just-playing
    -hard-to-get-38812856/?no-ist.
Žižek, Slavoj, ed. 2012. *Mapping Ideology*. London: Verso.
Zoellner, Tom. 2009. *Uranium: War, Energy, and the Rock That Shaped the World*. Lon-
    don: Penguin Books.
Zou, Weirong. 2014. "The Inner Mongolia Military Frontier Line Has the First Female
    Team." *Sina Military*, January 13. http://mil.news.sina.com.cn/2014-01-13
    /0833759769.html.
Zubok, Vladislav M. 2001. "The Mao-Khrushchev Conversations, 31 July–3 August 1958
    and 2 October 1959." *Cold War International History Project* 12/13 (Fall/Winter):
    254.
Zuo, Ping, ed. 2000. *Fifty Years of the Women's Movement in Baotou*. Baotou: Inner Mon-
    golia Baotou Municipality Women's Committee.

Index

Abraham, David S., 7, 41
Academy of Sciences (Russia), 218
accumulation, primitive, 108–109
activism: environmental, 18, 32, 122–123; political, garimpeiros, 177–180, 198
activism, indigenous: Cabeça do Cachorro, 170, 172–175, 186, 189–190; garimpeiros in, 178; policy reform efforts, 178–180; purpose of, 180
activists, indigenous: characterizations, 194; resistance, methods of, 188
Act on Greenland Self-Government, 148
Acuña, Cristóbal de, 182
Afghanistan, 19–20, 150–153, 162, 163, 164
Afghanistan Geological Survey, 150–151
Afzali, Ikram, 152
Agreement Governing the Activities of States on the Moon and Other Celestial Bodies (Moon Treaty), 31, 208–210, 223, 262n8
Agreement on General Principles for the Settlement of Questions between the Republic of China and the USSR, 255n15
Ak-Tyuz mine (Kyrgyzstan), 19, 51
Aldrin, Buzz, 262n1
Aluminum Corporation of China (Chinalco), 157
Amazon: centers of calculation and power, 29; China model, applying to, 235; frontier subjectivity, inculcating in, 16; gold, search for, 187; military concerns regarding, 173; rubber supplies, 183
Amazon frontier: appeal of, 56–57; biopolitics on the, 195; border-marking, 16; border security, 30; colonialism and control in the, 182–183, 191; contested hinterland, 191–197; foreign expropriation of resources, fears of, 184–185; geological survey, 184–185; hinterland, rationalizing as, 195; militarizing the, 188–191, 194–195; military dictatorship control over, 184–185; mineral exploration, jurisdiction over, 185; mining, constitutional prohibitions on, 185–186; mining, justifications for, 190, 197–198; nation-building on the, 181–191; necropolitics on the, 194–195; in the popular imaginary, 194; violence on the, 16, 187–188

Amazonian Surveillance System (Sistema de Vigilância dos Amazonas) (SIVAM), 190, 195–196
Amazon Military Command (Comando Militar da Amazônia) (CMA), 171–174, 171f, 193, 259n7
Ames National Laboratory, 47, 145
amethyst, 177
Andersen, Kai Holst, 152
Anderson, Eric, 225
Andersson, John Gunnar, 75
animals, pollution's impact on, 121, 124
antimony, 75, 254n7
apocalypse: environmental, 217, 226; narratives of, 241
Araxá, 57, 160, 167f, 168–169, 258n17
Arctic Circle, 25
Argentina, 182–183
Army Corps of Engineers (US), 183–184, 236, 260n25
arsenic, 1, 32, 122, 123–127
arsenical dermatosis, 123
asteroids, 223, 252n13, 263n13
ATECH (Brazil), 261n38
atomic age: aspirations of the, 54, 72, 93–95, 102, 219; rare earth extraction and the, 47, 73
Atomic Energy Act (India), 47, 52
Atomic Energy Commission (US), 79
Australia, 17–19
Austria, 47, 56, 209

Bai, Lina, 118
Bailingmiao, China, 77
Balasubramanian, S., 208
Baniwa, André, 179
Baniwa people, 186–188
Baogang. See Baotou Iron and Steel Company (Baogang)
Baotou (IMAR): building, hardships of, 88; Cabeça do Cachorro compared, 168; churches in, 254n2; Cold War, atomic aspirations of, 93–101; consolidation of production control, 133, 139, 158, 229, 232–233, 240; deposits, formation and location of, 27, 61–62; environmental activism at, 122–123; history

PCN. *See* Northern Trench Project (Projeto Calha Norte) (PCN)
Peixoto, Marechal Floriano, 260n23
People's Liberation Army (PLA), 78
periodic table, 41–42, 42*f*, 43*f*
Peru, 193
petroleum, 63, 138, 144, 257n13
Planetary Resources, 205, 208, 225, 263n13, 264n24
platinum, 49
Plescia, Jeff, 221
Policy on Mineral Resources (PRC), 130
pollution: animals, impact on, 121, 124; China's toxic mining practices producing, 32; on the frontier, 57, 111, 132, 162; racialization of, 117, 144. *See also* environment, effects of REMP on the
polonium, 120
Pomeranz, Kenneth, 243
power, spatiality of, 54
praseodymium, 44, 49, 100, 131, 140, 218
present world, creating the, 8
promethium, 44, 50, 55, 119, 252n3
prospecting: on borders, 63; costs of, 28–29; export crisis (2010), response to, 6, 229–231; for geopolitical power, 182; for gold, 59, 234; international, early sites of, 28–29; New Silk Road campaign, 241; for uranium, 50–52, 63, 70, 97–98, 148; US funded, 151. *See also* exploration
protactinium, 120
public/private divide, 22–23, 224
pyrophoric, 252n7

Quintão, Leonardo, 192

race and gender: in the division of labor, 84–93; on the frontier, 23, 88–89, 92, 97, 211; in resource use decision-making, 211; of sacrifice, 33–34; shaping distribution of benefits and hazards, 79; social mobilization programs, 84–93, 195–196
race to the bottom, 56, 110, 153
racism: colonial, 169; politics of, on the frontier, 236; in pollution, 18–19, 55, 117, 144; in waste disposal, 20, 24, 55
Radar of the Amazon (Radar da Amazônia) (RADAM), 184–186, 191–192, 261n26
Radioactive Pollution Prevention Law (PRC), 117
radium, 113–114, 120
radon, 55, 119–121, 123, 163, 237, 257n16
radon daughters, 120

rare earth alloys, 48, 96, 99
rare earth economy: black market, 23, 36, 134, 142, 158, 254n25; China in the, 59; consumption statistics, 4; demand for non-Chinese rare earths, 160–161; export crisis (2010) and, 4–5, 138–143, 141*f*, 155, 212; modernity, rare earths role in, 4, 45–46, 48–49, 140; Moon mining in the, 215–216; prices, 3*f*, 4–5, 52, 69, 139–141, 140*f*; technology investment in the, 70. *See also* rare earths: applications
rare earth politics: Cold War era, 52–54; geography of, 62; necessity vs. pollution in tension, 54–57; power and, 59; span of, 66; trade, sacrifice, and security, 57–60; understanding, importance of, 10–11; World War I and II, 49–52
rare earths: abundance, by element, 44*f*; applications, 44–50, 66, 75, 96, 114, 140, 216; as conflict minerals, 42, 66; consumption, global, 4; elements classified as, 41–42, 42*f*, 43*f*, 45*f*, 49, 50; fiction and nonfiction works on, 7, 262n1; geological occurrence, 44; incompatible, 61; knowledge about, 36; metaphors for, 46, 166, 239; mining of, factors in, 28; modern life's dependence on, 4, 45–46, 48–49, 140; mythology, 9; properties of, 1, 47–48, 60; research in, 48, 101; term usage, 65–66; value of, 49, 65, 234. *See also specific elements*
rare earths, history of: alloys, 48; industrial usage, 46–47; research on, 48; scarcity myth, 41; technological/commercial innovations, 46–47
rare earths mining/production (REMP): drivers of, 57; environmental effects of. *see* environment, effects of REMP on the; epidemiological effects. *see* epidemiological effects of REMP; externalization of, 232; geoeconomics of, 4–5, 139–143, 141*f*, 155, 212; globally, 1, 2*f*; green, 23, 160–162, 238; legal vs. illegal, 23–24; locating in relation to centers of power, 233; necessity vs. pollution in tension, 104; options, 237–240; processing, 60–61, 100–101, 119–120, 160–161, 234; research recommendations, 240–242; research sites, globally, 2*f*; risks and benefits, organizing, 23; spatial politics of, 11; sustainable, 3–4, 160–162, 238; western, demise of, 5–6
rare earths mining/production (REMP), geography of: to extend power, 19; hazards pressuring, 11; spatial allocation of sacrifice in, 19

Lighting Source UK Ltd.
Milton Keynes UK
UKHW010747110921
389953UK00012B/416